African International Relations:
An Annotated Bibliography

Other Titles in This Series

Southern Africa Since the Portuguese Coup, edited by John Seiler

South Africa into the 1980s, edited by Richard E. Bissell and Chester A. Crocker

An African Experiment in Nation Building: The Bilingual Cameroon Republic Since Reunification, edited by Ndiva Kofele-Kale

African Upheavals Since Independence, Grace Stuart Ibingira

Africa and the West, edited by Michael A. Samuels

Zambia's Foreign Policy: Studies in Diplomacy and Dependence, Douglas G. Anglin and Timothy M. Shaw

The Arab-African Connection: Political and Economic Realities, Victor T. Le Vine and Timothy W. Luke

Ethnicity in Modern Africa, edited by Brian M. du Toit

Regionalism Reconsidered: The Economic Commission for Africa, Isebill V. Gruhn

Botswana: An African Growth Economy, Penelope Hartland-Thunberg

Alternative Futures for Africa, edited by Timothy M. Shaw

The Angolan War: A Study in Soviet Policy in the Third World, Arthur Jay Klinghoffer

Westview Special Studies on Africa

African International Relations:
An Annotated Bibliography
Mark W. DeLancey

This bibliography contains more than 2,500 annotated entries for international books, journal articles, and pamphlets in the field of African international relations. Primarily covering works published from 1960 to 1978, it also includes a few earlier publications and a few works published in 1979 and 1980. References are organized under eleven broad subject headings, and all entries are cross-listed in a detailed index of place names and subjects. A substantial list of commonly used abbreviations and acronyms is also included.

Mark W. DeLancey is a Fulbright professor at the University of Yaounde, Cameroon, for the 1980-81 academic year, on leave from his position as associate professor and chairman of the Department of Government and International Studies at the University of South Carolina. Dr. DeLancey is also associate executive director of the International Studies Association and has previously spent six years teaching and conducting research in Nigeria and Cameroon.

African International Relations:
An Annotated Bibliography
Mark W. DeLancey

Westview Press / Boulder, Colorado

This is a Westview reprint edition, manufactured on our own premises using equipment and methods that allow us to keep even specialized books in stock. It is printed on acid-free paper and bound in softcovers that carry the highest rating of NASTA in consultation with the AAP and the BMI.

Westview Special Studies on Africa

All rights reserved. No part of this publication may be reproduced or transmitted in any form or by any means, electronic or mechanical, including photocopy, recording, or any information storage and retrieval system, without permission in writing from the publisher.

Copyright © 1981 by Westview Press, Inc.

Published in 1981 in the United States of America by
　　Westview Press, Inc.
　　5500 Central Avenue
　　Boulder, Colorado 80301
　　Frederick A. Praeger, Publisher

Library of Congress Cataloging in Publication Data
DeLancey, Mark.
　　African international relations, an annotated bibliography.
　　(Westview special studies on Africa)
　　Includes index.
　　1. Africa--Foreign relations--Bibliography. I. Title.
Z6465.A36D44　　(DT31)　　016.3276　　80-21254
ISBN 0-89158-680-6

Composition for this book was provided by the author
Printed and bound in the United States of America

10　9　8　7　6　5　4　3

Contents

Preface ix
Commonly Used Abbreviations and Acronyms,
  Elizabeth D. Thompson xi

African International Relations: General Works 1
African States Foreign Policies 15
Inter-African Conflicts, Borders, and Refugees 59
Sub-continental Regionalism 81
The OAU, Pan-Africanism, and African Unity 109
The UN and International Law 135
Southern Africa 161
The USSR, the PRC, the UK, and France:
  Relations with Africa 207
The USA: Relations with Africa 245
Other States: Relations with Africa 273
Economic Factors in African International Relations . . . 295

Subject Index 339

Preface

This bibliography of African international relations covers publications from 1960 through 1978 and, more sparingly, through 1980; a few earlier publications are also included. Books, journal articles, pamphlets, and a few documents have been included. The compilation is not exhaustive, however, as most brief notes, one to two page articles, and documents have been excluded.

The references have been gathered by examining the catalogues of several major libraries; by searching several subject indices, bibliographies, and abstract collections; by examining the contents of selected journals; and by contacting directly certain scholars known to be actively engaged in appropriate research.

The references are divided into broad subject categories, and a subject index is provided to enable cross listing of items and to enable listing of a number of subcategories. This index is based on the titles and brief annotations; it is not based on a thorough reading of each item. A complete reference to a publication will appear only once in this bibliography, under the most appropriate chapter heading. The index must be used to locate items of secondary importance to a subject.

Over the several years that I have compiled this bibliography, many persons have assisted me. There is a long list of librarians, scholars, clerks, and student assistants who have aided the project. I wish to thank all of them and to note especially the efforts of Michael Janos, Timothy Shank, and Elizabeth Normandy. Special gratitude must be expressed to Ms. Deborah Harrison, who has provided unusually dedicated service in typing the manuscript.

Financial support for this project has been received from the Institute of International Studies at the University of South Carolina, Dr. Richard Walker, director, and from the Earhart Foundation.

Mark W. DeLancey

Commonly Used Abbreviations and Acronyms

Elizabeth D. Thompson

AAASA	Association for the Advancement of Agricultural Sciences in Africa
AACB	Association of African Central Banks
AACC	All Africa Conference of Churches
AADFI	Association of African Development Finance Institutions
AAGS	Association of African Geological Surveys
AAHO	Afro-Asian Housing Organization
AAI	African-American Institute
AAPAM	African Association for Public Administration and Management
AAPC	All-African Peoples' Conference
AAPO	All-African Peoples' Organization
AAPS	African Association of Political Science
AAPSO	Afro-Asian Peoples' Solidarity Organization
AARRO	Afro-Asian Rural Reconstruction Organization
AASM	African Associated States and Madagascar
AATUF	All-African Trade Union Federation
AAU	Association of African Universities
AAWORD	Association of African Women for Research and Development
ABC	African Bibliographic Center
ABCA	See AACB
ACC	Administrative Committee on Co-ordination (OAU)
ACCT	Agence de Coopération Culturelle et Technique
ACORD	Association de Coopération et de Recherches pour le Développement
ACOSCA	Africa Co-operative Savings and Credit Association
ACP	African, Caribbean and Pacific Group
ADB	African Development Bank
AEF	Afrique Equatoriale Française
AELE	See EFTA
AFCAC	African Civil Aviation Commission
AFL-CIO	American Federation of Labor-Congress of Industrial Organization
AFPU	African Postal Union

AFRAA	African Airlines Association
AFRASEC	Afro-Asian Organization for Economic Cooperation
AFRO	African Regional Organization of the International Confederation of Free Unions
AGC	African Groundnut Council
AHG	Assembly of Heads of State and Government (OAU)
AIAFD	See AADFI
AID	Agency for International Development (US)
AID	See IDA
AIDA	See IALA
AIPLF	Association Internationale de Parlementaires de Langue Francaise
AISA	See IASA
AISS	See ISSA
AKP	See ACP
ALD	African Liberation Day
AMCO	See OCAM
ANC	African National Congress of South Africa
ANC	Armee Nationale Congolaise (Congolese National Army)
AOF	Afrique Occidental Française
APF	African Peace Force
APSA	African Political Science Association
APTU	African Postal and Telecommunications Union
ASECNA	Agency for the Security of Air Navigation in Africa and Madagascar
ASGA	See AAGS
ATUC	African Trade Union Confederation (now OATUU)
BAD	See ADB
BCEAEC	Banque Centrale des Etats de l'Afrique Equatoriale et du Cameroun
BCEAO	Banque Centrale des Etats de l'Afrique de l'Ouest
BEAC	Banque des Etats de l'Afrique Centrale
BIRD	See IBRD
BIS	See IASB
BLS	Botswana, Lesotho, Swaziland
BOAD	See WADB
BOSS	Bureau of State Security (South Africa)
BPEAR	Bureau for the Placement and Education of African Refugees
BWA	Baptist World Alliance
CAA	See AGC
CAE	See EAC
CAFAC	See AFCAC
CAFRAD	Centre Africain de Formation et de Recherche Administratives pour le Développement
CAMES	Conseil Africain et Malgache pour l'Enseignement Supérieur
CAPTAO	West African Conference of Posts and Telecommmunications Administrations
CBLT	Commission du Bassin du Lac Tchad
CCTA	Commission for Technical Cooperation in Africa South of the Sahara

CDC	Commonwealth Development Corporation
CEA	See ECA
CEAO	Communauté Economique de l'Afrique de l'Ouest
CECAS	Conference of East and Central African States
CEE	See EEC
CEPGL	Communauté Economique des Pays des Grand Lacs
CETA	See AACC
CFA	Communauté Financiere Africaine
CFA	Conférence des Femmes Africaines
CFAO	Compagnie Française de l'Afrique Occidentale
CFN	Commission du Fleuve Niger
CHEAM	Centre des Haute Etudes Administratives sur l'Afrique et l'Asie Modernes
CIA	Central Intelligence Agency (US)
CIAS	Conference of Independent African States
CICC	See ICCC
CIDA	Canadian International Development Agency
CIDESA	Centre International de Documentation Economique et Sociale Africaine
CIF	See ICW
CILSS	Comité Permanent Inter-Etats de Luttre contre la Sécheresse dans le Sahel
CIPEA	See ILCA
CISL	See ICFTU
CMCA	Commission of Mediation, Conciliation and Arbitration (OAU)
CMEA	Council for Mutual Economic Assistance
CMT	See WCL
COCLA	Coordinating Committee for the Liberation of Africa (OAU)
CODESRIA	Commission for the Development of Economic and Social Research in Africa
COE	See WCC
COI	See IOC
COMECON	See CMEA
COPAL	Cocoa Producers' Alliance
CPA	Commonwealth Parliamentary Association
CPCM	Comité Permanent Consultatif du Maghreb
DDR	See GDR
DLCOEA	Desert Locust Control Organization of Eastern Africa
DOLCEA	See DLOCEA
DTA	Democratic Turnhalle Alliance
EAA	East African Airways Corporation
EAAFRO	East African Agricultural and Forestry Research Organization
EAC	East African Community
EACM	East African Common Market
EACSO	East African Common Services Organization
EADB	East African Development Bank
EADC	East African Development Corporation
EADCA	East African Directorate of Civil Aviation
EAEC	East African Economic Community

EAFFRO	East African Freshwater Fisheries Research Organization
EAMD	East African Meteorological Department
EAMFRO	East African Marine Fisheries Research Organization
EAMRC	East Africa Medical Research Council
EAR	East African Railways Corporation
EAVRO	East African Veterinary Research Organization
ECA	Economic Commission for Africa (UN)
ECM	Council of Ministers (OAU)
ECOSOC	Economic and Social Council (UN)
ECOWAS	Economic Community of West African States
EDC	Educational and Cultural Commission
EDF	See FED
EEC	European Economic Community
EFTA	European Free Trade Association
ENA	See IPA
ESCHC	Educational, Scientific, Cultural and Health Commission (OAU)
EWG	See EEC
FAC	Fonds d'Aide et de Coopération
FAO	Food and Agricultural Organization (UN)
FED	Fonds Européen de Développement
FFHC	Freedom from Hunger Campaign (FAO)
FIDES	Fonds d'Investissement pour le Développement Economique et Social
FISE	See UNICEF
FLCS	National Front for the Liberation of the Somali Coast
FLIPG	Front for the Liberation and Independence of Portuguese Guinea
FMTS	See WFSW
FNLA	Frente Nacional de Libertação de Angola
FRELIMO	Frente de Libertação de Moçambique
FRG	Federal Republic of Germany
FROLINAT	Chad National Liberation Front
FROLIZI	Front for the Liberation of Zimbabwe
FSA	Fonds de Solidarité Africain
FSM	See WFTU
GATT	General Agreement on Tariffs and Trade
GDR	German Democratic Republic
HSNC	Health, Sanitation and Nutrition Commission (OAU)
IABAR	Inter-African Bureau for Animal Resources
IACO	Inter-African Coffee Organization
IAEA	International Atomic Energy Agency
IAI	International African Institute
IALA	International African Law Association
IAMD	Inter-African and Malagasy Organization
IAPB	Inter-African Phytosanitary Bureau (OAU)
IAPSB	See IAPB
IASA	International Air Safety Association
IASB	Inter-African Soils Bureau (OAU)
IATA	International Air Traffic Association
IBAH	Inter-African Bureau for Animal Health (OAU)

IBRD	International Bank for Reconstruction and Development (World Bank)
ICAM	Institut Culturel Africain et Mauricien
ICAO	International Civil Aviation Organization
ICCC	International Conference of Catholic Charities - CARITAS
ICFTU	International Confederation of Free Trade Unions
ICJ	International Court of Justice (UN)
ICVA	International Council of Voluntary Agencies
ICW	International Council of Women
IDA	International Development Association
IDAF	International Defence and Aid Fund for Southern Africa
IDEP	Institut pour le Développement Economique et la Planification
IFAN	Institut Fondamental d'Afrique Noire
IFC	International Finance Corporation
IITA	International Institute of Tropical Agriculture
ILCA	International Livestock Centre for Africa
ILO	International Labour Organization
IMCO	Inter-governmental Maritime Consultative Organization
IMF	International Monetary Fund
INTELSAT	International Telecommunications Satellite Organization
IOC	Intergovernmental Oceanographic Commission
IPA	Institute of Public Administration
IPD	See PAID
IPPF	International Planned Parenthood Federation
IPU	Inter-Parliamentary Union
IRC	International Rice Commission
IRLCO-CSA	International Red Locust Control Organization for Central and Southern Africa
ISSA	International Social Security Association
ITU	International Telecommunications Union
KANU	Kenya African National Union
LAS	League of Arab States
MLD	Djibouti Liberation Movement
MLSTP	Committee for Liberation of Sao Tome and Principe
MNC	Multi-national Corporation
MOLINACO	National Movement for the Liberation of the Comoro Islands
MPJ	See PYM
MPLA	Movimento Popular de Libertação de Angola
MPLT	Movement for the Total Liberation of Chad
NATO	North Atlantic Treaty Organization
NIEO	New International Economic Order
NLC	National Liberation Council (Ghana)
OAA	See FAO
OACI	See ICAO
OAMCE	Organisation Africaine et Malgache de Coopération Economique
OAPEC	Organization of Arab Petroleum Exporting Countries

OATUU	Organization of African Trade Union Unity
OAU	Organization of African Unity
OCAM	Organisation Commune Africaine et Mauricienne
OCAMM	Organisation Commune Africaine, Malgache, et Mauricienne
OCBN	Organisation Commune Bénin-Niger des Chemins de Fer et des Transports
OCCGE	Organisation de Coordination et de Coopération pour la Lutte contre les Grandes Endemies en Afrique de l'Oueste
OCDE	See OECD
OCDN	Niger-Dahomeyan Common Organization
OCEAC	Organisation de Coordination pour la Lutte contre les Endémies en Afrique Centrale
OCLALAV	Organisation Commune de Lutte Anticridienne et de Lutte Antiaviaire
ODTA	L'Organisation pour le Développement du Tourisme Africain
OECD	Organization of Economic Cooperation and Development
OERS	Organisation des Etats Riverains du Sénégal
OIAC	See IACO
OICMA	Organization International contre le Criquet Migrateur Africain
OIT	See ILO
OMCI	See IMCO
OMM	See WMO
OMS	See WHO
OMVS	Organisation pour le Mise en Valeur du Fleuve Sénégal
ONU	See UN
ONUC	Opération des Nations Unies au Congo
ONUDI	See UNIDO
OPAEP	See OAPEC
OPEC	Organization of Petroleum Exporting Countries
OSPAAAL	Organisaton de Solidarité des Peuples d'Afrique, d'Asie et d'Amerique Latine
OTAN	See NATO
OUA	See OAU
OXFAM	Oxford Committee for Famine Relief
PAC	Pan African Congress of South Africa
PAFMECA	Pan African Freedom Movement of East and Central Africa
PAFMECSA	Pan African Movement of Eastern, Central and Southern Africa
PAID	Pan African Institute for Development
PAIGC	Partido Africano da Independência da Guiné e Capo Verde
PAYM	See PYM
PNUD	See UNDP
PRC	People's Republic of China
PYM	Pan African Youth Movement
RFT	See FRG
ROC	Republic of China

RSA	Republic of South Africa
SADIAMIL	Société Africaine pour le Développement des Industries Alimentaires à Base de Millet et de Sorgho
SAMA	See AASM
SCOA	Société Commerciale de l'Oueste Africain
SCSA	Supreme Council for Sport in Africa
SEDES	Société d'Etudes pour le Développement Economique et Social
SERESA	Société d'Etudes et de Réalisations Economiques et Sociales dans l'Agriculture
SESAF	Société Anonyme pour les Echanges entre la Suisse et l'Afrique
SFI	See IFC
SIDA	Swedish International Development Authority
SIFIDA	Société Internationale Financière pour les Investissements et le Développement en Afrique
STRC	Scientific, Technical and Research Commission (OAU)
SWA	South West Africa
SWAPO	South West African Peoples' Organization
TAH	Trans-African Highway Co-ordinating Committee
TRC	Transport and Communications Commission (OAU)
UAM	Union Africaine et Malgache
UAMBO	Union Africaine et Malgache de Banques pour le Développement
UAMCE	Union Africaine et Malgache de Coopération Economique
UAMD	African and Malagasy Union for Defense
UAMPT	See UAPT
UAPT	L'Union Africaine des Postes et Télécommunications
UAR	United Arab Republic
UAS	Union of African States
UDAO	Union Douanière des Etats de l'Afrique Occidentale
UDE	Union Douanière Equatoriale
UDEAC	Union Douanière et Economique de l'Afrique Centrale
UDEAO	Union Douanière des Etats d'Afrique de l'Ouest
UDI	Unilateral Declaration of Independence
UEAC	Union des Etats d'Afrique Centrale
UGTAN	Union Générale des Travailleurs d'Afrique Noire
UIT	See ITU
UMOA	See WAMU
UN	United Nations
UNACAST	United Nations Advisory Committee on the Application of Science and Technology to Development
UNCHR	See UNHCR
UNCTAD	United Nations Conference on Trade and Development
UNDP	United Nations Development Programme
UNESCO	United Nations Educational, Scientific and Cultural Organization
UNFPA	United Nations Fund for Population Activities
UNHCR	United Nations High Commission for Refugees
UNICEF	United Nations Childrens' Fund
UNIDO	United Nations Industrial Development Organization
UNITA	Union for the Total Independence of Angola

UNITAR	United Nations Institute for Training and Research
UNOTC	United Nations Office for Technical Cooperation
UNSO	United Nations Sahelian Office
UNTAG	United Nations Transitional Assistance Group (Namibia)
UPAF	See AFPU
UPU	Universal Postal Union
URTNA	African National Radio and Television Union
USAID	See AID
USPA	See AATUF
UTA	Union Transportes Aériennes
WAAC	West African Airways Corporation
WACA	West African Court of Appeal
WADB	West African Development Bank
WAEC	West African Examinations Council
WAHS	West African Health Secretariat
WAMU	West African Monetary Union
WARDA	West African Rice Development Association
WASA	West African Science Association
WCC	World Council of Churches
WCL	World Confederation of Labour
WFP	World Food Programme
WFSW	World Federation of Scientific Workers
WFTU	World Federation of Trade Unions
WHO	World Health Organization
WMO	World Meteorological Organization
ZANU	Zimbabwe African National Union
ZAPU	Zimbabwe African Peoples' Union

African International Relations: General Works

1 ABDEL-MALEK, A. "La Vision du Problème Colonial par le Monde Afro-Asiatique." Cahiers Internationaux de Sociologie, 10, 35, July-December 1963, 145-56.
　　The desire of the recently colonized Afro-Arab states is to regain their identity and to elaborate a new world humanism.

2 ADELAJA, KOLA. "Africa's International Relationships." In: Reflections on the First Decade of Negro-African Independence. Paris: Présence Africaine, 1971. 17-34.
　　A general discussion of goals and actions in African diplomacy. Author concludes that the first decade of African independence has not been successful in respect of international affairs.

3 Africa South of the Sahara, 1978-1979. New York: Unipub, R. R. Bowker, 1978. 1,193 p. London: Europa, 1978. 1,253 p.
　　This reference volume, now in its 8th edition, contains chapters on each state, detailed information on regional organizations, a who's who, statistics, and directories of diplomatic corps.

4 AKE, C. "Explanatory Notes on the Political Economy of Africa." Journal of Modern African Studies, 14, 1, March 1976, 1-23.
　　The characteristics of Africa's political economy can best be analyzed in terms of dependence, an international division of labor which exploits Africa.

5 AKINDELE, R. A. "Reflections on the Preoccupation and Conduct of African Diplomacy." Journal of Modern African Studies, 14, 4, December 1976, 557-76, tabl.
　　Classifies and analyzes the resolutions of the OAU which bear on Africa's external relations in order to identify the major concerns of African diplomacy, 1963-1973.

6   ALEXANDRE, PIERRE. "Sur Quelques Aspects Internationaux de la Linguistique Négro-Africaine." _Afrique et Asie_, 58, 2, 1962, 14-22.
    The influence of the wide-spread use in Africa of European languages.

7   AMIN, SAMIR. [Transl. by Francis McDonagh.] _Neo-Colonialism in West Africa_. Harmondsworth: Penguin, 1973. 298 p., bibl., index, tabls.
    A major analysis based on the dependency concept.

8   _____. "Underdevelopment and Dependence in Black Africa-Origins and Contemporary Forms." _Journal of Modern African Studies_, 10, 4, December 1972, 503-24.
    Divides contemporary black Africa into three broad regions, and examines each through four periods of history. Concludes that the final and most recent period of colonialism destroyed the traditional societies and produced only dependent peripheral societies.

9   AMISSAH, AUSTIN N. E. "Air Transport in Africa and Its Importance for African Land-locked Countries." In: Cervenka, Z., (ed.). _Land-locked Countries of Africa_. Uppsala, 1973, 63-78.
    The role of air transport in Africa, especially in the land-locked states, and the need for international cooperation in air transport.

10 ANTOLA, E. "The Roots of Domestic Military Interventions in Black Africa." _Instant Research on Peace and Violence_, 5, 4, 1975, 207-21.
    African dependence in the international economic order is a major cause of military intervention in African domestic politics.

11 ARKHURST, FREDERICK S., (ed.). _Arms and African Development_. New York: Praeger, 1972. 158 p., tabls.
    Disarmament, science and technology, and regional integration are the major topics. Also see references by Gingyera-Pincywa and Aluko.

12 BAKER, ROSS K. "Tropical Africa's Nascent Navies." _United States Naval Institute Proceedings_, 95, 1, January 1969, 64-71.
    The colonial transfer of naval resources to African states was negligible. But for reasons of national pride, African states have built navies.

13 BERMAN, BRUCE J. "Clientelism and Neocolonialism: Center-periphery Relations and Political Development in African States." _Studies in Comparative International Development_, 9, 2, Summer 1974, 3-25, bibl.

14 BIELFELDT, M. "Aufgaben des Westens in Afrikca." Aussenpolitik, 11, 9, September 1960, 583-93.
    Before the African states become independent it is necessary to plan and to act to block the spread of communist influence.

15 BOURGUIBA, H. "The Outlook for Africa." International Affairs, 37, 4, October 1961, 425-31.
    The manner of decolonialization is an important determinant of which side the newly independent state will take in the Cold War.

16 BOZEMAN, ADDA B. Conflict in Africa: Concepts and Realities. Princeton, N. J.: Princeton University, 1976. 429 p., bibl., index, maps.
    This study based on anthropological materials attempts to determine African views of conflict and conflict resolution. The author then assesses the compatibility of African and European views.

17 BRETTON, HENRY L. Power and Politics in Africa. Chicago: Aldine, 1973. 402 p., bibl., figs., index, maps, tabls.
    This general survey of African politics includes a chapter titled "Sources and Uses of Power: International," pp. 55-92.

18 BRIONNE, B. "Nouveaux Facteurs de Déséquilibre en Afrique." Défense Nationale, April 1975, 65-79.
    Great power rivalry, African conflicts, and actions of the OAU have led to instability in African international relations.

19 BROOKS, ANGIE E. "Africa and the World Community." In: El-Ayouty, Y. and Brooks, H. C., (eds.). Africa and International Organization. Hague, 1974, 230-4.
    Author is former President of the United Nations General Assembly.

20 BROWN, R. W., et al. Africa and International Crises. Syracuse, N. Y.: Maxwell School of Citizenship and Public Affairs, Syracuse University, 1976. 106 p., maps, tabls.

21 BURCHETT, WILFRED and ROEBUCK, DEREK. The Whores of War: Mercenaries Today. Harmondsworth: Penguin, 1977. 240 p., map.
    The international mercenary business is described with some material on Angola. Burchett is a journalist and Roebuck is a law specialist.

22 CHAULEUR, P. "Afrique 1974." Etudes, June 1974, 831-49.
    The major development of the year has been the growth
    of Arab influence.

23 CLIFFORD, MARY L. and ROSS, EDWARD S. Africa: The Beginning of Tomorrow; Government, Statesmen, and African Unity.
    New York: Noble and Noble, 1971. 64 p., bibl., ill.
    One of a series of introductory readers on Africa.

24 COLA ALBERICH, JULIO. "Africa y Sus Problemas." Revista de
    Política Internacional, 157, May/June 1978, 91-104.

25 COPSON, RAYMOND W. "African International Politics: Underdevelopment and Conflict in the Seventies." Orbis, 22, 1,
    Spring 1978, 227-45.
    "An overview of the problems of African international
    conflict" from a systems analysis point of view. Argues
    that conflict is a product of African underdevelopment.

26 CROCKER, CHESTER A. "Military Dependence: The Colonial
    Legacy in Africa." Journal of Modern African Studies, 12,
    2, June 1974, 265-86.
    African dependency has an important military aspect and
    this must be considered in the analysis of African international relations.

27 DELANCEY, MARK W., (ed.). Aspects of International Relations in Africa. Bloomington: African Studies Program,
    Indiana University, 1979. 253 p., figs., index, tabls.
    A collection of six essays specifically designed for
    instructional purposes in university courses. See references under Boyd, DeLancey, Grundy, Liebenow, McGowan/
    Johnson, and Shaw/Newbury.

28 _____. "Current Studies in African International Relations." Africana Journal, 7, 3, 1976, 195-239.
    A bibliographical essay.

29 _____. "The Study of African International Relations."
    In: DeLancey, M. W., (ed.). Aspects of International Relations in Africa. Bloomington, Ind., 1979, 1-38.
    A bibliographical essay with 250 references examined.

30 _____. Teaching the International Relations of Africa:
    A Collection of Syllabi. Columbia, S. C.: Institute of
    International Studies, University of South Carolina, 1975.
    74 p., bibl.
    Syllabi from graduate and undergraduate courses taught
    at US and Canadian universities.

31 DOBERT, MARGARITA. "Who Invaded Guinea?" Africa Report, 16,
    3, March 1971, 16-8.

32 DOMINGUEZ, J. I. "Mice That Do Not Roar: Some Aspects of International Politics in the World's Peripheries." International Organization, 25, 2, Spring 1971, 175-208.
    A statement of the sub-system concept and its application to Africa, Asia, and Latin America.

33 DORO, MARION E. "Selected Bibliography on International and Continental African Affairs." A Current Bibliography on African Affairs, 4, 2, March 1971, 87-95.

34 DRABEK, A. G. and KNAPP, W. The Politics of African and Middle Eastern States: An Annotated Bibliography. Oxford: Pergamon, 1976. 192 p.

35 DU BOIS, W. E. B. The World and Africa. New York: International Publishers, 1965. 352 p., ill., maps. [Reprinted by Kraus-Thomson, Millwood, New York, 1976.]
    An analysis of Africa's world role in the past and the future.

36 DUMOGA, JOHN. Africa between East and West: A Background Book. Chester Springs: Dufour Editions, 1969. 142 p., index.
    The various chapter subjects include economic aid, African unity, African foreign policies, and communism in Africa. The author is Ghanaian.

37 EDMONDSON, LOCKSLEY. "Africa and the African Diaspora: Interactions, Linkages, and Racial Challenges in the Future World Order." In: Mazrui, A. A. and Patel, H., (eds.). Africa in World Affairs: The Next Thirty Years. New York, 1973, 1-21.
    The growing importance of race in international affairs with particular reference to Africa and a look at future developments.

38 ETINGER, YAKOV Y. Mezhgosudarstvennye Otnosheniya v Afrike: Politicheskiye Problemy Evolutsiya Organizatsionney Formey. Moscow: Nauka, 1972. 318 p., bibl.
    A general survey of African politics and foreign relations.

39 FOLTZ, WILLIAM J. "Military Influences." In: McKay, Vernon, (ed.). African Diplomacy: Studies in the Determinants of Foreign Policy. New York, 1966, 69-90.
    The conditions of African military establishments and their role in international politics.

40 GARDINER, ROBERT A.; ANSTEE, M. J.; and PATTERSON, C. L., (eds.). Africa and the World. Addis Ababa: Oxford University, 1970. 255 p.

41 GHAI, DHARANI P. "Africa, the Third World and the Strategy for International Development." In: Mazrui, A. A. and Patel, H., (eds.). Africa in World Affairs: The Next Thirty Years. New York, 1973, 235-56.
   The concept of a "Third World" hides important differences in conditions within the states included by that concept.

42 GLANTZ, MICHAEL H., (ed.). The Politics of Natural Disaster: The Case of the Sahel Drought. New York: Praeger, 1976. 340 p., figs., index, tabls.
   Of the numerous essays in this volume, three are directly relevant to African international relations. See references by El-Khawas, Sheets/Morris, and Wiseberg.

43 GOOD, ROBERT C. "Changing Patterns of African International Relations." American Political Science Review, 58, 3, September 1964, 632-41.
   Economic necessity is forcing most African states to adopt similiar foreign policies.

44 GRUNDY, K. W. "Africa in the World Arena." Current History, 52, 307, March 1967, 129-35.
   A brief review of events in a special issue on Africa.

45 _____. "On Machiavelli and the Mercenaries." Journal of Modern African Studies, 6, 3, October 1968, 295-310.
   Discusses two issues central to Machiavelli's The Prince, in light of recent African experience with mercenaries: the problems of force in institutionally weak societies and of control of various sources of armed force.

47 HARRELL-BOND, BARBARA. "Africa's Dependency and the Remedies: The 4th International Congress of African Studies." American Universities Field Staff Reports, Africa, 31, 1979, 1-8, ill.
   A discussion of the definition of dependence established at the Congress. Economic, technological, and cultural dependency are considered.

48 HARTWIG, GERALD W. and O'BARR, WILLIAM M. The Student Africanist's Handbook: A Guide to Resources. New York: John Wiley, 1974. 152 p., bibl.
   This multidisciplinary study guide includes sections on international relations and on each country.

49 HENTSCH, THIERRY. Face au Blocus: La Croix-Rouge Internationale dans le Nigeria en Guerre (1967-1970). Geneva: Institut Universitaire des Hautes Études Internationales, 1973. 307 p., ill., maps.
    The author examines the role of the International Committee of the Red Cross in the Nigerian Civil War, but only that portion of its activites played in Biafra.

50 HVEEM, H. "Afrika i Verdensøkonomien: Utvikling og Underutvikling." Internasjonal Politikk, 1976, 729-46.
    "Africa in the World Economy: Development and Underdevelopment." Africa remains dependent and economic development is not taking place. The author examines the role of sub-imperial states such as Nigeria and South Africa.

51 LE GAL, FRANCIS. "Le Néocolonialisme Face à l'Indépendance des Peuples d'Afrique Noire et de Madagascar." Cahiers du Communisme, 10, October 1970, 84-98.

52 _____. "Pour une Véritable Coopération avec les États d'Afrique Noire et Madagascar." Cahiers du Communisme, 5, May 1971, 92-102.

53 LEGUM, COLIN, (ed.). Africa Contemporary Record. London: Africa Research. New York: Africana, 1979. Index, maps, tabls.
    This annual publication includes essays on "Current Issues," a collection of documents, and a "Country-by-Country Review." There are usually essays on the USA, USSR, PRC, France, EEC, Commonwealth, and the OAU.

54 LIEBENOW, J. GUS. "Africa in World Affairs." In: Martin, Phyllis M. and O'Meara, Patrick, (eds.). Africa. Bloomington, Ind., 1977, 395-414.
    A general introduction to African international relations in a textbook designed as an introduction to African studies.

55 LONDON, KURT, (ed.). New Nations in a Divided World: The International Relations of the Afro-Asian States. New York: Praeger, 1963. 336 p., tabls.
    Contains essays by Blyden on neutralism, Cattell on Soviet policy, Li on China's economic impact, Nutter on Soviet economic policy, and Yakobsen on the USSR and Ethiopia.

56 MARCILIO, F. "Comunidade Internacional e Estados Africanos." Rivista de Direito Público e Ciênca Politica, 7, 1, January-April 1964, 5-28.
    The international community has increased in size with the independence of the African states. What are the effects of this expansion on international law?

57 MARCUM, JOHN. The Angolan Revolution. Cambridge, Mass.: M. I. T. Vol. 1, 1969, 380 p. Vol. 2, 1978, 473 p., index.
Chapters on US reactions to the rebellion against Portugal and NATO policy are in vol. 1. In vol. 2 Pan-Africanism and the problems of exile are considered.

58 MARVIN, DAVID K. Emerging Africa in World Affairs. San Francisco: Chandler, 1965. 314 p.
Sections titled "The African Economies and the Outside World" and "Africa in World Politics" contain numerous selections on international trade, neutralism, the UN, and the Katanga crisis. African integration is also included.

59 MATTHIES, VOLKER. Schwarzafrika: Politische Konflickte und Entwicklungskrisen. Opladen: Leske, 1971. 112 p., bibl., map.

60 MAZRUI, A. A. "African International Relations." In: Paden, John and Soja, Edward, (eds.). The African Experience, Vol. 1. Evanston, Ill., 1970, 532-45.
A brief survey in a textbook designed as an introduction to African studies.

61 _____. Africa's International Relations: The Diplomacy of Dependency and Change. Boulder, Col.: Westview. London: Heinemann, 1977. 310 p., index.
A major statement covering a wide variety of topics by one of the most published of African political scientists.

62 _____. On Heroes and Uhuru-Worship. London: Longmans, 1967. 264 p., index.
Contains several essays on African unity, East African integration, Egypt's Africa policy, Nkrumah, the UN, African relations with the Third World, and other subjects.

63 _____. Towards a Pax Africana. London: Weidenfeld and Nicholson. Chicago: University of Chicago, 1967. 287 p., index.
A collection of Mazrui's essays, some of which have been published elsewhere, and most of which are concerned with African international relations. Among the topics included are human rights, Pan-Africanism, non-alignment, the OAU, and Ghana's role in African diplomacy.

64 _____ and Patel, Hasu H., (eds.). Africa in World Affairs: The Next Thirty Years. New York: Third Press, 1973. 265 p., figs., index, tabls.
A collection of papers presented in 1969 at a conference funded by the World Law Fund and the Carnegie Endowment. See references under Edmundson, Shaw, Tandon, Gordenker, Mazrui, Nsibambi, Seidman, and Ghai.

65 McGOWAN, PATRICK J. and SMITH, DALE L. "Economic Dependency in Black Africa: An Analysis of Competing Theories." International Organization, 32, 1, Winter 1978, 179-235.
    The authors conduct quantitative analyses of three models of the role of Third World states in international relations.

66 McKAY, VERNON. Africa in World Politics. New York: Harper and Row, 1963. 468 p., bibl., index, tabls.
    "This book . . . . highlights the major African issues in international relations." Major sections are "(1) the impact of the United Nations on Africa, (2) Pan-African, Afro-Asian, and Eurafrican movements, (3) the policies of India and the Soviet Union on Africa, and (4) American interests and policies in Africa."

67 _____ , (ed.). African Diplomacy: Studies in the Determinants of Foreign Policy. New York: Praeger for the Johns Hopkins University. London: Pall Mall, 1966. 210 p.
    Contains eight essays by outstanding US Africa experts of that time. Conflict, ideology, trade, military considerations are included.

68 _____ . "Research Needs." In: McKay, Vernon, (ed.). African Diplomacy: Studies in the Determinants of Foreign Policy. New York, 1966, 177-210.
    The editor discusses topics in need of further study.

69 MELADY, THOMAS P. "The Impact of Africa on Recent Developments in the Roman Catholic Church." Race, 1, 2, October 1965, 147-56.
    "Africa has forced Catholicism to face courageously the consequences of something it has preached for centuries- the Universal Church."

70 The Middle East and Africa, 1978-1979. New York: Unipub, R. R. Bowker, 1978. 936 p. London: Europa, 1978. 951 p.
    This reference volume, now in its 25th edition, contains descriptions of individual states, statistics, information on regional organizations, directories of diplomatic corps, etc.

71 MOORE, CLARK D. and DUNBAR, ANN, (eds.). Africa: Yesterday and Today. New York: Bantam, 1968.
    Contains sections on Pan-Africanism, US-Africa relations, and the UN. Chapters consist of excerpts from sources published elsewhere.

72 MORRISON, D. G., et al. Black Africa: A Comparative Handbook. New York: Free Press, 1972. 483 p., figs., maps, tabls.

73 NEUBERGER, BENYAMIN. "The African Concept of Balkanisation." *Journal of Modern African Studies*, 14, 3, September 1976, 523-29.

74 NICOL, D. "Toward a World Order: An African Viewpoint." *Daedalus*, 95, 2, Spring 1966, 674-93.
    Because of their colonial past, the African states have a unique bicultural view which allows them to make a significant contribution to international affairs.

75 NKRUMAH, KWAME. *Neo-Colonialism, the Last Stage of Imperialism*. New York: International Publishers, 1965. 280 p., index.
    A statement on the continuing control of Africa by external, capitalist powers.

76 OKUMA, J. J. "The Place of African States in International Relations." In: Schou, A. and Brundtland, A., (eds.). *Small States in International Relations*. Stockholm, 1971, 147-65.
    "To investigate the place of African states in international politics it is necessary to look into the general nature of . . . the colonial situtation."

77 OLA, OPEYEMI. "The New Africa: Beyond the Nation-State." *Civilisations*, 26, 3/4, 1976, 216-346.

78 PADELFORD, NORMAN J. and EMERSON, RUPERT, (eds.). *Africa and World Order*. New York: Praeger, 1963. 153 p., bibl., figs., map, tabl.
    Contains essays by Emerson on Pan-Africanism, Holmes on the Commonwealth, Hoffman on the Congo Operation, Henry on the UN role in African development, African behavior at the UN by Spencer, and integration movements in Africa by Kloman. All of the essays appeared originally in a special edition of *International Organization*.

79 PAOLOZZI, URSULA. *Communism in Sub-Saharan Africa: An Essay with Bibliographic Supplement*. Washington: Center for Research in Social Systems, 1969. 44 p.

80 RIVKIN, ARNOLD. *Africa and the West: Elements of Free World Policy*. New York: Praeger, 1962. 241 p., index.
    African economic development and international relations. EEC, USA, and Israeli relations and interests are considered. Arms sales, the Congo crisis, and trust territories are given special attention.

81 _____. *The African Presence in World Affairs: National Development and Its Role in Foreign Policy*. New York: Free Press, 1963. 304 p., index.
    The concept of neutralism is given particular attention.

82 ROBERTSON, CHARLES L. "Africa: Independence and Ordeal."
In: Robertson, C. L. International Politics Since World
War II: A Short History. 2nd. ed., New York, 1975, 347-74.
Largely a review of post-1945 events.

83 RUBIN, LESLIE and WEINSTEIN, BRIAN. Introduction to African
Politics: A Continental Approach. New York: Praeger, 1977.
[First edition, 1974.]. 337 p., bibl., index, tabls.
This introduction to the study of African politics contains chapters on inter-African relations and Africa in world politics.

84 SATHYAMURTHY, T. V. "New States and the International Order.
The Experience of East African States: Theoretical and
Methodological Considerations." Genève-Afrique, 12, 1, 1973, 63-82.
Depicts the process by which the new states of East
Africa have defined themselves as states in the international environment.

85 SHAW, ROBERT B. and SKLAR, RICHARD L. A Bibliography for
the Study of African Politics, Vol. 1. Los Angeles:
African Studies Center, University of California, 1973.
206 p., index.
An excellent collection of references. Volume II compiled by A. Solomon.

86 SHAW, T. M. "Discontinuities and Inequalities in African
International Politics." International Journal, 30, 3,
Summer 1975, 369-90.
An overview of the changing pattern of African international relations. Africa is today a complex, mixed actor system.

87 _____. "Inequalities and Interdependence in Africa and
Latin America." Cultures et Développement, 10, 2, 1978, 231-63.

88 _____ and GRIEVE, MALCOLM. "Dependence or Development:
International and Internal Inequalities in Africa." Development and Change, 8, 3, July 1977, 377-408.
A review article discussing the various models of relations between rich and poor states and the relevance of these for development plans.

89 _____ and NEWBURY, M. CATHARINE. "Dependence or Interdependence: Africa in the Global Political Economy." In:
DeLancey, M. W., (ed.). Aspects of International Relations
in Africa. Bloomington, Ind., 1979, 39-89.
"This chapter will examine Africa's place in world
politics and in the global political economy."

90 SHEPHERD, GEORGE W., JR. The Politics of African Nationalism: Challenge to American Policy. New York: Praeger, 1962. 244 p., bibl., index, map.

Although primarily a collection of case studies of selected countries, there are chapters on the role of communist states in Africa, African activities at the UN, and US Africa policy.

91 SKURNIK, W. A. E. Sub-Saharan Africa: A Guide to Information Services. Detroit: Gale Research, 1977. 130 p.

Each chapter begins with a bibliographic essay and concludes with a bibliographic list. Chapter titles are "Pan-Africanism, Unity, and Foreign Policy;" "Western Europe and Africa;" "The United States and Africa;" "The Socialist Countries and Africa;" "African Liberation Movements;" and "Reference Works."

92 SOGSTAD, K. "Utviklingen av Militaere Styrker og Muligheter for Rustningskontroll i Afrika." Internasjonal Politikk, 2, 1967, 103-39.

"Military Power Development and the Possibilities of Armament Control in Africa." Arms control will be difficult in Africa because prestige is represented by military forces.

93 SOLOMON, ALAN C. A Bibliography for the Study of African Politics, Vol. 2. Waltham, Mass.: Crossroads, 1977. 193 p., index.

An excellent collection of references. Volume I compiled by R. Shaw and R. Sklar.

94 SPIRO, HERBERT J. Politics in Africa: Prospects South of the Sahara. Englewood Cliffs, N. J.: Prentice-Hall, 1962. 183 p., bibl., index, map, tabls.

Although largely a study of domestic politics in selected states, there are chapters on US Africa policy, African states in the UN, and Africa's role in world politics.

95 Die Staaten Afrikas und Asiens: Innere Entwicklung, Aussenpolitik. Frankfurt: Marxistische Blätter, 1971. 255 p., bibl.

96 TANDON, Y., (ed.). Readings in African International Relations. Nairobi: East African Literature Bureau, 1972. Vol. I, 380 p. 1974. Vol. II, 270 p.

Contains a large number of previously published essays on a wide range of topics related to Africa in the world from the colonial through the modern era.

97 TETZLAFF, RAINER. "Staat und Klasse in Peripher-kapitalistischen Gesellschaftsformationen: Die Entwicklung des Abhängigen Staatskapitalismus in Afrika." Verfassung und Recht in Übersee, 10, 1, 1977, 43-77.
   Class conflict in Africa and the international ties of capitalists in the developed and underdeveloped states. "State and Class in Peripheral Capitalist Social Formations: The Development of Dependent State Capitalism in Africa."

98 THOBHANI, AKBARALI H. "The Mercenary Menace." Africa Today, 23, 3, July-September 1976, 61-8.
   The increasing number of US nationals involved is poisoning the African-American relationship and leading to further internationalization of conflict in Africa.

99 THORNE, C. T., JR. "External Political Pressures." In: McKay, Vernon, (ed.). African Diplomacy: Studies in the Determinants of Foreign Policy. New York, 1966, 145-76.
   Africa in the international arena - the western states the communist states, the UN, and Afro-Asia.

100 VENGROFF, RICHARD. "Dependency, Development, and Inequality in Black Africa." African Studies Review, 20, 2, September 1977, 17-26.

101 WALLERSTEIN, I. "Semi-peripheral Countries and the Contemporary World Crisis." Theory and Society, 3, 4, Winter 1976, 461-84.
   Nigeria, Zaire, and South Africa are among the semi-peripheral states considered.

102 ZARTMAN, I. WILLIAM. "Africa." In: Rosenau, James N.; Thompson, K. W.; and Boyd, G., (eds.). World Politics. New York, 1976, 569-94, map, tabls.
   An introduction to the African international system. Written for use in introductory international relations courses.

103 _____. "Africa as a Subordinate State System in International Relations." International Organization, 21, 3, Summer 1967, 545-64. Also in: Doro, M. E. and Stultz, N. M., (eds.). Governing in Black Africa. Englewood Cliffs, N. J., 1970, 324-41. Also in: Falk, R. A. and Mendlovitz, S. H., (eds.). Regional Politics and World Order. San Francisco, 1973, 384-97.
   An important application of systems analysis to the study and description of African international relations.

104 ZARTMAN, I. WILLIAM. "Coming Political Problems in Africa." In: Whitaker, J. S., (ed.). <u>Africa and the United States: Vital Interests</u>. New York, 1978, 87-119.

    Domestic and international problems today and into the 1980s are considered.

105 _____. <u>International Relations in the New Africa</u>. Englewood, Cliffs, N. J.: Prentice Hall, 1966. 175 p., index, maps, tabls.

    An examination of "the development of foreign relations among the new states of North and West Africa, 1956-1965."

African States Foreign Policies

106 'ABBĀS, FU'ĀD IBRĀHĪM. "Ifrīqīyah wa-al-Thawrah al-Filastīnīyah." Al-Kātib, 169, April 1975, 8-17.
   What factors have led to the increased interest of the African states in the Palestinian question?

107 ADAMOLEKUN, 'LADIPO. "The Foreign Policy of Guinea." In: Aluko, O., (ed.). The Foreign Policies of African States. London, 1977, 98-117.
   One of eleven case studies in this volume.

108 ADEBISI, B. "Nigeria's Relations with South Africa, 1960-1975." Africa Quarterly, 16, 3, 1977, 67-89.

109 ADELAJA, KOLA, (ed.). Perspectives on African Foreign Policy; Proceedings of the Inaugural Conference of the Nigerian Society of International Affairs, Held at the University of Ibadan, September 1971. Ile-Ife: University of Ife, 1973. 240 p.

110 "Africa Speaks to the United Nations: A Symposium of Aspirations and Concerns Voiced by Representative Leaders at the United Nations." International Organization, 16, 2, Spring 1962, 303-30. Also in: Padelford, N. J. and Emerson, R., (eds.). Africa and World Order. New York, 1963, 35-62.
   Interviews with 18 African representatives at the UN.

111 AGUDA, OLUWADARE. "Arabism and Pan-Arabism in Sudanese Politics." Journal of Modern African Studies, 11, 2, June 1973, 177-200.
   Sudan's policies have given it recognition as a true Arab state.

112 AGYEMAN, OPOKU. "Kwame Nkrumah and the Congo (Zaire) Revisited." African Review, 4, 4, 1976, 531-42.

113   AHIDJO, AHMADOU. In Defense of Peace, Justice and Solidarity in International Society. Yaounde: c. 1971. 74 p., ill.
A collection of speeches on international politics by the President of Cameroon. Included are essays on the OAU and the UN.

114   AHMAD, NĀZLĪ MU'AWWAD. "Diblūmāsīyat Nayjiryā 'alā al-Masrah al-Ifrīqī." Al-Siyāsah al-Duwalīyah, 37, July 1974, 141-52.
Nigeria's African diplomacy and Nigerian-Arab relations.

115   AHMED, JAMEL MOHAMMED. "The Sudan in World Economy." Studia Diplomatica, 29, 3, 1976, 365-70.

116   AKINDELE, R. A. "The Conduct of Nigeria's Foreign Relations, 1960-1975." International Problems, 12, 3/4, October 1973, 46-65.
"An analytical review of the conduct of Nigeria's foreign relations in the first decade after independence suggests the existence of a new direction in the interpretation of Nigeria's national interest and a new style in the conduct of foreign policy since 1967."

117   _____. "Kwame Nkrumah - A Man and His Foreign Policy." Queen's Quarterly, 79, 3, Autumn 1972, 379-87.
Nkrumah was a true African nationalist and his ideas were ahead of his time. But, his means were not as appropriate as his ends.

118   _____. "Nigerian Parliament and Foreign Policy, 1960-1966." Quarterly Journal of Administration, 9, 3, April 1975, 270-91.
"This article examines the hypothesis that the House of Representatives . . . acted more as an ineffective rubber stamp manipulated at will by the executive."

119   _____. "On the Operational Linkage of External and Internal Dimensions of Balewa's Foreign Policy: A Review Article." Odu: A Journal of West African Studies, N. S. 12, July 1975, 110-22.
A review of the major works by Idang and Akinyemi on Nigerian foreign policy.

120   AKINSANYA, ADEOYE. "The Afro-Arab Alliance: Dream or Reality." African Affairs, 30, October 1976, 511-29.
The Afro-Arab alliance could be a powerful political and economic factor. The Entebbe raid stressed to African states the view of Israel as a threat to African territorial integrity.

121 AKINYEMI, A. B. Foreign Policy and Federalism: The Nigerian Experience. Ibadan: Ibadan University, 1974. 217 p., bibl., index.
   Case study of Nigerian foreign policy, 1958-1966. Chapters on the Congo crisis, relations with Ghana, Afro-Arab relations, Pan-Africanism, and Fernando Poo/Cameroon are included.

122 _____, (ed.). Nigeria and the World: Readings in Nigerian Foreign Policy. Ibadan: Oxford University, 1978. 152 p., bibl.
   Papers from a conference conducted in 1976.

123 AKPAN, M. E. "African Goals and Strategies toward Southern Africa." African Studies Review, 14, 2, September 1971, 243-63.
   African goals have not been reached in spite of various strategies employed. Material power, and the African lack of it, is the cause of this failure.

124 ALALADE, F. O. "French-speaking Africa - France Relations: A Critical Bibliographical Survey with Particular Reference to Ivory Coast." A Current Bibliography on African Affairs, 9, 1, 1976/1977, 84-93. Repeated: 9, 4, 1976/1977, 325-34.
   A survey of the literature with suggestions for further research.

125 ALUKO, OLAJIDE. "The Civil War and Nigerian Foreign Policy." Political Quarterly, 42, 2, April-June 1971, 177-90.
   Discusses the effect of the civil war on the three main components of Nigeria's foreign policy: nonalignment, African affairs, and West African relations.

126 _____. "The Determinants of the Foreign Policies of African States." In: Aluko, O., (ed.). The Foreign Policies of African States. London, 1977, 1-23.
   A summary based on the findings of the eleven case studies included in the volume.

127 _____, (ed.). The Foreign Policies of African States. London: Hodder and Stoughton, 1977. 243 p.
   Contains essays on Algeria, Egypt, Ethiopia, Ghana, Guinea, Ivory Coast, Kenya, Nigeria, Tanzania, Zambia, and Zaire. A summary essay by the editor is included.

128 _____. "The Foreign Service." Quarterly Journal of Administration, 5, October 1970, 33-52.
   "We shall trace the development of the Service and identify some of the administrative problems which it has had to cope with." A Nigerian case study.

129   ALUKO, O. "Ghana's Foreign Policy." In: Aluko, O., (ed.).
      The Foreign Policies of African States. London, 1977, 72-97.
            One of three essays contributed by Aluko to this volume.
      He analyzes both internal and external influences.

130   _____. "Ghana's Foreign Policy under the National Liberation Council." Africa Quarterly, 10, 4, January-March 1971, 312-28.
            The NLC had different foreign policy objectives and utilized different methods than the Nkrumah regime.

131   _____. "The 'New' Nigerian Foreign Policy: Developments Since the Downfall of General Gowon." Round Table, 264, October 1977, 405-14.
            The post-Gowon government has a more dynamic foreign policy.

132   _____. "Nigeria and Britain after Gowon." African Affairs, 304, July 1977, 303-20.
            Anglo-Nigerian relations have basically improved since the end of the Gowon regime.

133   _____. "Nigeria and the European Economic Community." International Studies, 13, 3, July-September 1974, 465-74.
            Nigeria served as a spokesman for Africa in the negociations.

134   _____. "Nigeria and the Superpowers." Millenium: Journal of International Studies, 5, 2, Autumn 1976, 127-41.
            "The two superpowers have significant interests to promote in Nigeria which neither could easily ignore."

135   _____. "Nigerian Foreign Policy." In: Aluko, O., (ed.). The Foreign Policies of African States. London, 1977, 163-95, tabls.
            The author, a lecturer at the University of Ife, stresses Nigeria's relations with African states and its non-alignment policy.

136   _____. "Nigeria's Role in Inter-African Relations with Special Reference to the Organisation of African Unity." African Affairs, 72, 287, April 1973, 145-62.
            Since the completion of the civil war Nigeria has played a more active and leading role in continental and OAU affairs.

137   _____. "Public Opinion and Nigerian Foreign Policy under the Military." Quarterly Journal of Administration, 7, 3, April 1973, 253-69.
            Public opinion plays only a minor role in Nigerian foreign policy. Compare with the findings of M. Peil as reported in her book on Nigerian public opinion.

138 ALUKO, O. "The Role of Ghana in the Rhodesian Question, 1961-66." Quarterly Journal of Administration, 6, 3. April 1972, 301-21.
Ghana's initiatives at the UN and in the Commonwealth have caused Britain to put pressure on Rhodesia.

139 ANABTAWI, S. N. "The Afro-Asian States and the Hungarian Question." International Organization, 17, 4, Autumn 1963, 872-900.
An analysis of UN voting patterns helps to delimit the characteristics of the Afro-Asian neutral bloc.

140 ANGLIN, DOUGLAS G. "Confrontation in Southern Africa: Zambia and Portugal." International Journal, 25, 3, Summer 1970, 497-517.
A description of the dynamics of Zambia's relations and objectives in Southern Africa.

141 _____. "Nigeria: Political Non-alignment and Economic Alignment." Journal of Modern African Studies, 2, 2, July 1964, 247-64.
Economic alignment with the West has not meant political alignment, for Nigeria has clearly shown her independence in this respect.

142 _____. "Zambia and Southern African 'Détente'." International Journal, 30, 3, Summer 1975, 471-503.
Kaunda of Zambia prefers a negotiated settlement in Rhodesia and he sought to test the validity of Vorster's policy of detente.

143 _____. "Zambia and the Recognition of Biafra." African Review, 1, 2, September 1971, 102-36.
Recognition was a symbolic, humanitarian measure, not an approval of secession. Zambia did provide a variety of supports to Biafra.

144 _____ and SHAW, TIMOTHY M. Zambia's Foreign Policy: Studies in Diplomacy and Dependence. Boulder, Col.: Westview, 1979. 300 p., bibl., figs., index, tabls.

145 ANSPRENGER, FRANZ. "Communism in Tropical Africa." In: Hamrell, Sven and Widstrand, Carl G., (eds.). The Soviet Bloc, China and Africa. Uppsala, 1964, 75-100, figs.
Communism is of only secondary concern to Africans who are mainly worried by imperialism and neo-colonialism.

146 _____. "Les Etats Africains et le Pouvoir Blanc en Afrique." Études Internationales, 1, 4, December 1970, 17-34.
Describes the campaign of black African states to bring changes to white-dominated Southern Africa.

147 ARIKPO, OKOI. "Nigeria and the Organisation of African Unity." Nigerian Journal of International Affairs, 1, 1, July 1975, 1-11. Also in: Quarterly Journal of Administration, 9, 1, October 1974, 49-59.
   The text of a lecture presented on 31 July 1974 by the Commissioner for External Affairs of Nigeria.

148 ARNOLD, GUY. Modern Nigeria. London: Longman, 1977. 192 p., ill., index, maps.
   See "Foreign Affairs," pp. 133-43, and "Neocolonialism", pp. 152-62. The author is a journalist.

149 AUDA, ABDEL M. "Pan-Africanism in Egyptian National Culture." L'Egypte Contemporaine, 343, January 1971, 17-40.

150 AYELE, NEGUSSAY. "The Foreign Policy of Ethiopia." In: Aluko, O., (ed.). The Foreign Policies of African States. London, 1977, 46-71.
   One of eleven case studies in this volume. Covers the period until the end of Selassie's reign.

151 BAILEY, MARTIN. "Les Relations Extérieures de Zanzibar." Revue Française d'Études Politiques Africaines, 75, March 1972, 65-84.
   Mainly concerned with the union of Tanganyika and Zanzibar.

152 BAKRI, A. K. "The Economic Factor in African-Arab Relations." Current Bibliography on African Affairs, 9, 3, 1976/1977, 213-27.
   Since the 1973 October War the Africans and Arabs have come closer to establishing an international sub-system based on cooperation in Africa and the Middle East.

153 BALEWA, ABUBAKAR T. "Nigeria Looks Ahead." Foreign Affairs, 41, 1, October 1962, 131-41. Also in: Quigg, P. Africa. New York, 1964, 302-13.
   A statement by Nigeria's Prime Minister on Nigeria's domestic and foreign policies.

154 BALOGUN, KOLAWOLE. Mission to Ghana: Memoir of a Diplomat. New York: Vantage, 1963. 73 p.

155 BALTA, PAUL. "La Politique Africaine de l'Algérie." Revue Française d'Etudes Politiques Africaines, 132, December 1976, 54-73.
   Algeria stands in firm support of the liberation of Southern Africa, for Afro-Arab cooperation, and for the OAU.

156 BALTA, P. "La Politique Extérieure de l'Algérie. Sur la Brèche Africaine." Défense Nationale, June 1977, 87-98.
   Algeria is developing strong relations with "progessive" African states.

157 BAYART, J. F. "La Politique Extérieure du Cameroun, 1960-1971." Revue Française d'Études Politiques Africaines, 75, March 1972, 47-64.
   Diversification of relations, closer ties with the Arab world, and coolness to Pan-Africanism characterize the foreign policies of President Ahidjo.

158 PAYNE, E. A. "The Issue of Greater Somalia." American Universities Field Staff Reports, Northeast Africa, 13, 1 and 2, 1-6, 1-14, ill., map.
   Part I examines relations between the PRC and Somalia and the second part examines relations between Ethiopia and Somalia.

159 BELL, J. BOWYER. "The Sudan's African Policy: Problems and Perspectives." Africa Today, 20, 3, Summer 1973, 3-12.
   The government has "turned from Arab matters toward a new involvement in Africa."

160 BENEDICT, BURTON, (ed.). Problems of Smaller Territories. London: Athlone for the Institute of Commonwealth Studies, 1967. 153 p., bibl., maps, tabls.
   In additon to essays on general aspects of the problem, there are case studies. See the essay by J. E. Spence on the High Commission territories.

161 BENGU, S. M. E. Chasing Gods Not Our Own. Petermaritzburg, S. Africa: Shuter and Shooter, 1975. 170 p., bibl.
   The original title of this dissertation was African Cultural Identity and International Relations: Analysis of Ghanaian and Nigerian Sources, 1958-1974. "The purpose of our research is to analyse post-independence African writings with a view to determining what independence has meant to the Africans up to now, how African identity issues influence relations between the so-called independent African states and their former metropoles, and how these relations, in turn, are influencing and will influence the entire international scene."

162 BLACK, J. E. and THOMPSON, K. W., (eds.). Foreign Policies in a World of Change. New York: Harper and Row, 1963. 756 p., maps.
   Contains essays by Coleman on Nigeria, Kiano on East Africa, and Webb on the Union of South Africa.

163 BLYDEN, EDWARD W. III. "The Idea of African 'Neutralism' and 'Non-alignment': An Exploratory Survey." In: London, Kurt, (ed.). New Nations in a Divided World: The International Relations of the Afro-Asian States. New York, 1963, 146-60.
Although "neutralism" and "non-alignment" are borrowed by Africans, they have added a distinctly African meaning to these concepts.

164 BONE, MARION. "The Foreign Policy of Zambia." In: Boorston, Ronald P., (ed.). The Other Powers: Studies in the Foreign Policies of Small States. London, 1973, 121-53, tabls.
"Seven years is a short period over which to study Zambia's foreign policy, but the outlines have already been traced with remarkable clarity as the country has made considerable impact on African and world affairs."

165 "Botswana between Black and White Africa." Swiss Review of World Affairs, 22, 5, August 1972, 20-2.
In spite of economic dependence on South Africa, Botswana follows an independent foreign policy. In part, this is possible because of South Africa's desire to make the "outward policy" work.

166 BOUTROS-GHALI, BOUTROS. "The Foreign Policy of Egypt." In: Aluko, O., (ed.). The Foreign Policies of African States. London, 1977, 41-5.
A brief note on Egypt's relations with Africa and the Arab world.

167 _____. "Réflexions sur le Dialogue Afro-Arabe (Reflections on the Afro- Arabian Dialogue)." Revue Egyptienne de Droit International, 32, 1976, 155-64.

168 BOUVIER, P. "L'Afrique à l'Heure Algérienne: Les Chances du Rapprochement Arabo-Africain." Studia Diplomatica, 28, 3, 1975, 305-25.
The chances for an Afro-Arab alliance are analyzed.

169 BRAUN, D. "The Indian Ocean in Afro-Asian Perspective." In: University of Southampton, The Indian Ocean in International Politics. Southampton, 1973, 179-94.
The littoral states distrust the major powers and dislike interference in their affairs.

170 BREYTENBACH, W. J. "Black Africa and the Arabs." Bulletin of the Africa Institute of South Africa, 12, 10, 1974, 432-35.
Discussion of the possible rifts in Afro-Arab solidarity.

171 BRIONNE, BERNARD. "Le Nigéria, Grande Puissance Africaine." Défense Nationale, 29, December 1973, 63-78.
   A brief note on Nigeria's economic and demographic strength as an African power.

172 BROWN, IRENE. "Studies on Non-alignment." Journal of Modern African Studies, 4, 4, December 1966, 517-27.
   Review of two books on non-alignment, Cecil Crabb's The Elephants and the Grass (1965), and G. H. Jansen's Nonalignment and the Afro-Asian States (1966).

173 BUHLMAN, MARGARET A. and TOMLIN, BRIAN W. "Relative Status and Foreign Policy: Status Partitioning and the Analysis of Relations in Black Africa." Journal of Conflict Resolution, 21, 1, March 1977, 187-216.

174 CARRUTHERS, OLIVER. "The Algebra of African Foreign Policy: A Need for Regular Review." Round Table, 270, April 1978, 182-90.

175 CERVENKA, ZDENEK. "The Afro-Arab Alliance." Africa, 31, March 1974, 76-9.
   Africa has allied with the Arab states because of disillusionment with the western states.

176 _____. "The Emergence and Significance of the African-Arab Solidarity." Instant Research on Peace and Violence, 4, 2, 1974, 102-9.
   The 1973 OAU Heads of State Meeting marks an important swing to the Arab side by the African states.

177 _____. "Swaziland's Links with the Outside World." In: Cervenka, Z., (ed.). Land-locked Countries of Africa. Uppsala, 1973, 263-70.

178 CHAUDHRI, MOHAMMED AHSEN. "Foreign Policy of Sudan under President Numeiri." Pakistan Horizon, 28, 4, 1975, 19-52.

179 CHAULEUR, PIERRE. "La Guinée de M. Sekou Touré." Etudes, November 1977, 437-53.
   The return of good relations between Guinea and France is caused by economic factors.

180 _____. "Les Problèmes du Nigeria." Etudes, June 1978, 737-51.
   By population and economic power Nigeria should play an important role in West African and international politics, but domestic problems have prevented this.

181 CHERNYSHOV, A. "Peace and Security for the Indian Ocean." International Affairs (USSR), 12, December 1976, 42-51.

182 CHIKH, S. "L'Algérie et l'Afrique 1954-1962." Revue Algérienne des Sciences Juridiques, Politiques et Economiques, 5, 3, September 1968, 703-46.
  Algeria's Africa policy is a product of her long war of liberation.

183 CHIPEMBERE, HARRY B. M. "Malawi's Growing Links with South Africa: A Necessity or a Virtue?" Africa Today, 18, 2, March/April 1971, 27-47.

184 CHOURCRI, N. "The Nonalignment of Afro-Asian States: Policy, Perception, and Behavior." Canadian Journal of Political Science, 2, 1, March 1969, 1-17.

185 CLAPHAM, C. "Sub-Saharan Africa." In: Clapham, C., (ed.). Foreign Policy Making in Developing States: A Comparative Approach. Farnborough, 1977, 75-109, bibl., map, tabls.

186 CLAUSEN, U. "Die Afro-Arabische Komponente in der Mauretanischen Aussenpolitik." Orient, 18, 2, June 1977, 67-86.
  "The Afro-Arab Component in Mauritania's Foreign Policy." The Africa policy of Mauretania and the effects of the claims on Western Sahara are examined.

187 COLA ALBERICH, J. "La Conferencia de Lusaka." Revue Politica Internacional, 112, November/December 1970, 139-52.
  Analysis of the agreements reached at the Conference and the means of their implementation.

188 COLEMAN, JAMES S. "The Foreign Policy of Nigeria." In: Black, Joseph E. and Thompson, Kenneth W., (eds.). Foreign Policies in a World of Change. New York, 1963, 379-406, tabl.
  After discussing the domestic political background, the author describes the main features of Nigeria's foreign policy.

189 COLLINS, JOHN N. Foreign Conflict Behavior and Domestic Disorder in Africa. Syracuse, N. Y.: Program of Eastern African Studies, Syracuse University, 1971. 128 p., bibl.
  A quantitative analysis of the hypothesis that there is a relationship between domestic disorder and foreign conflict behavior.

190 CONSTANTIN, FRANÇOIS and COULON, CHRISTIAN. "Des Casernes aux Chancelleries: La Variable Militaire dans la Politique Extérieure de Trois États Africains: Haute-Volta, Togo, Mali." Canadian Journal of African Studies, 9, 1, 1975, 17-49, tabls.
  Do military take-overs change the international position of the states in which they occur? Authors conclude that no change occurs.

191 COPSON, RAYMOND W. "East Africa and the Indian Ocean - A 'Zone of Peace'?" African Affairs, 304, July 1977, 339-58.
    The East African states concept of a zone of peace in the Indian Ocean is no longer realistic. Those states must reexamine and adjust their policies.

192 CORNEVIN, ROGER. "La Politique Etrangère du Gouvernement de Bangui." Revue Française d'Etudes Politiques Africaines, 10, 117, September 1975, 105-13.

193 _____. "La Politique Extérieure du Togo." Revue Française d'Etudes Politiques Africaines, 82, October 1972, 50-79.
    Comparison of Togo's foreign policy under three regimes.

194 _____. "La Politique Extérieure du Zaire." Revue Française d'Etudes Politiques Africaines, 79, July 1972, 58-70.
    The foreign policy of General Mobutu, dynamic and unclear.

195 COULON, C. "Les Relations entre l'Afrique Noire et le Reste du Monde." Année Afrique 1973, 1973, 172-223; 1974, 139-62; 1975, 97-115.
    African foreign relations in 1973 showed a marked tendency to broaden ties beyond the ex-colonial power. This annual survey examines relations with ex-colonial powers and the communist states in particular.

196 COWAN, L. GRAY. "Nigerian Foreign Policy." In: Tilman, Robert O. and Cole, Taylor, (eds.). The Nigerian Political Scene. Durham, N. C., 1962, 115-43.
    "Concerned with the basic considerations that play a part in foreign policy in Nigeria . . . the principal positions adopted . . . [and] to what extent these positions are reflections of . . . parties and of press and public opinion."

197 _____. "Political Determinants." In: McKay, Vernon, (ed.). African Diplomacy: Studies in the Determinants of Foreign Policy. New York, 1966, 119-44.
    An analysis of foreign policy influences from African domestic and inter-African politics.

198 CRABB, CECIL V. The Elephants and the Grass: A Study of Non-alignment. New York: Praeger, 1965. 237 p., bibl.

199 DALE, RICHARD. Botswana and Its Southern Neighbor: The Patterns of Linkage and the Options in Statecraft. Athens, Ohio: Ohio University, Center for International Studies, 1970. 22p.
    Although dependent upon South Africa, Botswana exerts an independent foreign policy.

200 DALE, R. "Botswana's First Decennium. An Inquiry into the Continuity and Change in the Symbols and Substance of African Independence, 1966-1976." Cultures et Développement, 9, 3, 1977, 455-76.
   Domestic and international policies of Botswana are greatly influenced by the existence of neighboring apartheid states.

201 _____. "The Challenges and Restraints of White Power for a Small African State: Botswana and Its Neighbors." Africa Today, 25, 3, July-September 1978, 7-23.
   Discussion of Botswana's place in an international hierarchy and its role of clientage - from a historical perspective.

202 _____. "President Sir Seretse Khama, Botswana's Foreign Policy, and the Southern African Subordinate State System." Plural Societies, 7, 1, Spring 1976, 69-87.

203 _____. The Racial Component of Botswana's Foreign Policy. Denver: Center on International Race Relations, University of Denver, 1971. 32 p., bibl.

204 DAVIS, M. Interpreters for Nigeria: The Third World and International Public Relations. Urbana: University of Illinois, 1977. 197 p., bibl., index, tabls.

205 DECHERF, D. "Du Non-alignement au Pan-socialisme: L'Evolution de la Politique Etrangère de la Tanzanie Expliquée par Son Contexte Régional." Politique Etrangère, 40, 5, 1975, 493-523.
   Tanzanian socialism is seen as a major influence on her East African and Chinese foreign policies.

206 _____. "Socialisme Ujamaa et Relations Extérieures de la Tanzanie." Revue Juridique et Politique Indépendance et Coopération, 30, 1, January-March 1976, 44-57.
   The influence of Tanzanian socialism on the country's foreign policy.

207 _____. "La Tanzanie entre l'Océan Indien et le Continent Sud-Africain." Défense Nationale, February 1976, 115-24.
   Analysis of the foreign policy and strategic implications for Tanzania of the Tanzam Railroad,

208 DECRAENE, P. "Esquisse d'une Nouvelle Politique Etrangère Gabonaise." Revue Française d Etudes Politiques Africaines, 90, June 1973, 58-66.
   The foreign policy of President Bongo places Gabon's national interests before any ideological matters.

209 DEI-ANANG, MICHAEL. The Administration of Ghana's Foreign Relations, 1957-1965: A Personal Memoir. London: Athlone for the Institute of Commonwealth Studies, 1975. 88 p., bibl., index.
Dr. Dei-Anang was Principal Secretary at the Ghana Ministry of Foreign Affairs (1959-1961) and Head of the African Affairs Secretariat (1961-1966).

210 _____. Ghana Resurgent. Accra: Waterville Publishing Houses, 1964. 248 p.
A personal account by an important member of Nkrumah's foreign affairs establishment.

211 DELANCEY, MARK W. "Nigerian Views of the Angolan Conflict." Bulletin of the Southern Association of Africanists, 4, 2, Summer 1976, 39-54.
Analysis of various Nigerian writers' views of the Angolan Civil War.

212 DELORME, NICOLE. "The Foreign Policy of the Ivory Coast." In: Aluko, O., (ed.). The Foreign Policies of African States. London, 1977, 118-35.
A tribute to the policy of Houphouët-Boigny that concludes that Ivory Coast policy is based on neutralism.

213 DIA, MAMADOU. African Nations and World Solidarity. New York: Praeger, 1961. 145 p.
An important statement of the Eurafrique philosophy by a Senegalese politician.

214 DIAKHITE, LAMINE. "l'Afrique Noire et le Monde Arabe; Ou les Bases Millénaires d'une Coopération." Remarques Africaines, 458/459, February/March 1975, 78-81.
Author was Senegalese ambassador to Morocco.

215 DJALILI, MOHAMMED REZA and KAPPELER, DIETRICH. "La Situation Militaire des Pays de l'Océan Indien." Politique Etrangère, 42, 5, 1977, 517-27

216 "Dr. Munyua Waiyaki, Kenyan Minister of Foreign Affairs." Africa Report, 22, 2, March/April 1977, 37-40.

217 EAST, MAURICE A. "Foreign Policy-making in Small States: Some Theoretical Observations Based on a Study of the Uganda Ministry of Foreign Affairs." Policy Sciences, 4, 4, December 1973, 491-508, tabls.
The effects of small states on world politics.

218 ELEWAINY, M. "Africa and the Fourth Arab-Israeli War." Majallat Ma'had al-Buḥūth Wa-al-Dirāsāt al-'Arabīyah, 5, June 1974, 65-79.

219 EL-KHAWAS, M. A. "Africa and the Middle Eastern Crisis." Issue, 5, 1, Spring 1975, 33-42.
Cites the increasing importance of African views on the crisis as part of the emergence of a Third World approach to international problems. Looks at Egyptian and Israeli efforts to woo African support.

220 ———. "African-Arab Solidarity: The Emergence of a New Alliance." Current Bibliography on African Affairs, 8, 2, 1975, 134-45.
Analyzes African-Arab relations since 1955 with emphasis on the implications of the Arab-Israeli conflict and the liberation of Southern Africa.

221 ERIKSEN, T. L. "Zambia: Den Vanskelige Veien mot Motokonomisk Uavhengighet." Internasjonal Politikk, 4, October-December 1975, 557-74.
Zambia's strategy for and difficulties in reducing dependence on South Africa.

222 ESSEKS, JOHN D. "Political Independence and Economic Decolonization: The Case of Ghana under Nkrumah." Western Political Quarterly, 24, 1, March 1971, 59-64.

223 ETINGER, Y. "African Countries: Anti-imperialist Foreign Policies." International Affairs (USSR), 8, August 1973, 75-81.

224 FAUJAS, ALAIN. "M. Houphouët-Boigny et la Diplomatie Ivoirienne." Revue Française d'Etudes Politiques Africaines, 68, August 1971, 23-36.
Dominant position of Houphouët-Boigny in the foreign policy of the Ivory Coast.

225 ———. "La Politique Etrangère du Niger." Revue Française d'Etudes Politiques Africaines, 72, December 1971, 41-60.
The dynamism of Diori's foreign policy means that Niger plays a surprisingly large role in international affairs.

226 ———. "La Politique Extérieure de la Haute-Volta." Revue Française d'Etudes Politiques Africaines, 83, November 1972, 59-73.
A comparison of the policies of President Yameogo and General Lamizana.

227 FISCHER, G. "La Zambie et la British South Africa Company." Revue Française de Science Politique, 17, 2, April 1967, 286-93.
In a special edition titled "Les Conflicts entre Etats et Compagnies Privées," edited by J. B. Duroselle.

228 FLORY, M. "Les Conférences Islamiques." Annuaire Français de Droit International, 16, 1970, 233-43.
     Black African states played only a minor role in the Rabat Conference.

229 "Francis Deng, Sudanese Minister of State for Foreign Affairs." Africa Report, 22, 2, March/April 1977, 13-20.

230 FROËLICH, J.-C. "L'Egypte et les Peuples Noirs." Orient, 32/33, 1964/1965, 13-28.
     Although Egypt has tried to develop good relations with black states, she has underestimated several negative factors and has not met with full success.

231 GALLAGHER, CHARLES F. "Morocco and Its Neighbors: Part III, Morocco and Mauritania." American Universities Field Staff Reports, North Africa, 13, 4, 1967, 1-12.
     What to do with Spanish Sahara? The major issue in Morocco's relations with Mauritania.

232 GAMBARI, I. A. "Nigeria and the World: A Growing Internal Stability, Wealth, and External Influence." Journal of International Affairs, 29, 2, Fall 1975, 155-69.
     Raises the questions: What are the internal and external factors responsible for the economic and political transformation of Nigeria since independence? What are the likely political and diplomatic manifestations of its new economic standing in Africa and the world?

233 GARBA, J. N. "The New Nigerian Foreign Policy." Quarterly Journal of Administration, 11, 3, April 1977, 135-46.
     A description of Nigeria's foreign policy under the military government of General Obasanjo.

235 GARDINER, R. K. A. "Race and Color in International Relations." Daedalus, 96, Spring 1967, 296-311.
     African states must reconcile their loyalties to the OAU, francophonie or the Commonwealth, and to humanity.

236 GAVRANOV, V. "Trêca Konferencija Nesvrstanih." Socijalizam, 13, 10, October 1970, 1324-31.
     Review of the Lusaka Conference of the non-aligned states.

237 GELDENHUYS, D. "Lesotho en Suid-Afrika - Studie in Waagpolitick." Politikon, 1, 1, June 1974, 28-43.
     Lesotho is heavily dependent upon South Africa but its foreign policy shows the force of the attraction of a pro-OAU orientation.

238   GHANA. INFORMATION SERVICES. <u>Nkrumah's Subversion in Africa</u>. Accra: State Publishing, 1969. 91 p., ill., maps.
      Documentary and other evidence suggesting that the Nkrumah government actively sought the overthrow of various African leaders.

239   GHEZZI, C. "La Politica Estera della Zambia." <u>Rivista di Studi Politici Internazionali</u>, <u>39</u>, 3, July-September 1972, 372-92.
      "Zambia's Foreign Policy." Geographic location causes problems for Kaunda's policy of non-alignment.

240   GINIEWSKI, PAUL. "Deux Nouvelles Indépendances Africaines: Lesotho et Botswana." <u>Politique Etrangère</u>, <u>31</u>, 4, 1966, 382-90.
      Their location next to South Africa will be a major influence on their foreign policies.

241   GITELSON, SUSAN A. "Major Shifts in Recent Ugandian Foreign Policy." <u>African Affairs</u>, 304, July 1977, 359-80.
      Uganda's foreign policy in the Amin era based on interviews and analysis of relevant literature.

242   _____. "Why Do Small States Break Diplomatic Relations with Outside Powers? Lessons from the African Experience." <u>International Studies Quarterly</u>, <u>18</u>, 4, December 1974, 451-84.
      External causes are more important than domestic factors in explaining the breaking of diplomatic relations.

243   GOGUEL, A. M. "La Diplomatie Malgache." <u>Revue Française d'Etudes Politiques Africaines</u>, 78, June 1972, 78-103.
      The very pro-West policy of Tsiranana will be relaxed by the military government.

244   GRETTON, GEORGE. "The Sudan: Caught between Asia and Africa." <u>New Middle East</u>, 7, April 1969, 26-30.

245   GRUHN, ISEBILL. "British Arms Sales to South Africa: The Limits of African Diplomacy." <u>Studies in Race and Nations</u>, <u>3</u>, 3, 1971/1972, 1-30.
      Considers the ability of small states to influence major powers through diplomacy with the attempts of African states to influence Britain as a case study.

246   GRUNDY, K. W. and FARLOW, R. L. "Internal Sources of External Behaviour: Ghana's New Foreign Policy." <u>Journal of Asian and African Studies</u>, <u>4</u>, 3, July 1969, 300-14.
      Ghana's foreign policy under the National Liberation Council was determined by four general variables - societal, idiosyncratic, systemic, and role-governmental.

247 GUELLAL, CHERIF. "Africa vis-à-vis the Western Powers." Annals of the American Academy of Political and Social Sciences, 354, July 1964, 9-21.
A general survey of the goals of African foreign policies.

248 GUITON, R. J. "Von Bandung bis Kairo. Das Zustandekommen der Afro-asiatischen Zusammenarbeit." Europa-Archiv, 16, 12, June 1961, 301-12.
The Cairo Conference of 1957 was the first opportunity for black African states to play a major role in Afro-Asian solidarity. It also was the first public display of the conflict between Nasser and Nkrumah.

249 HADSEL, F. L. "Africa and the World: Nonalignment Reconsidered." Annals of the American Academy of Political and Social Sciences, 372, July 1967, 93-104. Also in: Doro, M. E. and Stultz, N. M., (eds.). Governing in Black Africa. Englewood Cliffs, New Jersey, 1970, 342-52. Also in: Markovitz, I. L., (ed.). African Politics and Society. New York, 1970, 431-431.
Early acceptance of non-alignment was reconsidered at the end of the 1960s.

250 HAGOPIAN, ELAINE C. "Arab Nationalism, the Arab-Israeli Crisis and the African Response." Pan-African Journal, 3, Spring 1970, 78-92.

251 HALL, RICHARD. The High Price of Principles. Kaunda and the White South. Harmondsworth: Penguin, 1973. 286 p. New York: Africana, 1970. 256 p., index.
This study of Zambia's foreign policy places major emphasis on the role of the Rhodesian question.

252 HANISCH, R. "The Political Image of Ghanaian Civil Servants." Verfassung und Recht in Ubersee, 9, 3, 1976, 343-55.
A study of the international outlook of Ghanaian civil servants. Data was collected through survey research.

253 HASSAN, A. D. "Big Power Rivalry in Indian Ocean - A Tanzanian View." Africa Quarterly, 15, 3, 1976, 80-6.
The author, the High Commissioner for Tanzania in India, makes a political statement.

254 HENDERSON, ROBERT D'A. "Principles and Practice in Mozambique's Foreign Policy." World Today, 34, 7, July 1978, 276-86.

255 _____. "Relations of Neighborliness - Malawi and Portugal, 1964-74." Journal of Modern African Studies, 15, 3, September 1977, 425-55, map, tabls.
Conflict and cooperation between Malawi and Portugal.

256 HENDERSON, WILLIE. "Independent Botswana: A Reappraisal of Foreign Policy Options." African Affairs, 73, 290, January 1974, 37-49.
In spite of dire predictions, Botswana has an independent foreign policy.

257 HERMANS, H. C. L. "Botswana's Options for Independent Existence." In: Cervenka, Z., (ed.). Land-locked Countries of Africa. Uppsala, 1973, 197-211.
Botswana cannot solve alone the problems caused by its location.

258 HERSKOVITS, JEAN. "Dateline Nigeria: A Black Power." Foreign Policy, 29, Winter 1977/1978, 167-88.
Domestic and international politics of Nigeria.

259 ———. "One Nigeria." Foreign Affairs, 51, 2, January 1973, 392-407, maps.
Optimistic appraisal of internal political developments in post-civil war Nigeria. Comments on Nigeria's role in international politics.

260 HESS, ROBERT L. Ethiopia: The Modernization of Autocracy. Ithaca: Cornell University, 1970. 272 p., bibl., ill., index, maps.
See "The End of Ethiopian Isolation," pp. 197-215 on foreign relations.

261 HILL, CHRISTOPHER R. "Botswana and the Bantustans." In: Shaw, T. M. and Heard, K. A., (eds.). Cooperation and Conflict in Southern Africa: Papers on a Regional Subsystem. Washington, D. C., 1977, 339-50.
Botswana's internal problems, her foreign relations, and some comparisons between Botswana and the Bantustans.

262 ———. "Independent Botswana: Myth or Reality?" Round Table, 245, January 1972, 55-62.
Consideration of linkages between domestic and foreign policies.

263 HINZ, MANFRED O. "Guinea: Zur Rechtlich - Politischen Gestalt eines Westafrikanischen Staates." Jahrbuch de Öffentlichen Rechts der Gegenlichin Rechts der Gegenwart, 19, 1970, 501-19.

264 HIPPOLYTE, M. "Guinea's Foreign Policy." Africa Quarterly, 11, 4, January-March 1972, 302-6.

265 HODGES, TONY. "Mozambique: The Politics of Liberation." In: Carter, G. M. and O'Meara, P., (eds.). Southern Africa in Crisis. Bloomington, Ind., 1977. 48-88, figs.
 This essay analyzes Mozambique's foreign policy and the interactions of this state in the Southern Africa sub-system.

266 _____. "Omar Hadrami, Polisario Representative." Africa Report, 23, 2, March/April 1978, 39-43.

267 HOOKER, JAMES R. "The Unpopular Art of Survival." American Universities Field Staff Reports, Central and Southern Africa Series, 14, 1, 1970, 1-9.
 President Banda of Malawi has often been more friendly with white Southern African regimes than most African leaders.

268 HOSKYNS, CATHERINE. "Africa's Foreign Relations: The Case of Tanzania." International Affairs, 44, 3, July 1968, 446-62. Also in: Rweyamamu, Anthony, (ed.). Nation Building in Tanzania. Nairobi, 1970, 90-109.
 Initially Tanzania followed a conservative foreign policy, but the events of 1964 led to a radical foreign policy.

269 _____. Case Studies in African Diplomacy. Dar-es-Salaam: Oxford University, 1969. Vol. 1. The Organization of African Unity and the Congo Crisis. 75 p. Vol. 2. The Ethiopia-Somali-Kenya Dispute, 1960-1967. 91 p., map, tabl.
 Each volume contains a number of documents and brief commentary by Hoskyns.

270 _____. "The Part Played by the Independent African States in the Congo Crisis 1960-December 1961." In: Austin, D. and Weiler, H., (eds.). Inter-State Relations in Africa. Freiburg, 1965, 30-50.
 A description of African reactions and actions in the Congo crisis and an analysis of the effects of these actions on the crisis and on Pan-Africanism.

271 HOUPHOUËT-BOIGNY, FÉLIX. "On Dialogue with South Africa." Bulletin of the Institute of South Africa, 9, 6, June 1971. Also in: Roland, Joan G., (ed.). Africa: The Heritage and the Challenge. Greenwich, Conn., 1974, 487-92.
 A message sent by the President of the Ivory Coast to the Summit Meeting of the OAU in 1971 in which he defends dialogue.

272 HOWELL, JOHN. "An Analysis of Kenyan Foreign Policy." Journal of Modern African Studies, 6, 1, May 1968, 29-48.
 Kenya's foreign policy is described as radical and as a focus of national unity.

273 HOWELL, J. and HAMID, M. B. "Sudan and the Outside World, 1964-1968." African Affairs, 68, 273, October 1969, 299-315.
   Sudan's foreign policy swings from radical to conservative, but it is a direct function of the domestic political process.

274 HUGHES, ANTHONY. "J. W. Garba, Nigerian Commissioner for External Affairs." Africa Report, 21, 6, November/December 1976, 2-5.

275 _____. "Malawi and South Africa's Co-prosperity Sphere." In: Cervenka, Z., (ed.). Land-locked Countries of Africa. Uppsala, 1973, 212-32, tabls.
   Malawi's foreign policy has accepted the existence of apartheid.

276 _____ and SCHULTZ, BONNIE J. "Ambassador Shirley Yema Gbujama." Africa Report, 22, 1, January/February 1977, 37-40.

277 IDANG, GORDON J. Nigeria: Internal Politics and Foreign Policy, 1960-1966. Ibadan: Ibadan University, 1973. 171 p., bibl., index, tabls.
   Description of Nigeria's foreign policy (1960-1966) and analysis of the policy process. The roles of parliament, parties, public opinion, and the mass media are examined.

278 _____. "The Politics of Nigerian Foreign Policy: The Ratification and Renunciation of the Anglo-Nigerian Defence Agreement." African Studies Review, 13, 2, September 1970, 227-51.
   Author examines the relationship between domestic politics, the external environment, and Nigerian foreign policy in this case study.

279 INGHAM, K., (ed.). Foreign Relations of African States. London: Butterworths, 1974. 344 p., figs, maps, tabls.
   Although many of the essays in this volume are concerned with diplomatic history, there are several items which are relevant to this bibliography. See references under Allot, Cervenka, Kirk-Greene, Person, Shaw, Suret-Canale, and Tandon.

280 ISMAEL, T. Y. "The Sudan's Foreign Policy Today." International Journal, 25, 3, Summer 1970, 565-75.
   Although initially western-oriented, Sudanese foreign policy became pro-Soviet after 1967. The effects of Soviet penetration of Sudan are discussed.

281 JACOMY-MILLETTE, A.-M. "Anatomie d'un Pays en Voie de Développement à la Lumière de Ses Engagements Internationaux: Le Cas de l'Ethiopie." Revue Générale de Droit International Public, 78, 4, 1974, 1017-45.
Ethiopian foreign policy was directly related to two major domestic concerns - concentration of power in the hands of the Emperor and economic development.

282 JANSEN, G. H. Non-alignment and the Afro-Asian States. New York: Praeger, 1966. 432 p., index.
The author, an Indian, states that "this work is not an impartial objective account .... It is a tract for the times which has been written principally for Afro-Asian readers."

283 JOHNS, DAVID H. "Exchanges of Diplomats within Africa." Africa Quarterly, 11, 3, December 1971, 203-15.
The analysis of diplomatic exchanges is one indication of the status of inter-African international relations.

284 _____. "The Foreign Policy of Tanzania." In: Aluko, O., (ed.). The Foreign Policies of African States. London, 1977, 196-219.
Argues that Tanzania has followed a remarkably consistent policy since independence and forecasts a continuation of that consistency.

285 _____. "The 'Normalisation' of Intra-African Diplomatic Activity." Journal of Modern African Studies, 10, 4, December 1972, 597-610.
Various quantitative measures of diplomatic exchange are used in this study of inter-African relations.

286 JOHNS, SHERIDAN. "Botswana's Strategy for Development: An Assessment of Dependence in the Southern African Context." Journal of Commonwealth Political Studies, 11, 3, November 1973, 214-30.
Analyzes Botswana's attempts to determine its future in the face of South African domination.

287 JUDD, PETER. "The Attitudes of the African States toward the Katanga Secession, July 1960-January 1963." Columbia Essays in International Affairs. New York, 1966, 239-54.

288 KAMARCK, ANDREW M. "Economic Determinants." In: McKay, Vernon, (ed.). African Diplomacy: Studies in the Determinants of Foreign Policy. New York, 1966, 55-68.
A discussion of the influence of economic factors on African foreign policy.

289 KANZA, THOMAS. Conflict in the Congo. Baltimore: Penguin, 1972. 346 p.
    Domestic and international politics of the Congo (Zaire) during the Lumumba regime. The author served in various capacities - including ambassador to the UN - in the Lumumba regime.

290 _____. "Zaire's Foreign Policy." In: Aluko, O., (ed.). The Foreign Policies of Africa. London, 1977, 235-43.
    "In Zaire it is President Mobutu that largely decides the content and direction of the country's foreign policy."

291 KATOND, D. "Los Estados Africanos y el Conflicto del Oriente Medio: El Caso de la República del Zaire." Revista de Política Internacional, 137, January/February 1975, 47-57.
    Israel's expansionist policy and its close relations with South Africa have caused Zaire to break relations with Israel.

292 KEITA, M. "The Foreign Policy of Mali." International Affairs, 37, 4, October 1961, 432-39.
    Positive neutralism and African political unity are the cornerstones of the foreign policy of Mali. Author was President of Mali.

293 KHAPOYA, VINCENT B. "Determinants of African Support for African Liberation Movements: A Comparative Analysis." Journal of African Studies, 3, 4, Winter 1976, 469-89, tabls.
    Critical analysis of the extent and kinds of support that African states have been extending to liberation movements, and of those factors which have in fact affected support for liberation movements.

294 _____. The Politics of Decision: A Comparative Study of African Policy toward the Liberation Movements. Denver: Graduate School of International Studies, University of Denver, 1975. 88 p. [Monograph Series in World Affairs, 12, 3].

295 KIMCHE, DAVID. The Afro-Asian Movement: Ideology and Foreign Policy of the Third World. Jerusalem: Israel Universities. New York: Halsted, 1973. 296 p., bibl., ill., index.
    "I have sought to trace the history of the Afro-Asian Movement per se, analysing the causes for its rise and decline, and endeavouring to place it within its proper context in contemporary history." India, Indonesia, and Tanzania are case studies.

296 KOCHAN, R. "An African Peace Mission in the Middle East: One-man Initiative of President Senghor." African Affairs, 287, April 1973, 186-96.
Description and analysis of the failure of Senghor's mission.

297 KOTSOKOANE, J. R. L. "Lesotho and Her Neighbors." African Affairs, 68, 271, April 1969, 135-38.
Supports Lesotho's non-hostile stance towards South Africa as a matter of survival, economic necessity, and preference for peace rather than war.

298 KOWET, DONALD. "Lesotho and the Customs Union with the Republic of South Africa." In: Cervenka, Z., (ed.). Landlocked Countries of Africa. Uppsala, 1973, 250-57.
An analysis of the dependence of Lesotho upon the Republic of South Africa.

299 LEGUM, COLIN. "The Growth of Africa's Foreign Policy: From Illusion to Reality." In: Gardiner, Robert K. A., (ed.). Africa and the World. Addis Ababa, 1970. 48-65.
Examines in the field of foreign policy what Africa's experience has been thus far in trying to place itself firmly in the modern world.

300 LEISTNER, GERHARD M. E. The Economic Problems and Policies of South Africa's Neighbouring Black African States. Johannesburg: South African Institute of International Affairs, 1973. 15 p., bibl.

301 LE VINE, VICTOR T. The Cameroon Federal Republic. Ithaca: Cornell University, 1971. 205 p., bibl. essay, index, maps, tabls.
See "External Relations," pp. 173-78.

302 LEYMARIE, PHILIPPE. "La Nouvelle Diplomatie Malgache." Revue Française d'Etudes Politiques Africaines, 97, January 1974, 29-40.

303 LIBBY, ROBERT T. "External Co-optation of a Less Developed Country's Policy Making: The Case of Ghana, 1969-1972." World Politics, 29, 1, October 1976, 67-89.

304 LIEBENOW, J. GUS. Liberia: Evolution of Privilege. Ithaca, New York: Cornell University, 1969. 247 p., bibl., figs., map.
See "A Place in the Sun: Liberia in World Affairs," pp. 189-205.

305 LIMAGNE, JOSEPH. "La Politique Etrangère de la République Islamique de Mauritanie." Revue Française d'Etudes Politiques Africaines, 75, March 1972, 34-46.
Pro-France, Pro-Arab, and faithful to African unity, these are the major themes of Mauritania's foreign policy.

306 _____. "Seize Ans de Diplomatie Mauritanienne." Revue Française d'Etudes Politiques Africaines, 87, March 1973, 82-92.

307 LUNDAHL, FREDERICK. "Thumbing Your Nose at the Giant: The Case of Botswana in Southern African Interstate Relations." Journal of International and Comparative Studies, 5, 3, Fall 1972, 27-50.

308 LYSTAD, ROBERT A. "Cultural and Psychological Factors." In: McKay, Vernon, (ed.). African Diplomacy: Studies in the Determinants of Foreign Policy. New York, 1966, 91-118.
Although he notes that great difficulties exist, the author attempts to describe "the relationship between African 'social values' and foreign policy."

309 MACHEL, SAMORA. "Address to the OAU." Africa Today, 23, 1, January-March 1976, 12-4.
The President of Mozambique calls for condemnation of South Africa's aggression, and support of Angola.

310 MACKINTOSH, JOHN P. Nigerian Government and Politics. London: George Allen and Unwin, 1966. 651 p., bibl., index, maps, tabls.
See chapter VI, "Nigerian External Relations," pp. 268-85. This is similar to the journal article by this author.

311 _____. "Nigeria's External Relations." Journal of Commonwealth Political Studies, 2, 3, November 1964, 207-18.
Nigeria's foreign policy is mainly the work of the Prime Minister, and under pressure from militant nationalists he has followed a policy of anti-colonialism.

312 MAGRINI, L. "Il Rinnovo delle Convenzione di Yaoundé: Oltre l'Idea Eurafricana." Politica Internazionale, 3, March 1975, 39-51.
Analysis of the Yaoundé agreement between the EEC and the African associated states.

313 Malawi: Dialogue and Development. London: Africa Publications Trust, 1973. 24 p., bibl., map.
Economic and political aspects of Malawi's domestic and international situation, particularly in respect of South Africa, are considered.

314 MALÉCOT, G. R. "La Politique Etrangère de L'Ethiopie."
Revue Française d'Etudes Politiques Africaines, 79, July
1972, 39-57.
Ethiopia considers international organizations, especially the UN, as a major source of protection for its fragile independence.

315 MARCUM, JOHN A. "Angola: Division or Unity?" In: Carter, G. M. and O'Meara, P., (eds.). Southern Africa in Crisis. Bloomington, Ind., 1977, 136-62.
Foreign intervention in Angola is discussed.

316 _____. "The Exile Condition and Revolutionary Effectiveness: Southern African Liberation Movements." In: Potholm, C. P. and Dale, R., (eds.). Southern Africa in Perspective: Essays in Regional Politics. New York, 1972, 262-75, ill.
External aid to liberation movements has had many negative effects for those movements. "An overreliance on external aid has undercut the effectiveness of Southern African liberation movements."

318 MARIÑAS OTERO, LUIS. "Seychelles: Bases y Coordenadas de la Política Exterior del Estado mas Pequeño de Africa." Revista de Política Internacional, 158, July/August 1978, 99-116.

319 MARKOVITZ, IRVING L. Leopold Sédar Senghor and the Politics of Negritude. New York: Atheneum, 1969. 300 p., bibl., index.
This study of negritude includes a chapter on Senghor's views of appropriate relations between France and Africa.

320 MARTIN, DENIS. "L'Occident, l'Océan et le Kenya." Annuaire des Pays de l'Océan Indien, 2, 1975, 229-42.
Because of its location, Kenya plays an important role in the international politics of East Africa and the Indian Ocean. Kenya's foreign policy is discussed in this perspective.

321 MASCHERONI, A. M. "Le Conferenze dei Paesi Afro-Asiatici Impegnati." Rivista Internazionale di Filosofia Politica e Sociale, 9, 1, January-March 1965, 59-77.
Africa's role and views within the non-aligned states are analyzed.

322 MATES, L. "The Non-aligned as a Pressure Group?" Pacific Community, 2, 3, April 1971, 512-22.
Analyzes the significance of the Lusaka Conference.

323 MAYALL, JAMES. "Malawi's Foreign Policy." World Today, 26, 10, October 1970, 435-45.
   This examination of Malawi's foreign policy centers on relations between Malawi and the Republic of South Africa.

324 _____. "Oil and Nigerian Foreign Policy." African Affairs, 75, 300, July 1976, 317-30.
   Sketches those traditional areas of external concern where oil revenue seems to have had an effect on policy and raises some preliminary questions about the effects of oil on the country's external relations.

325 MAZRUI, ALI A. "Africa and Egypt's Four Circles." In: Mazrui, A. A. On Heroes and Uhuru Worship. London, 1967, 96-112.
   This essay on Africa's position in Egyptian foreign policy was previously published in African Affairs, 1964.

326 _____. "Africa in World Affairs: Conflict and Change in the Next Thirty Years." Nigerian Journal of International Affairs, 1, 1, July 1975, 71-82.

327 _____. "Africa, the West and the World." Australian Outlook, 26, 2, August 1972, 115-36.
   Culture and race are important aspects of the international relations between Africa and the world.

328 _____. "African Diplomatic Thought and the Principle of Legitimacy." In: Gappert, G. and Garry, T., (eds.). The Congo, Africa and America. Syracuse, 1965, 46-62.
   "Can we now discuss a general philosophy which has helped to guide and influence the diplomatic behavior of African states?"

329 _____. "Black Africa and the Arabs." Foreign Affairs, 53, 4, July 1975, 725-42.
   Arab relations with Africa are linked to the problems of Israel and the Republic of South Africa, but the manner in which the Arab states assist in African economic problems is a key factor in the future of Afro-Arab relations.

330 MBADINUIJU, C. CHINWOKE. "The Ideological Framework of African Opposition to South Africa's Apartheid Policy." Third World Review, 1, 1, Fall 1974, 52-8.

331 McGEE, GALE W. "The U. S. Congress and the Rhodesian Chrome Issue." Issue, 2, 2, Summer 1972, 2-7.
   Decries the action by the Senate in late 1971, allowing the US to unilaterally breach the UN sanctions against Southern Rhodesia, and urges a speedy reversal of the action.

342 McGOWAN, PATRICK J. "Africa and Non-alignment: A Comparative Study of Foreign Policy." International Studies Quarterly, 12, 3, September 1968, 262-95.
A variety of quantitative techniques are utilized to test hypotheses about interaction between African states and the Communist bloc. Conclusions about non-alignment are drawn from this analysis.

343 _____. "The Pattern of African Diplomacy: A Quantitative Comparison." Journal of Asian and African Studies, 4, 3, July 1969, 202-21.
Several statistical measures are used in this effort to measure aspects of African foreign policy. Included are decolonization, relations with the Communist bloc, and participation in the international system.

344 _____ and GOTTWALD, K. P. "Small State Foreign Policies: A Comparative Study of Participation, Conflict, and Political and Economic Dependence in Black Africa." International Studies Quarterly, 19, 4, December 1975, 469-500.
The authors present a typology of states and foreign policy patterns, draw hypotheses, and test these using events data.

346 _____ and JOHNSON, THOMAS H. "The AFRICA Project and the Comparative Study of African Foreign Policy." In: DeLancey, M. W., (ed.). Aspects of International Relations in Africa. Bloomington, Ind., 1979, 190-241.
"We wish to describe the methods and some findings of the African Foreign Relations and International Conflict Analysis (AFRICA) Project as they relate to the comparative study of African foreign policy."

347 McHENRY, DONALD F. "Captive of No Group." Foreign Policy, 15, Summer 1974, 142-9.
The US State Department has tried without success to develop American blacks as a constituency for US Africa policy.

348 McKAY, VERNON. "The African Operations of United States Government Agencies." In: Goldschmidt, Walter, (ed.). The United States and Africa. New York, 1963, 273-95, tabls.
After a brief description of the Africa activites of several branches of the US government, McKay concludes that there is a "vast problem of organization and coordination in the [US] Africa policy machine."

349 _____. "Changing External Pressures on Africa." In: Goldschmidt, Walter, (ed.). The United States and Africa. New York, 1963, 74-112, tabls.
The UN, India, Egypt and Islam, the Soviet Union, the US, and Europe as external pressures on Africa.

350   McKAY, VERNON. "The Propaganda Battle for Zambia." Africa Today, 18, 2, April 1971, 18-26.

351   McMASTER, CAROLYN. Malawi: Foreign Policy and Development. London: Julian Friedmann. New York: St. Martins, 1975. 270 p., bibl., index, maps, tabls.
      "The determinants and development of the foreign policy of one small, land-locked member of the international system, Malawi, will be examined. One question is central to this study: Namely, how wide is the range of foreign policy options open to Malawi?"

352   "Mediniyut Mitzrayim Be-Afrika." Hamizrab Hebadasb, 12, 1/2, 1962, 19-27.
      "Egypt's Policy in Africa." An examination of Egypt's goals in Africa and the means whereby the government attempts to attain these goals.

353   MEERSSCHE, P. VAN DE. "Algerije en Afrika in de Politiek van De Gaulle." Internationale Spectator, 15, 9, May 1961, 243-77.
      "Algeria and Africa in De Gaulle's Policy."

354   _____. "Hoe Evolueert de Franse Gemeenschap?" Internationale Spectator, 14, 11, June 1960, 255-70.
      "How Is the French Community Evolving?" Discusses the views of Toure, Senghor, and Houphouët-Boigny on Franco-African relations."

355   MELADY, T. P. "Nonalignment in Africa." Annals of the American Academy of Political and Social Sciences, 362, November 1965, 52-61.
      The African states see domestic economic problems of primary interest to them. These states have their own values and their own way of doing things. These are major roots of African nonalignment.

356   _____ and MELADY, M. "The Expulsion of the Asians from Uganda." Orbis, 19, 4, Winter 1976, 1600-20.
      A description of the process of expulsion and comment upon the general lack of African reaction.

357   MILLAR, T. B. and MILLER, J. D. B. "Afro-Asian Disunity: Algiers 1965." Australian Outlook, 19, 3, December 1965, 306-21.
      The Algiers Conference shows that the Afro-Asian states are not stooges of the Chinese and they are sceptical about the good faith of the western states.

358 MILLER, J. D. B. "The Intrusion of Afro-Asia." *International Affairs*, November 1970, 46-55.
   Non-alignment, neo-colonialism, and renewal of spirit are concepts uniting the African and Asian states. Their common effort is impressive in appearance but not in substance.

359 _____. *Survey of Commonwealth Affairs: Problems of Expansion and Attrition, 1953-1969.* London: Oxford University for the Royal Institute of International Affairs, 1974. 550 p., bibl., index, tabls.
   Part III, "Africa," pp. 101-267 contains sections on South Africa, Rhodesia, and black-ruled members of the Commonwealth. Foreign policy matters are discussed in some detail.

360 MINTER, WILLIAM. "Major Themes in Mozambican Foreign Relations, 1975-1977." *Issue*, 8, 1, Spring 1978, 43-9.
   Mozambique's foreign policy has clear direction and purpose. The state is the pawn of no foreign power.

362 MISRA, K. P. "Indian Ocean as a Zone of Peace: The Concept and Alternatives." *India Quarterly*, 33, 1, January-March 1977, 19-32.

363 MITCHELL, C. "The Horn of Africa and the Indian Ocean." In: University of Southampton, *The Indian Ocean in International Politics*. Southampton, 1973, 153-78.
   Focuses on African reactions to external pressure.

364 MITTELMAN, J. H. "The State of Research on African Politics: Contribution on Uganda." *Journal of Asian and African Studies*, 11, 3/4, July-October 1976, 152-65.
   Among other comments, the author notes that there has been insufficient study of African foreign policies.

365 MODERNE, F. "Le Panafricanisme et la Politique Extérieure de la Tanzanie." *Revue Française d'Etudes Politiques Africaines*, 62, February 1971, 81-99.
   Nyerere considers African unity as the only solution to Africa's economic and political problems. This belief is the basic element in Tanzania's foreign policy.

366 MOHAN, JITENDRA. "African Liberation Struggle, in Continental and International Perspective." Economic and Political Weekly, 11, 4, January 1976, 105-16.
Imperialism and anti-imperialism as the major factors in African international relations.

367 _____. "Ghana, the Congo and the United Nations." Journal of Modern African Studies, 7, 3, October 1969, 369-406.
Ghana's foreign policy and the Congo crisis, 1960-1963.

368 MORTIMER, ROBERT A. "The Algerian Revolution in Search of the African Revolution." Journal of Modern African Studies, 8, 3, October 1970, 363-87.
Algeria views its role in Africa as being largely to radicalize the thinking of African leaders and to promote African unity on the basis of a common heritage of colonial exploitation.

369 _____. "From Federalism to Francophonia: Senghor's African Policy." African Studies Review, 15, 2, September 1972, 283-306.
This general review of Senegal's Africa policy concentrates on the issues of Gambia, Guinea-Bissau, francophonia, West Africa as a region, and the Senegal River basin.

370 MPAKATI, ATTATI. "Malawi: The Birth of a Neo-colonial State." African Review, 3, 1, 1973, 33-68, tabls.
Traces the history of Malawi from early society, through colonization and on to independence and neo-colonial status. Puts greater emphasis on the latter.

371 MTSHALI, BENEDICT V. "Zambia and the White South." In: Cervenka, Z., (ed.). Land-locked Countries of Africa. Uppsala, 1973, 188-93.
Zambia's attempts to decrease dependence on the white states of Southern Africa may cause serious internal problems.

372 _____. "Zambia's Foreign Policy." Current History, 58, 343, March 1970, 148-53, 177-79.
Disengagement from South Africa has raised major problems for the Zambian government.

373 MUNGAI, NJOROGE. "Kenya's Foreign Policy on Southern Africa." Pan-African Journal, 4, 2, Spring 1971, 218-22.

374 MUNGER, EDWIN S. "President Kenneth Kuanda of Zambia." American Universities Field Staff Reports, Central and Southern Africa Series, 14, 2, 1970, 1-22.
Foreign policy stance of President Kuanda, who "seems to be assuming the mantle of Third World leadership."

375 MUNGER, E. S. "Trading with the Devil: Malawi's Economic Relations with Portugal, Rhodesia and South Africa." American Universities Field Staff Reports, Central and Southern Africa Series, 13, 2, 1969, 1-17.
Effects of its land-locked status on Malawi's foreign policy, especially in Southern African relations.

376 MWAANGA, VERNON J. "Zambia's Policy toward Southern Africa." In: Potholm, C. P. and Dale, R., (eds.). Southern Africa in Perspective: Essays in Regional Politics. New York, 1972, 234-44.
The author was Permanent Representative of Zambia at the UN. In this essay he briefly describes Zambian policy toward each of the states in Southern Africa.

377 NATUFE, O. IGHO. "Nigeria and Soviet Policy towards African Regimes, 1965-1970." Survey, 22, 1, Winter 1976, 93-111.

378 NDESHYO, R. "Note sur la Politique Africaine de Nonalignement." Cahiers Economiques et Sociaux, 11, 1/2, 1973, 107-20.
After a general synopsis of non-alignment policy, the author concentrates on specific African contributions on views of non-alignment.

379 N'DONGO, SALLY. Voyage Forcé: Itinéraire d'un Militant. Paris: F. Maspero, 1975. 224 p.
Senegalese politics and relations with France.

380 NIBLOCK, T. C. "Tanzanian Foreign Policy: An Analysis." African Review, 1, 2, September 1971, 91-101.
Tanzania's clash with the western powers in 1964/1965 was due to changes in environment, not changes in Tanzania's foreign policy.

381 NJOYA, ADAMOU N. Le Cameroun dans les Relations Internationales. Paris: Librairie Générale de Droit et de Jurisprudence, 1976. 414 p., bibl., maps.

382 _____. "Sociology of African Diplomacy." Mawazo, 4, 3, 1975, 7-23.

383 NKRUMAH, KWAME. "African Prospect." Foreign Affairs, 37, 1, October 1958, 45-53. Also in: Quigg, P. Africa. New York, 1964, 274-82.
Nkrumah's views on the appropriate relationship between Africa and Europe.

384 _____. Rhodesia File. London: Panaf, 1976. 168 p.
Nkrumah's views and aspects of Ghanaian foreign policy for Rhodesian matters are examined in this collection of Nkrumah's writings and speeches and government documents.

385 NNOLI, OKWUDIBA. "The Place of Morals in African Foreign Policy." Africa Quarterly, 14, 1/2, April-September 1975, 48-71.

386 _____. "Political Will and the Margin of Autonomy in Tanzanian Foreign Policy." Africa Quarterly, 17, 1, July 1977, 5-26.

387 _____. Self Reliance and Foreign Policy in Tanzania: The Dynamics of the Diplomacy of a New State, 1961 to 1971. New York, 1978. 340 p., index, tabls.
"Treats Tanzanian foreign policy in its totality as the nation's reaction to the external environment."

388 NUNN, GRADY M. "Nigerian Foreign Relations." In: Blitz, L. F., (ed.). The Politics and Administration of Nigerian Government. New York, 1965, 249-64.
The chapter is divided into three sections, "Factors Affecting Nigerian Foreign Policy," "Organisation of Nigerian Foreign Policy," and "Nigerian Foreign Policy."

389 NWORAH, DIKE. "Nationalism versus Coexistence: Neo-African Attitudes to Classical Neutralism." Journal of Modern African Studies, 15, 2, June 1977, 213-37.
African emphasis on regional economic cooperation and national welfare may be changing the concept of traditional coexistence (i.e., neutralism).

390 NWUNELI, O. E. and DARE, O. "The Nigerian Press and the Civil War in Angola." A Current Bibliography on African Affairs, 9, 4, 1976/1977, 302-16, tabls.
Analysis of press coverage of the war and the effects on press coverage of a major shift in Nigerian foreign policy. Includes important evidence of the dependence on foreign news services of Nigeria's press.

391 NYERERE, JULIUS K. Freedom and Unity: A Selection from Writings and Speeches, 1952-1965. London: Oxford University, 1967. 366 p., index.
Contains several speeches on the OAU, African unity, the Commonwealth, the UN, and Tanzania's foreign relations.

392 _____. "Nonalignment in the 1970s." In: Legum, Colin and Hughes, Anthony, (eds.). Africa Contemporary Record. London, 3, 1971, c34-c37.

393 ODIER, J. "La Politique Etrangère de Mobutu." Revue Française d'Etudes Politiques Africaines, 120, December 1975, 25-41.
A positive view of the foreign policy of the Mobutu regime.

394 OFOEGBU, MAZI RAY. "The Relations between Nigeria and Its Neighbours." *Nigerian Journal of International Studies*, 1, 1, July 1975, 28-40. Also in: *Odu: A Journal of West African Studies*, n.s. 12, July 1975, 3-24, tabls.
 This essay provides some background information on the feasibility of the ECOWAS proposal.

395 _____ and OGBUAGU, CHIBUZO. "Towards a New Philosophy of Foreign Policy for Nigeria." In: Akinyemi, A. B., (ed.). *Nigeria and the World*. Ibadan, 1978, 116-37.
 "We are now in a new era of detente which demands from us more action and greater involvement in world affairs."

396 OGUNBADEJO, OYE. "General Gowon's African Policy." *International Studies*, 16, 1, January-March 1977, 35-50.
 Description of Nigerian foreign policy in respect of continental issues and regional issues.

397 _____. "Nigeria and the Great Powers: The Impact of the Civil War on Nigerian Foreign Relations." *African Affairs*, 75, 298, January 1976, 14-32.
 Assesses the impact in three sections: Nigerian relations with the western powers, with the Soviet Union and China, and an overall evaluation of the new "non-alignment" policy. Confirms his initial thesis that Nigeria's stance is far from being truly independent, and while it has been evolving in the right direction, still requires a big jolt to the left if it is to be truly non-aligned.

398 OHAEGBULAM, FESTUS U. "The Democratic Republic of the Congo and the International Politics of Non-Alignment." *Africa Today*, 15, December/January 1969, 8-11.

399 OJEDOKUN, OLASAPO. "The Anglo-Nigerian Entente and Its Demise, 1960-1962." *Journal of Commonwealth Political Studies*, 9, 3, November 1971, 210-33.
 Internal and external factors led to dramatic changes in Anglo-Nigerian relations in 1962.

400 _____. "The Changing Pattern of Nigeria's International Economic Relations: The Decline of the Colonial Nexus, 1960-1966." *Journal of Developing Areas*, 6, 4, July 1972, 535-53, tabls.
 Analyzes factors, economic and political, domestic and foreign, which shaped the character of the country's international economic ties.

401 OJO, OLATUNDE J. B. "Nigeria-Soviet Relations: Retrospect and Prospects." *African Studies Review*, 19, 3, December 1976, 43-63, bibl., tabls.
 There has been continuity in relations between the two, but no sharp reversal of Nigerian policy in favor of the Soviet Union as is commonly thought to have occurred in 1967.

402 OKOLO, JULIUS E. and LANGLEY, WINSTON E. "The Changing Nigerian Foreign Policy." *World Affairs*, 135, 4, Spring 1973, 309-27.

403 OKUMU, JOHN K. "Kenya's Foreign Policy." In: Aluko, O., (ed.). *The Foreign Policies of African States*. London, 1977, 136-62, tabls.
The author stresses the relationships between foreign policy and economic development.

404 _____. "Some Thoughts on Kenya's Foreign Policy." *African Review*, 3, 2, June 1973, 263-90.

405 OLYMPIO, SYLVANUS E. "African Problems and the Cold War." *Foreign Affairs*, 40, 1, October 1961, 50-7. Also in: Quigg, P., (ed.). *Africa*. New York, 1964, 294-303.
One of several statements by African heads-of-state to be published by this journal in the early 1960s.

406 OSTHEIMER, JOHN M. *Nigerian Politics*. New York: Harper and Row, 1973. 200 p., bibl., index, maps, tabls.
See Chapter 9, "Foreign Policy," pp. 161-85. Author compares pre- and post-civil war foreign policies.

407 OTHMAN, H. M. "The Arusha Declaration and the Triangle Principles of Tanzania Foreign Policy." *East African Journal*, 7, 5, May 1970, 35-42.

408 OYEBODE, AKINDELE B. "Towards a New Policy on Decolonization." In: Akinyemi, A. B., (ed.). *Nigeria and the World*. Ibadan, 1978, 97-115.
"We intend to focus attention on the present stage of the national liberation process and . . . attempt to draw the parameters of what should be a new Nigerian policy."

409 PAUKER, G. "The Rise and Fall of Afro-Asian Solidarity." *Asian Survey*, 5, 9, September 1965, 425-32.
Contains limited information on African attitudes concerning the second Afro-Asian Conference of 1965.

410 PETTMAN, JAN. *Zambia: Security and Conflict*. London: Julian Friedmann, 1974. 284 p., bibl., index, maps.
Analysis of Zambia's internal and external politics. The search for security has become identified with the survival of the government.

411 PHILLIPS, CLAUDE S., JR. *The Development of Nigerian Foreign Policy*. Evanston, Ill.: Northwestern University, 1964. 154 p., index, tabl.
"Foreign policy developments in the formative period between 1959-63 are studied." Interviews, parliamentary debates and published materials are the primary sources of data.

412 POLHEMUS, JAMES H. "Nigeria and Southern Africa: Interest, Policy, and Means." Revue Canadienne des Études Africaines, 11, 1, 1977, 43-68.

413 POTHOLM, C. P. "Remembrance of Things Past? The Process of Institutional Change in Swaziland." Africa Quarterly, 13, 1, April-June 1973, 1-22.
   Since independence Swaziland has attempted to decrease dependence on South Africa and to increase its contacts with black Africa.

414 PRATT, R. C. "African Reactions to the Rhodesian Crisis." International Journal, 21, 2, Spring 1966, 186-98.
   The author offers explanations for the style of African reactions to the Rhodesian situation, using Tanzania as an example. Options for British policy are examined.

415 _____. "Foreign Policy Issues and the Emergence of Socialism in Tanzania, 1961-1968." International Journal, 30, 3, Summer 1975, 445-60.
   Analyzes major factors in foreign affairs that led to a decision to adopt a policy of socialism in Tanzania.

416 AL-QAR'Ī, AHMAD YŪSUF. "Al-Taharruk al-Misrī fī Ifrīqiyā." Al-Siyāsah al-Duwalīyah, 31, January 1973, 166-74.
   Egypt's Africa policy.

418 QUAISON-SACKEY, ALEX. Africa Unbound: Reflections of an African Statesman. London: Andre Deutsch. New York: Praeger, 1963. 174 p.
   A diplomat from Ghana writes on African integration, positive neutralism, Africa and the UN, and other matters.

419 RAINERO, R. "L'Evolution du Neutralisme et Ses Caractéristiques Africaines." Genève-Afrique, 4, 1, 1965, 35-47,
   The meaning of neutralism and African views of it.

420 _____. "L'Idea del Neutralismo e le Sue Caracterizzazione Africane." Rivista di Studi Politici Internazionali, 32, 4, October-December 1965, 493-508.
   The true basis of African neutrality is the conflict between rich and poor states and the need for the poor to work together in the struggle against the rich.

421 RIVIÈRE, C. "La Politique Etrangère de la Guinée." Revue Française d'Etudes Politiques Africaines, 68, August 1971, 37-68.
   The positive neutralism which is the core of Guinea's foreign policy is based on a political and ideological foundation, but it is also a result of the conditions under which the state became independent.

422 RIVKIN, ARNOLD. "Nigeria: A Unique Nation." Current History, 45, 268, December 1963, 329-34.
Nigeria's role in African inter-state relations.

423 RONDOT, P. "Die Neuen Staaten Nordafrikas und Ihre Stellung in der Welt." Europa-Archiv, 18, 14, July 1963, 521-32.
Africa represents one of four major areas of attention in the foreign policies of Morocco, Algeria, Tunisia, and Libya.

424 ———. "La Politique Extérieure de la République Arabe Unie." Revue de Défense Nationale, December 1965, 1918-31.
The author included a section on relations between the UAR and the African and Third World states.

425 ROTHCHILD, DONALD and CURRY, ROBERT L. "Expanding Botswana's International Options." In: Shaw, T. M. and Heard, K. A., (eds.). Cooperation and Conflict in Southern Africa: Papers on a Regional Subsystem. Washington, D. C., 1977, 312-28, tabls.
"We hope that this policy-oriented line of analysis will further the examination of what problem-solving initiatives may be available to statesmen in late developing countries."

426 SAXENA, S. C. "Malawi's Relations with South Africa and Their Implications for African Unity." Indian Journal of Politics, 10, 2, July-December 1976, 39-52.
Malawi's Southern Africa policies are examined and their effects on African unity are considered.

427 SCHNEIDMAN, WITNEY J. "FRELIMO's Foreign Policy and the Process of Liberation." Africa Today, 25, 1, January-March 1978, 56-67.
Major goal of FRELIMO's foreign policy was to win diplomatic support.

428 SCHWARZ, F. A. O., JR. Nigeria. Cambridge, Mass.: M.I.T., 1965. 316 p., bibl., ill., index.
"Foreign Policy," pp. 213-37, a descriptive chapter.

429 SEATON, EARLE E. and MALITI, SOSTHENES T. Tanzania Treaty Practice. Nairobi: Oxford University, 1973. 200 p., index.
The Nyerere Doctrine and "the succession to the rights and obligations arising out of treaties concluded" by the preceeding colonial powers are the major topics. Authors are officials in the government of Tanzania.

430 SERRE, F. DE LA. "Les Revendications Marocaines sur la Mauritanie." Revue Française de Science Politique, 16, 2, April 1966, 320-31.
Relations between Morocco and Mauritania improved in 1966 after a period of conflict.

431 SHAMUYARIRA, N. M. "Commentary on the Lusaka Manifesto."
African Review, 1, 1, March 1971, 73-8.
Description and analysis of the Lusaka Manifesto.

432 _____. "The Lusaka Manifesto." East African Journal, 6, November 1969, 23-8.

433 SHAW, TIMOTHY M. "African States and International Stratification: The Adaptive Foreign Policy of Tanzania." In: Ingham, K., (ed.). Foreign Relations of African States. London, 1974, 213-36, figs.
"This paper attempts to construct a synthetic model based on the concepts of subsystem and adaptation . . . . a small contribution to the analyses of inherited structural dependence and of alternative adaptive strategies to erode the impact of international stratification."

434 _____. Dependence and Underdevelopment: The Development and Foreign Policies of Zambia. Athens: Ohio University, 1976. 60 p., map, tabls. [Papers on International Studies, Africa Series, No. 28].
"The policies and problems of Zambia cannot be explained without reference to its inheritance and acceptance of dependence."

435 _____. The Foreign Policy of Tanzania, 1961-69. Nairobi: East African Publishing House, 1975.

436 _____. "The Foreign Policy of Zambia: An Events Analysis of a New State." Comparative Political Studies, 11, 2, July 1978, 181-210.

437 _____. "The Foreign Policy of Zambia: Ideology and Interests." Journal of Modern African Studies, 14, 1, March 1976, 79-105.
Zambian concepts of humanism play an important role in the formulation and content of Zambia's foreign policy.

438 _____. "The Foreign Policy System of Zambia." African Studies Review, 19, 1, April 1976, 31-63.
Structure and process are examined. Author argues that a great diversity of actors is involved in Zambia's foreign policy process.

439 _____. "Kenya and South Africa: 'Subimperialist' States." Orbis, 21, 2, Summer 1977, 375-94.
Focuses on the understanding of the phenomenon of subimperialism in Africa, and on linkages among national, regional, and global "imperialisms" in an age of inequalities.

440 SHAW, T. M. "Uganda under Amin: The Costs of Confronting Dependence." Africa Today, 20, 2, Spring 1973, 32-45.
Amin's unplanned demolition of dependence has led to a new dependence on the Arab states.

441 _____. "Zambia: Dependence and Underdevelopment." Revue Canadienne des Études Africaines, 10, 1, 1976, 3-22.

442 _____. "Zambia's Foreign Policy." In: Aluko, O., (ed.). The Foreign Policies of African States. London, 1977, 220-34.
An analysis of Zambia's role in the politics of the Southern Africa region.

442 a _____ and MUGOMBA, A. T. "The Political Economy of Regional Detente: Zambia and Southern Africa." Journal of African Studies, 4, 4, Winter 1977, 392-413, tabl.

443 SHEPHERD, GEORGE W., JR. "Counterracism in New African Policies." In: Shepherd, George W., Jr., (ed.). Racial Influences on American Foreign Policy. New York, 1970, 150-64.
Race is a factor in many aspects of the relations between African states and between Africa and the rest of the world.

444 _____. Nonaligned Black Africa: An International Subsystem. Lexington, Mass.: Heath Lexington, 1970. 151 p., index, tabls.
The concept of non-alignment is defined in chapters on African relations with the USA, the Commonwealth, and the Communist states. There is emphasis on Southern African problems.

445 SILLA, O. "L'Afrique - le Maghreb - le Proche-Orient." Etude Internationales, 1, 4, December 1970, 35-49.
A study of the various common interests of North Africa and the rest of the continent - anti-colonialism, Islam, economic development, etc.

446 SINGH, L. P. "Das Fiasko der Zweiten Konferenz der Afro-Asiatischen Staaten in Algier." Europa-Archiv, 21, 11, June 1966, 399-408.
"The Failure of the Second Conference of the Afro-Asiatic States in Algiers." The failure was due to Sino-Soviet competition and domestic problems in the Afro-Asian states.

447 SKURNIK, W. A. E. The Foreign Policy of Senegal. Evanston, Ill.: Northwestern University, 1972. 308 p., bibl., charts, map, tabls.
An "attempt to tell the story of the foreign policy behavior of a 'moderate' African state."

448 "South Africa and Lesotho." Bulletin of the Africa Institute of South Africa, 12, 4, 1974, 141-6.
"Traces South Africa policy of Lesotho under Jonathan regime, from friendlier stance . . . to more militant posture by the end of 1974."

449 SPECK, SAMUEL W., JR. "Malawi and the Southern African Comlex." In: Potholm, C. P. and Dale, R., (eds.). Southern Africa in Perspective. New York, 1972, 207-18, tabls.
"Malawi can be expected to disengage from Southern Africa only when disengagement would no longer jeopardize it economically or threaten it politically."

450 SPENCE, J. E. "The High Commission Territories with Special Reference to Swaziland." In: Benedict, Burton, (ed.). Problems of Smaller Territories. London, 1967, 97-111, figs.
External relations are considered.

451 _____. "The Implications of the Rhodesian Issue for the Former High Commission Territories." Journal of Commonwealth Political Studies, 7, 2, July 1969, 104-12.
The Rhodesia policies of Botswana, Lesotho, and Swaziland are similar; each desires a rapid settlement of the conflict to prevent an increase of pressure on them from the Republic of South Africa.

452 _____. Lesotho: The Politics of Dependence. London: Oxford University, 1968. 88 p., map.
Although the larger portion of the volume is on domestic political and economic aspects, Lesotho's place in world affairs and especially relations between Lesotho and South Africa are discussed.

453 STEVENS, RICHARD P. "The 1972 Addis Ababa Agreement and the Sudan's Afro-Arab Policy." Journal of Modern African Studies, 14, 2, 1976, 247-74.
Sudan's Africa policy takes on special significance in view of the country's Afro-Arab heritage. Some analysis of Sudan-Nigeria relations is included.

454 STOJKOVIĆ, N. "Azijsko-Afrička Konferencija Solidarnosti." Socijalizam, 8, 6, June 1965, 843-51.
"The Afro-Asian Solidarity Conference." A statement on the significance of the 1965 Conference held in Ghana.

455 STREMLAU, JOHN J. The International Politics of the Nigerian Civil War, 1967-1970. Princeton: Princeton University, 1977. 425 p., bibl., index, maps.
Nigerian foreign policy, Biafran foreign policy, the OAU, and the policies of Britain and the USA are considered in detail.

456 SURET-CANALE, J. "Les Relations Internationales de la République de Guinée." In: Ingham, K., (ed.). Foreign Relations of African States. London, 1974, 259-76.
Analysis of the effects of France's rejection of Guinea in the early years of independence and the influence of Guinea as a model for other francophone states.

457 SUTTON, F. X. "Authority and Authoritarianism in the New Africa." Journal of International Affairs, 15, 1, January 1961, 7-17.
African foreign policies are related to the sources and character of authority in these states. Nyasaland (Malawi) is used as a case study.

458 TANDON, YASHPAL. "An Analysis of the Foreign Policy of African States: A Case Study of Uganda." In: Ingham, K., (ed.). Foreign Relations of African States. London, 1974, 191-211.
After admitting the dangers of generalizing from a case study, Tandon suggests "that the factors that gave rise to the peculiarities of Uganda's foreign policy may well be common for the entire continent."

459 _____. "Juristic Dis-imperialism: A Case Study of Tanzania's Dispute with the United Kingdom over the Pensions Issue." In: Tandon, Y., (ed.). Readings in African International Relations, Vol. I. Nairobi, 1972, 222-5.

460 THIAM, DOUDOU. "Les Fondements Idéoligiques de la Politique Internationale dans les Etats Indépendants d'Afrique Noire." Revue Française d'Etudes Politiques Africaines, 37, January 1969, 70-86.
African foreign policies are based on two main themes - nationalism and socialism. African unity is a third theme.

461 _____. The Foreign Policy of African States: Ideological Bases, Present Realities and Future Prospects. New York: Praeger. London: Phoenix House, 1965. 134 p., index.
This volume was originally published in France in 1963. The work is in two parts - the ideological bases of international policies and international policy in practice.

462 THOMPSON, W. SCOTT. Ghana's Foreign Policy, 1957-1966: Diplomacy, Ideology and the New State. Princeton: Princeton University, 1969. 462 p., figs., index, tabls.
An analysis of the foreign policy of a small power that attempted to expand its international influence not through the use of force or power but through a moral factor.

463 _____. "Ghana's Foreign Policy under Military Rule." Africa Report, 14, 5/6, May/June 1969, 8-15.

464 THOMPSON, W. S. "Nonalignment in the Third World: The Record of Ghana." Orbis, 11, 4, Winter 1968, 1233-55.
Evolution of diplomatic contacts, UN voting, and relations with East and West are examined for three time periods during the Nkrumah regime.

465 TIMMLER, M. "Die Afrika-Konferenzen von Tunis, Accra und Conakry." Aussenpolitik, 11, 5, May 1960, 328-35.
"The African Conferences of Tunis, Accra, and Conakry." The first steps toward an independent and united African foreign policy.

467 TOURÉ, SEKOU. "African Independence as an International Issue." In: Tandon, Y., (ed.). Readings in African International Relations, Vol. I. Nairobi, 1972, 226-30.
An accusation by Guinea's President against the imperialist powers.

468 _____. "Africa's Future and the World." Foreign Affairs, 41, 1, October 1962, 141-57. Also in: Quigg, P., (ed.). Africa. New York, 1964, 316-28.
An early statement by the President of Guinea. Africa's main objective is total emancipation of the continent.

469 TUNTENG, KIVEN. From Pan Negroism to African Paramountcy: The Role of Kwame Nkrumah. Montreal: Vanier College, 1977. 179 p., bibl.

470 UKPABI, S. C. "Military Considerations in African Foreign Policies." Transition, 6, 31, June/July, 1967, 35-40.

471 AL-'UWAYNĪ, MUHAMMAD 'ALĪ. "Africa and the Fourth Arab-Israeli War." Ma'had al-Buhūth wa-al-Dirāsāt al-'Arabīyah. Majallah, 5, June 1974, 65-80.

472 VELLAS, PIERRE. "La Diplomatie Marocaine dans l'Affaire du Sahara Occidental." Politique Etrangère, 43, 4, 1978, 417-28.
What is Morocco's position on the Western Saharan question and what actions are being taken by the Moroccan government to achieve its goals?

473 VELLUT, J. L. "Congolese Foreign Policy and African 'Middle Powers,' 1960-1964." Australian Outlook, 19, 3, December 1965, 287-305.
The UN intervention in the Congo has important effects on the structure of African international relations. Lumumba did not understand this pattern of relations.

474 VENGROFF, R. "Instability and Foreign Policy Behavior: Black Africa in the UN." American Journal of Political Science, 20, 3, August 1976, 425-38.
   An effort to operationalize and test James Rosenau's pretheory of foreign policy. Results indicate that systematic variables are more important then individual in explaining variations in African foreign policy behavior.

475 WACHUKU, JAJA. "Nigeria's Foreign Policy." In: MacLure, Millar and Anglin, Douglas, (eds.). Africa: The Political Pattern. Toronto, 1961, 62-73.

476 WALLENSTEEN, P. "Dealing with the Devil. Five African States and South Africa." Instant Research on Peace and Violence, 3, 1971, 85-99.
   A comparative study of the South Africa policies of Botswana, Lesotho, Swaziland, Malawi, and Zambia.

477 WATERBURY, JOHN. "The Sudan in Quest of a Surplus: Part II, Domestic and Regional Politics." American Universities Field Staff Reports, Northeast Africa, 21, 9, 1976, 1-22, ill., maps.
   "Numeiry must perform both a domestic and an international balancing act of no small proportions."

478 WAYAS, JOSEPH. Nigeria's Leadership Role in Africa. London: Macmillan, 1979. 122 p.

479 WEINSTEIN, F. B. "The Second Asian-African Conference: Preliminary Bouts." Asian Survey, 5, 7, July 1965, 359-73.
   The inability to conduct a second conference indicates that the future of Afro-Asian cooperation is not promising.

480 WEISFELDER, RICHARD F. "Lesotho and South Africa: Diverse Linkages." Africa Today, 18, 2, April 1971, 48-55.

481 WINAI-STROM, G. "The Influence of Multinational Corporations on Lesotho's Politics and Economics." African Review, 5, 4, 1975, 473-91.

482 WOLDE-MARIAM, MESFIN. "Ethiopia and the Indian Ocean." In: Cottrell, A. J. and Burrell, R. M., (eds.). The Indian Ocean: Its Political, Economic, and Military Importance. New York, 1972, 181-92.
   "It is necessary to have a clear idea of the context in which Ethiopia finds itself, especially in terms of three major forces - Islam, political orientation, and East-West power blocs."

483 ZARTMAN, I. WILLIAM. "Characteristics of Developing Foreign Policies." In: Lewis, William H., (ed.). French-speaking Africa: The Search for Identity. New York, 1965, 179-93.
Discusses "a number of salient characteristics of foreign policy in the developing states of former French West Africa (AOF) and the Maghreb."

484 _____. "National Interest and Ideology." In: McKay, Vernon, (ed.). African Diplomacy: Studies in the Determinants of Foreign Policy. New York, 1966, 25-54.
A motivational study, a study "of reasons for acting, not a study of policy but of decision-making."

Inter-African Conflicts, Borders, and Refugees

485 AALL, CATO. "Refugee Problems in Southern Africa." In: Hamrell, S., (ed.). Refugee Problems in Africa. Uppsala, 1967, 26-44.
General comments on the situation of refugees and examination of their conditions in Southern Africa.

486 ABDALLA, I. H. "The 1959 Nile Waters Agreement in Sudanese-Egyptian Relations." Middle East Studies, 7, 3, October 1971, 329-41.
Sudanese agreements were needed before the High Dam on the Nile could be undertaken. This essay describes the negotiations and analyzes the agreement.

487 ABDI, SAID YASEF. "Independence for the Afars and Issas: Complex Background; Uncertain Future." Africa Today, 24, 1, January-March 1977, 61-7.

488 ADAMOLEKUN, LADIPO. "L'Aggression du 22 Novembre 1970: Faits et Commentaires." Revue Française d'Etudes Politiques Africaines, 14, June 1975, 79-114.
Description and analysis of the 1970 invasion of Sekou Toure's Guinea by a mixture of people from Portugal, Guine-Bissau, and Guinea.

489 ADANALIAN, ALICE A. "The Horn of Africa." World Affairs, 131, 1, April-June 1968, 38-42.
The major focus of tension in this area of growing superpower interest is the Kenya-Somalia-Ethiopia conflict.

490 ADDO, N. O. "Foreign African Workers in Ghana." International Labour Review, 109, 1, January 1974, 47-68, tabls.
An important domestic and international factor in African politics is the large numbers of foreign Africans in most African states.

491 ADDO, N. O. "Immigration into Ghana: Some Social and Economic Implications of the Aliens Compliance Order of 18 November 1969." Ghana Journal of Sociology, 6, 1, February 1970, 20-42.
Ghana's government has put pressure on foreign African nationals for domestic and international reasons.

492 AKAKPO, AMOUZOUVIE MAURICE. "La Délimitation des Frontières Togolaises." In: Deutschen UNESCO-Kommission, (ed.). Symposium Leo Frobenius. Köln, 1974, 92-109, maps.
A historical account of the delimitation of Togo's borders.

493 AKINYEMI, A. B. "Nigeria and Fernando Poo, 1958-1966: The Politics of Irredentism." African Affairs, 69, 276, July 1970, 236-49.
For many years some Nigerians have desired the annexation of this island.

494 ALEXANDRE, PIERRE. "The Land-locked Countries of Afrique Occidentale Française (AOF): Mali, Upper Volta and Niger." In: Cervenka, Z., (ed.). Land-locked Countries of Africa. Uppsala, 1973, 137-45, tabl.
Three stages in developments - from federal regulations, to international agreements, to compromise.

495 ALLOT, ANTHONY N. "Boundries and the Law in Africa." In: Widstrand, Carl G., (ed.). African Boundry Problems. New York, 1969, 9-21.
"My object . . . [is] to point out some of the main legal considerations which have affected the creation of boundaries, the legal context within which they have been imposed, the problems . . . and the extent to which they relate to pre-existing indigenous norms."

496 _____. "The Changing Legal Status of Boundaries in Africa: A Diachronic View." In: Ingham, K., (ed.). Foreign Relations of African States. London, 1974, 111-28, bibl.
"This paper is no more than a first attempt to unravel some of the knotty legal problems posed by African boundaries."

497 ALUKO, OLAJIDE. Ghana and Nigeria 1957-60: A Study in Inter-African Discord. New York: Barnes and Noble. London: Rex Collings, 1976. 275 p., bibl., index, map.
An observer might assume that Ghana and Nigeria would cooperate in African affairs, but "very little co-operation did really take place between the two countries between March 1957-1970."

498 AMIN, SAMIR, (ed.). Modern Migrations in Western Africa. London: Oxford University for the International African Institute, 1974. 426 p., bibl., ill.
   Contains seventeen essays on internal and international migration. See references by Adomako-Sarfoh, Ahooja-Patel, Deniel, Dussauze-Ingrand, and Kumekpor/Looky.

499 ANENE, J. C. The International Boundaries of Nigeria, 1885-1960; The Framework of an Emergent African Nation. London: Longman. New York: Humanities, 1970. 331 p., bibl., maps.
   Documentary sources, ethnographic material, and oral history are the sources for this study.

500 ANKOMAH, KOFI. "The Colonial Legacy and African Unrest." Science and Society, 34, 2, Summer 1970, 129-45.

501 ANKPRAH, KODWO E. "The Stranger within the Gates: The Case of Refugees in Africa." In: Parsons, R. T., (ed.). Windows on Africa. 1971, 107-32.

502 ASHFORD, D. E. "The Irredentist Appeal in Morocco and Mauritania." Western Political Quarterly, 15, 4, December 1962, 641-51.
   Irredentism is seen as a replacement for nationalist solidarity.

503 AUSTIN, DENNIS. "The Uncertain Frontier: Ghana-Togo." Journal of Modern African Studies, 1, 2, 1963, 139-45, tabl.
   Discusses problems created by the colonially-inherited boundary between the two countries and possible solutions.

504 _____ and WEILER, HANS N., (eds.). Inter-State Relations in Africa. Freiburg: Arnold-Bergstraesser Institut, 1965. 105 p., tabl.
   Contains five papers presented at a conference in 1963. See references under D. Austin, C. Hoskyns, E. Brett, K. Panter-Brick, and M. McWilliam.

505 BADR, G. M. "The Nile Waters Question. Background and Recent Development." Revue Égyptienne de Droit International, 15, 1959, 94-117.
   International law and the concerns of Egypt and Sudan with respect to the waters of the Nile.

506 BARBIER, M. "L'Avenir du Sahara Espagnol." Politique Etrangère, 40, 4, 1975, 353-80.
   Background and analysis of the conflict over the decolonization of Spanish Sahara.

507 BOUTROS-GHALI, BOUTROS. Les Conflicts de Frontières en Afrique. Paris: Editions Techniques et Economiques, 1972. 158 p., bibl., index, maps.
   General considerations with case studies of the Algeria-Morocco, Somalia-Ethiopia, and Somalia-Kenya disputes. Thirty-one relevant documents are included.

508 BOUVIER, P. "Un Problème de Sociologie Politique: Les Frontières des Etats Africains." Revue de l'Institut de Sociologie, 4, 1972, 685-720.
   The origins of African boundaries and the post-independence conflicts over boundaries are discussed.

509 BOYCE, FRANK. "The Internationalizing of Internal War: Ethiopia, the Arabs, and the Case of Eritrea." Journal of International and Comparative Studies, 5, 3, Fall 1972, 51-73.

510 BOYD, J. BARRON, JR. "The Origins of Boundary Conflict in Africa." In: DeLancey, M. W., (ed.). Aspects of International Relations in Africa. Bloomington, Ind., 1979, 159-89.
   "The factors which make the role of boundaries in African foreign policy particularly worthy of note and investigation derive from the process by which the contemporary African borders were defined and the rather unique set of pressures which are consequently present in the foreign policy environment of the African states."

511 BROOKS, HUGH C. and EL-AYOUTY, YASSIN, (eds.). Refugees South of the Sahara: An African Dilemma. Westport: Negro Universities, 1970. 307 p., ill., index, maps.
   Contains 16 chapters plus several appendices. See references by T. Hovet, G. L. Metcalf, G.-I. Smith, and J. Wine.

512 BROWNLIE, IAN. African Boundaries: A Legal and Diplomatic Encyclopedia. Los Angeles: University of California, 1979. 1488 p., maps.
   A comprehensive study of the boundaries of 48 African states, with a separate section and map of each of the 105 boundary alignments.

513 CAMPBELL, J. F. "Rumblings along the Red Sea: The Eritrean Question." Foreign Affairs, 48, 3, April 1970, 537-48.
   Background and a somewhat optimistic view of the future.

514 CERVENKA, ZDENEK, (ed.). Land-locked Countries of Africa. Uppsala: Scandinavian Institute of African Studies, 1973. 360 p., bibl., maps, tabls.
   Contains the papers presented in 1972 at a seminar at the Scandinavian Institute.

515 CERVENKA, Z. "The Limitations Imposed on African Land-locked Countries." In: Cervenka, Z., (ed.). Land-locked Countries of Africa. Uppsala, 1973, 17-33, map, tabls.
A general review as an introduction to the essays contained in the volume.

516 _____. "The Need for a Continental Approach to the Problems of African Land-locked States." In: Cervenka, Z., (ed.). Land-locked Countries of Africa. Uppsala, 1973, 316-28.
Bilateral and regional approaches to the problems of land-lockedness are insufficient. A continental approach is needed.

517 _____. "The Right of Access to the Sea of African Land-locked Countries." Verfassung und Recht in Übersee, 6, 3, 1973, 299-310.
The land-locked states of Botswana, Lesotho, Swaziland, and Zambia face special problems because of the politics of race in Southern Africa.

518 CHALIAND, GÉRARD. "The Horn of Africa's Dilemma." Foreign Policy, 30, Spring 1978, 116-31.

519 CHARLIER, T. "A Propos des Conflits de Frontières entre la Somalie, l'Ethiopie et le Kenya." Revue Française de Science Politique, 16, 2, April 1966, 310-19.
Background to the continual conflict between Ethiopia, Somalia, and Kenya.

520 CHARTRAND, P. E. "The Organization of African Unity and African Refugees: A Progress Report." World Affairs, 137, 4, Spring 1975, 265-85.
Analysis of the legal status of refugees and the obligations of host countries under African international law.

521 CHAULEUR, P. "L'Afrique Divisée mais Solidaire." Études, February 1976, 189-202.
Relations between African states and between them and the great powers are related to problems of economic development.

522 CHIME, SAMUEL. "The Organization of African Unity and African Boundaries." In: Widstrand, Carl G., (ed.). African Boundary Problems. New York, 1969, 65-78.
The structures within the OAU utilized for the settlement of disputes and some brief case studies of the structures in action are presented.

523 CHUKWURA, A. O. "Organization of African Unity and African Territorial and Boundary Problems 1963-1973." Nigerian Journal of International Studies, 1, 1, July 1975, 56-81.

524 CLAPHAM, CHRISTOPHER. "Ethiopia and Somalia." In: International Institute of Strategic Studies. Conflicts in Africa. London, 1972, 1-24, ill., maps, tabls.

525 COLA ALBERICH, JULIO. "Africa, Humillada." Revista de Política Internacional, 150, March/April 1977, 157-182; 151, May/June 1977, 125-44; 153, September/October 1977, 169-82.
    A review and analysis of events in Africa during 1976 and 1977. Africa has been "humiliated" by the actions of its tyrannical rulers and by the Soviet-Cuban takeover of much of the continent.

526 COLLINS, J. N. "Foreign Conflict Behavior and Domestic Disorder in Africa." In: Wilkenfeld, J., (ed.). Conflict Behavior and Linkage Politics. New York, 1973, 251-93, tabls.
    "It is the intent of this research to examine one widely hypothesized sufficient condition of foreign conflict behavior - the condition of domestic disorder as it applies in the African International System."

527 CONSTANTIN, F. "Les Etats Africains et la Représentation de la Chine aux Nations Unies." Année Afrique 1971, 1971, 273-308.
    This study of UN voting indicates that divisions between the African states have prevented them from effectively using their bloc voting power at the UN.

528 _____. "Les Etats Africains Face à la Guerre du Nigéria." Année Afrique 1969, 1969, 114-39.
    African governments are faced with a dilemma when asked to recognize a secessionist government.

529 _____. "Frontières Douteuses et Leaders Incertains dans l'Afrique Noire de 1974." Année Africaine 1974, 1974, 183-205.
    Description of continental relationships during 1974.

530 COPSON, RAYMOND W. "Foreign Policy Conflict among African States, 1964-1969." In: McGowan, P. J., (ed.). Sage International Yearbook of Foreign Policy Studies, 1, 1973, 189-217, figs., tabls.
    Event data analysis is used in an attempt to find answers to two questions. "Are there variables in several social domains which enable us to explain or predict conflict?" "Can any trend toward the resolution or termination of conflict be discerned?"

531 CUÉNOD, JACQUES. "The Problem of Rwandese and Sudanese Refugees." In: Hamrell, S., (ed.). Refugee Problems in Africa. Uppsala, 1967, 45-54.
   A case study of problems and efforts to resolve them. The role of the UNHCR is stressed.

532 DALE, RICHARD. "Refugees from South Africa to Botswana: The External Ramifications of Internal Security Policy." Cultures et Développement, 6, 2, 1974, 305-29.
   A quantitative analysis of the relationships between two nations in terms of domestic political system and the production of refugees.

533 DECRAENE, R. "Problèmes et Tensions entre Etats d'Afrique Noire." Etudes Internationales, 1, 4, December 1970, 12-6.
   Economic differences, ideology, and personality conflicts as causes of international tension.

534 DESCHAMPS, HUBERT. "Les Frontières de la Sénégambie." Revue Française d'Études Politiques Africaines, 80, August 1972, 44-57.

535 DIABATE, M. "Problèmes de l'Appartenance des Etats Africains Francophones à la 'Communauté' et des Etats Africains Anglophones au 'Commonwealth'." Présence Africaine, 96, 1975, 595-628, tabls.
   The development of relations between anglophone and francophone states between 1960 and 1975 has not been extensive.

536 DOOB, LEONARD W., (ed.). Resolving Conflict in Africa: The Fermeda Workshop. New Haven: Yale University, 1970. 209 p.
   An attempt to promote the solution of international conflict through informal conversation between citizens of the conflicting states - Kenya, Somalia, and Ethiopia in this instance.

537 DU BOIS, VICTOR D. "The Economic, Social, and Political Implications of Voltaic Migration to the Ivory Coast." American Universities Field Staff Reports, West Africa, 14, 1, 1972, 1-9.
   Migration from Upper Volta has important domestic and international political and economic effects.

538 ELDRIDGE, JOHN. "Education and Training of Refugees and Their Potential Contribution to Development." In: Hamrell, S., (ed.). Refugee Problems in Africa. Uppsala, 1967, 65-84.
   The author describes the various programs for education and training of refugees in Southern Africa and exposes some of the problems and benefits of these programs.

539 FARAH, NURUDDIN. "I Negus, Menghistu e la Via del Mare." Politica Internazionale, 4, April 1978, 18-26.
"The Negus, Mengistu and the Road to the Sea." Present conflicts in the Horn of Africa are a continuation of Ethiopia's ancient and unfulfilled desire for access to the Red Sea.

540 FARER, TOM J. "Dilemmas on the Horn." Africa Report, 22, 2, March/April 1977, 2-6.

541 FORSYTH, FREDERICK. The Making of an African Legend: The Biafra Story. Harmondsworth: Penguin, 1977. 286 p., maps.
International aspects of the Nigerian Civil War are included in this report by a newspaper correspondent who covered Biafra.

542 FRADE, FERNANDO. "El Conflicto del Cuerno de Africa." Revista de Política Internacional, 156, March/April 1978, 161-76.

543 FRANCK, THOMAS M. "Afference, Efference and Legitimacy in Africa." In: El-Ayouty, Y. and Brooks, H. C., (eds.). Africa and International Organization. The Hague, 1974, 3-10.
The author takes issue with the often repeated statements that African boundaries are artificial and that African states suffer from a lack or weakness of nationalism.

544 _____. "The Stealing of the Sahara." American Journal of International Law, 70, 4, October 1976, 694-721.
The settlement of the Spanish Sahara issue is not in keeping with past UN decolonization decisions. The use of force by Morocco and Mauritania is a bad precedent.

545 GLAGOW, R. "Das Rote Meer: Eine Neue Konfliktregion? Politik und Sicherheit im Afro-Arabischen Grenzbereich." Orient, 18, 2, June 1977, 16-50.
"The Red Sea: A New Region of Conflict? Politics and Security in an Afro-Arab Border Region." Egypt and Saudi Arabia are competing for influence in the Red Sea, and there are great power complications. The Ethiopian and Somalian problems are involved.

546 GROMYKO, ANATOLII A. "Colonialism and Territorial Conflicts in Africa: Some Comments." In: Widstrand, Carl G., (ed.). African Boundary Problems. New York, 1969, 165-67.
"Africa's territorial problems have a father, whose name is colonialism." The author is a member of the Africa Institute, Moscow.

547 HAMRELL, SVEN, (ed.). Refugee Problems in Africa. Uppsala: Scandinavian Institute of African Studies, 1967. 123 p., tabls.
Contains case studies and discussion of several general aspects of the refugee problem. See references under Aall, Cuénod, Legum, Omari, Matthews, George Ivan Smith, and Eldridge.

548 HANSEN, ART and OLIVER-SMITH, ANTHONY, (eds.). Involuntary Migration and Resettlement: The Problems and Responses of Dislocated Peoples. Boulder, Col.: Westview, 1981.
Contains several essays based on research in Africa.

549 HENDERSON, ROBERT D'A. "Relations of Neighbourliness - Malawi and Portugal, 1964-74." Journal of Modern African Studies, 15, 3, September 1977, 425-55.
An "issue areas" approach to bilateral relations.

550 HERTSLET, E. The Map of Africa by Treaty. London: Frank Cass, 1967. 1354 p., index, maps. [Reprint of 1908 edition].
This is an important reference for the study of African boundaries and conflicts concerning these boundaries.

551 HILL, CHRISTOPHER R. "The Botswana-Zambia Boundary Question." Round Table, October 1973, 535-41, map.
Boundary problems must be resolved and new road construction completed to alleviate Botswana'a dependence on routes to the south.

552 HIYĀLĪ, SULAYMĀN YŪSUF. "Ittijahat al-Diblūmāsīyah al-Lībī-yah fī Ifrīqiyā." Al-Siyāsah al-Duwalīyah, 33, July 1973, 176-80.
Libya's relations with black African states since 1969.

553 HOLBORN, LOUISE W. Refugees: A Problem of Our Time: The Work of the United Nations High Commission for Refugees, 1951-1972. Metuchen, N. J.: Scarecrow, 1975. 1525 p., bibl., ill., index, maps, tabls.
This general study contains several sections discussing refugees in Africa, their status and rights, and the efforts to develop programs and protection for them. See especially, "Refugees in Africa" and "Programs in the Major African Countries of Asylum," pages 823-1397.

554 HOVET, THOMAS, JR. "Boundary Disputes and Tensions as a Cause of Refugees." In: Brooks, Hugh C. and El-Ayouty, Y., (eds.). Refugees South of the Sahara. Westport, 1970, 21-32.

555 HUGHES, ANTHONY J. "Dennis Akumu, Secretary General of the Organization of African Trade Union Unity." Africa Report, 22, 6, November/December 1977, 49-55.

556 HULL, GALEN. "Internationalizing the Shaba Conflict." Africa Report, 22, 4, July/August 1977, 4-9.

557 HUNTER, FREDERICK. "Arabs and Africans." Worldview, 15, 12, December 1972, 25-8.

558 International Institute for Strategic Studies. Conflicts in Africa. London: The Institute, 1972. 52 p. [Adelphi Papers, 93].
   Contains essays by Clapham on the Ethiopia-Somalia conflict, Abir on Red Sea politics, and Chad.

559 IRVINE, K. "Sahel, Africa's Great Divide: Friction between Arab and Black Africa." Current History, 64, March 1973, 118-20.

560 KAPIL, R. L. "On the Conflict Potential of Inherited Boundaries in Africa." World Politics, 18, 4, July 1966, 656-73.
   The author establishes a typology of boundaries and analyzes the conflict potential of each type.

561 KATOND, DIUR. "Los Obstaculos al Encuadramiento de los Refugiados Africanos por la OUA." Revista de Politique Internacional, 140, July/August 1975, 183-90.

562 KIMMINICH, O. "Der Schutz der Politischen Flüchtlinge in Afrika." Verfassung und Rechtin Übersee, 3, 4, 1970, 443-59.
   "The Protection of Political Refugees in Africa." Analysis of UN and OAU resolutions on refugee rights and state responsibilities.

563 KIRK-GREENE, A. H. M. "Diplomacy and Diplomats: The Formation of Foreign Service Cadres in Black Africa." In: Ingham, K., (ed.). Foreign Relations of African States. London, 1974, 279-322, figs., tabls.
   "This relatively pioneer attempt to examine the formation of foreign service cadres in anglophone Black Africa" makes use of quantitative data.

564 LABOUZ, M. F. "Le Reglement du Contentieux Frontalier de l'Ouest Maghrébin: Aspects Politiques et Juridiques." Maghreb, 53, September/October 1972, 50-4.
   An optimistic view preceeding the outbreak of war in this region.

565 LANNE, BERNARD. "Les Frontières du Tchad et de la Libye." Revue Juridique et Politique, Indépendance et Coopération, 31, 3, July-September 1977, 953-66.
   The present conflict between Chad and Libya is placed in historical perspective. Agreements made in 1899, 1919, and 1935 by France, Britain, Turkey, and Italy cause confusion over border locations.

566 LEGUM, COLIN and LEE, BILL. Conflict in the Horn of Africa. London: Rex Collings, 1977. 95 p., map, tabl.
   Foreign intervention in the conflict is discussed in some detail.

567 LEGUM, MARGARET. "Problems of Asylum for Southern African Refugees." In: Hamrell, S., (ed.). Refugee Problems in Africa. Uppsala, 1967, 54-64.
   A summary of the needs of the refugees, host country attitudes concerning refugees, and proposals for future steps to be taken on behalf of the refugees.

568 LEMARCHAND, R. "The Limits of Self-Determination: The Case of the Katanga Secession." American Political Science Review, 56, 2, June 1962, 404-16.
   In spite of external and settler support, there is no genuine basis for Katangese secession.

569 LEWIS, I. M. "Culture and Conflict in Africa." Millenium: Journal of International Studies, 6, 2, Autumn 1977, 175-81.

570 LEWIS, WILLIAM H. "Islam and Nationalism in Africa." In: Kerekes, T., (ed.). The Arab Middle East and Muslim Africa. New York, 1961, 63-84.
   The author discusses the spread of Islam across state boundaries and the possible effects of this spread on international relations.

571 LIEBENOW, J. GUS. "The Caucus Race: International Conflict in East Africa and the Horn." American Universities Field Staff Reports, General, 11, 1, 1977, 1-9.
   "Conflict in East Africa and the Horn has already passed from the potential to actual. In terms of complexity of issues, it rivals if not surpasses the situation in Southern Africa, and raises the possibility of direct Great Power confrontation."

572 LINIGER-GOUMAZ, MAX. "Eléments de Bibliographie: Gabon-Guinée Équatoriale - Problème Frontalier." Genève-Afrique, 12, 1, 1973, 99-102.
   Bibliography of articles and books concerned with the Gabon-Equatorial Guinea border. References date from 1860 to 1968.

573 LIPPERT, ANNE. "Emergence or Submergence of a Potential State: The Struggle in Western Sahara." Africa Today, 24, 1, January-March 1977, 41-60, map.

574 LOBBAN, RICHARD. "The Eritrean War: Issues and Implications." Revue Canadienne des Études Africaines, 10, 2, 1976, 335-46.

575 MARAIS, BEN. "Islam: Political Factor in Africa." Bulletin of the Africa Institute of South Africa, 9, 2, March 1971, 51-64.

576 MARIAM, M. W. "Background to the Ethio-Somalian Boundary Dispute." Journal of Modern African Studies, 2, 2, July 1964, 189-220.
    The genesis of the boundaries of this area and its relation to the present conflict between Ethiopia and Somalia.

577 MARKS, T. A. "Spanish Sahara - Background to Conflict." African Affairs, 75, 298, January 1976, 3-13, map.
    International and domestic factors made it difficult for Spain to give up the colony of Spanish Sahara.

578 MATTHEWS, ROBERT O. "Domestic and Interstate Conflict in Africa." International Journal, 25, 3, Summer 1970, 459-85.

579 _____. "Interstate Conflicts in Africa: A Review." International Organization, 24, 2, Spring 1970, 335-60.
    A review of recent publications on interstate conflict in Africa. The author concludes that African states are unlikely to engage in external adventures and will tend to support the international status quo.

580 _____. "Refugees and Stability in Africa." International Organization, 26, 1, Winter 1972, 62-83, tabls.
    Describes the international ramifications of African refugees in order to learn to what extent they contributed to interstate tensions in Africa, and to discover what means the African states have devised for resolving this problem.

581 MATTHEWS, Z. K. "The Role of Voluntary Organisations in the Refugee Situation in Africa." In: Hamrell, S., (ed.). Refugee Problems in Africa. Uppsala, 1967, 97-109.
    Voluntary organizations have various advantages in respect of government organizations in refugee work. Examples from Africa are presented.

582 MAYALL, JAMES. "The Battle for the Horn: Somali Irredentism and International Diplomacy." World Today, 34, 9, September 1978, 336-45.

583 MAYALL, J. "The Malawi-Tanzania Boundary Dispute." Journal of Modern African Studies, 11, 4, December 1973, 611-28.
 The dispute is related to the different policies the two states have for South Africa.

584 MAZRUI, ALI A. "Inter-African Migration: A Case Study of the East African Community." Round Table, 242, April 1971, 293-300.

585 _____. "Violent Contiguity and the Politics of Retribalization in Africa." Journal of International Affairs, 23, 1, Winter 1969, 89-105.

586 _____. "Who Are the Afro-Saxons? A Preface." Indian Journal of Politics, 7, 2, July-December 1973, 158-64.
 Further consideration of the role of the English language in Africa. See the Mazruis' earlier article in Political Quarterly (1967).

587 _____ and MAZRUI, M. "The Impact of the English Language on African International Relations." Political Quarterly, 38, 2, April-June 1967, 140-55.
 The English language as a factor in African domestic and international politics is the subject of this article. Also see A. Mazrui's article in the Indian Journal of Politics (1973).

588 McEWEN, A. C. International Boundaries of East Africa. New York: Oxford University, 1971. 321 p., bibl., index, ill., maps.
 Kenya, Ethiopia, Somalia, Sudan, Tanzania, Burundi, Rwanda, Congo Republic, Malawi, Mozambique, Zambia, and Uganda are included in this volume.

589 McKAY, VERNON. "The Impact of Islam on Relations among the New African States." In: Proctor, J. H., (ed.). Islam and International Relations. New York, 1965, 158-93, map, tabls.
 General observations with extended comments on Guinea, Nigeria, Senegal, Cameroon, Chad, and Sudan as states with large Muslim populations and Ghana, Sierra Leone, Upper Volta, and Tanganyika as states with small Muslim populations. North Africa is treated separately.

590 _____. "International Conflict Patterns." In: McKay, Vernon, (ed.). African Diplomacy: Studies in the Determinants of Foreign Policy. New York, 1966, 1-23.
 Analyzes conflict potential between African states, between the great powers over Africa, and between African states and the great powers.

591 MELANDER, G. and NOBEL, P., (eds.). African Refugees and the Law. Uppsala: Scandinavian Institute of African Studies, 1978. 98 p.

592 MERCER, JOHN. "Confrontation in the Western Sahara." World Today, 32, 6, June 1976, 230-35.

593 MÉRIC, E. "Le Conflit Algéro-Marocain." Revue Française de Science Politique, 15, 4, August 1965, 743-52.
The persistent conflict between these two states appears to have its cause in boundary disagreements, but a variety of personality, historical, and other factors are involved, too.

594 MESTRE, T. "Balance de las Relaciones Interafricanas del Africa Independiente." Revista Política Internacional, 100, November/December 1968, 187-211.
"A Survey of Inter-African Relations in Independent Africa." The author notes the tendency for various African states to group together in ideological blocs, both prior to and after the formation of the OAU.

595 METCALFE, GEORGE L. "Effects of Refugees on the National State." In: Brooks, Hugh C. and El-Ayouty, Y., (eds.). Refugees South of the Sahara: An African Dilemma. Westport, 1970, 73-88.
The existence of so many refugees represents "the most outstanding plight of a continent seeking refuge from the problems of transition."

596 MEYRIAT, J. "Minorités Ethniques et Conflits Internationaux: Note Introductive." Revue Française de Science Politique, 17, 4, August 1967, 713-17.
This special issue contains five essays, one of which is concerned with Dahomey. See the reference under S. Bonzon.

597 MITTELMAN, JAMES H. "The Uganda Coup and the Internationalization of Political Violence." Munger Africana Library Notes, 14, 1972, 1-32, maps.
Analyzes the impact of the Ugandan coup of January 25, 1971 on African international relations.

598 MORGAN, EDWARD. "A Geographic Evaluation of the Ethiopia-Eritrea Conflict." Journal of Modern African Studies, 15, 4, December 1977, 667-74, map.

599 MTSHALI, B. V. "The Zambian Foreign Service, 1964-1972." African Review, 5, 3, 1975, 303-16, tabls.
The Foreign Service was initially staffed by transfers from other parts of the civil service, and thus the Foreign Service lacked experience. The author discusses other problems and the functioning of the organization.

600 MUSHKAT, MARION. "L'Afrique, le Tiers Monde et le Système Collectif de Sécurité Économique et Politique Internationale." Africa, 33, 1, March 1978, 1-22.
Tribal factors are the essence of the constant warfare in Africa. Suggestions to bring peace are presented.

601 MUSTAFA, ZUBEIDA. "The Eritrean Problem: Its International Implications." Pakistan Horizon, 28, 2, 1975, 67-70.

602 NOLDE, A. "Djibouti: Indépendance, Oui, Mais." Défense Nationale, February 1976, 69-77.
Summarizes the precarious international position of Djibouti, located between Ethiopia and Somalia and suffering internally from severe ethnic conflict.

603 OMARI, T. PETER. "From Refugee to Emigré: African Solutions to the Refugee Problem." In: Hamrell, S., (ed.). Refugee Problems in Africa. Uppsala, 1967, 85-96.
African problems should be solved by African governments with assistance from other parts of the world.

604 OSUNTOKUN, JIDE. "Relations between Nigeria and Fernando Po from Colonial Times to the Present." In: Akinyemi, A. B., (ed.). Nigeria and the World. Ibadan, 1978, 1-17.
"The emergence of a nationalist government in Nigeria and the closure of an American radio relay station in Kaduna, has led Fernando Po into the warm embrace of the United States and to the establishment of a strong American presence on the island."

605 PAYNE, R. M. "Divided Tribes: A Discussion of African Boundary Problems." New York University Journal of International Law and Politics, 2, 2, Winter 1969, 243-66.

606 PEDERSON, OLE KARUP and LEYS, ROGER. "A Theoretical Approach to the Problems of African Land-locked States." In: Cervenka, Z., (ed.). Land-locked Countries of Africa. Uppsala, 1973, 288-92.

607 PEIL, MARGARET. "The Expulsion of West African Aliens." Journal of Modern African Studies, 9, 2, August 1971, 205-29.
The presence of non-citizen Africans in many African states is both a problem for and a tool of African international politics.

608 PERSON, YVES. "L'Afrique Noire et Ses Frontières." Revue Française d'Etudes Politiques Africaines, 80, August 1972, 18-48.

609 POTHOLM, CHRISTIAN P. "Wanderers on the Face of Africa: Refugees in Kenya, Tanzania, Zambia, and Botswana." Round Table, 261, January 1976, 85-92.

610 PRESCOTT, J. R. V. "Africa's Major Boundary Problems." The Australian Geographer, 9, 1, March 1963, 3-12.

611 PRICE, DAVID L. "Morocco and the Sahara: Conflict and Development." Conflict Studies, 88, October 1977, 3-16.
A brief analysis of the causes of the conflict is followed by suggestions that a solution may be found through OAU and Arab League mediation and a regional economic program.

612 "The Problem of African Refugees." In: Hamrell, S., (ed.). Refugee Problems in Africa. Uppsala, 1967, 9-25, tabls.
Refugees are one of Africa's most acute problems and Africa's refugees are one of the world's most serious refugee problems. This introductory essay gives definitions and outlines the scope of the problem.

613 RAMCHANDANI, R. R. "Conflicts in the Horn of Africa and Western Sahara." IDSA Journal, 9, 4, April-June 1977, 449-73.
After describing the two conflict situations the author considers the problem of causation. Are the causes internal or a by-product of East-West relations?

614 REINER, ANTHONY S. "The Length and Status of International Boundaries in Africa: A List." In: Widstrand, Carl G., (ed.). African Boundry Problems. New York, 1969, 186-89.

615 REISMAN, W. MICHAEL. "African Imperialism." American Journal of International Law, 70, 4, October 1976, 801-2.

616 REYNER, ANTHONY S. Current Boundary Problems in Africa. Pittsburg: Institute of African Affairs, Duquesne University, 1964. [Africa Reprint, 15].

617 _____. "Length and Status of International Boundaries in Africa." African Studies Bulletin, 10, 2, September 1967, 6-9.

618 _____. "Morocco's International Boundaries: A Factual Background." Journal of Modern African Studies, 1, 3, September 1963, 313-26, maps.

619 _____. The Republic of the Congo: Development of Its International Boundaries. Pittsburg: Duquesne University, 1961. 8 p.

620 RONDOT, PHILIPPE. "La Mer Rouge Peut-elle Devenir un 'Lac de Paix' Arabe?" Défense Nationale, October 1977, 71-84.
The reopening of the Suez Canal means that the Red Sea has become a focus of international tension. The author examines Arab attitudes on this situation.

621 RUBIN, NEVILLE. "Africa and Refugees." African Affairs, 73, 292, July 1974, 290-311, tabls.
An introduction to the various aspects and problems of, as well as statistics on, refugees in Africa.

622 RYAN, SELWYN D. "Civil Conflict and External Involvement in Eastern Africa." International Journal, 28, 3, Summer 1973, 465-510.

623 SABOURIN, LOUIS. "Problems and Prospects of the Seven Landlocked Countries of French-speaking Africa: Central African Republic, Chad, Mali, Niger, Upper Volta, Rwanda, and Burundi." In: Cervenka, S., (ed.). Land-locked Countries of Africa. Uppsala, 1973, 146-57, tabl.

624 SCHACK, A. VON. "Schwarzafrika und Sein 'Weisser' Norden." Aussenpolitik, 28, 1, 1977, 102-13.
"Black Africa and Its 'White' North." Contains some discussion of relations between North Africa and the other African states.

625 SCHRÖDER, D. "Inter-African Diplomatic Communication." International Problems, 13, 1-3, January 1974, 75-87.

626 SHAMS, B. Conflict in the African Horn." Current History, 73, 432, December 1977, 199-204.

627 SHAW, TIMOTHY M. "Inter-State Rivalries: Another Barrier to Unity." Africa Today, 24, 3, July-September 1977, 69-71.
Review of Ghana and Nigeria 1957-70 by Olajide Aluko.

628 SHEIK-ABDI, ABDI. "Somali Nationalism: Its Origins and Future." Journal of Modern African Studies, 15, 4, December 1977, 657-65.
The force of Somali nationalism will continue to cause conflict in the Horn of Africa until all Somali people are reunited. The OAU must resolve this and all other situations caused by illogical colonial borders. The author is Somali.

629 SKURNIK, W. A. E. "Ghana and Guinea, 1966: A Case Study in Inter-African Relations." Journal of Modern African Studies, 5, 3, November 1967, 369-84.
Recent events in relations between Ghana and Guinea suggest that there is a fundamental conflict of ideologies in African international relations.

630 SMITH, GEORGE I. "The Role of the United Nations." In: Hamrell, S., (ed.). Refugee Problems in Africa. Uppsala, 1967, 110-21.
A very general note on various aspects of the UN's role, voluntary associations, and other factors involved with African refugees.

631 SMITH, G. I. "Some Aspects of the Refugee Problem South of the Sahara." In: Brooks, Hugh C. and El-Ayouty, Y., (eds.). Refugees South of the Sahara. Westport, 1970, 33-44.
   Africa's major refugee problems will arise in Southern Africa.

632 SUDARKASA, N. "Commercial Migration in West Africa, with Special Reference to the Yoruba in Ghana." African Urban Notes, Ser. B, 1, Winter 1974/1975, 61-103, bibl., tabls.
   Outlines some of the research problems involved in the study of commercial migration and presents descriptive data on Yoruba migrants as an important example of this phenomenon.

633 SVENDSEN, KNUD E. "The Economics of the Boundaries in West, Central and East Africa." In: Widstrand, Carl G., (ed.). African Boundary Problems. New York, 1969, 33-64, tabls.
   Discusses the interaction between the need for economic development and the boundary frontiers in Africa.

634 TÄGIL, SVEN. "The Study of Boundaries and Boundary Disputes." In: Widstrand, Carl G., (ed.). African Boundary Problems. New York, 1969, 22-32.
   "The foremost purpose of this lecture is to draw attention to some general features of international boundaries." Author is a historian.

635 TANDON, YASH. "Military Coups and Inter-African Diplomacy." Africa Quarterly, 6, 4, January-March 1967, 278-84.

636 THAHANE, T. T. "Lesotho, an Island Country: The Problems of Being Land-locked." African Review, 4, 2, 1974, 279-90.
   A theoretical foreward is followed by a case-study of Lesotho.

637 _____. "Lesotho, the Realities of Land-lockedness." In: Cervenka, Z., (ed.). Land-locked Countries of Africa. Uppsala, 1973, 239-49.

638 THOLOMIER, R. "La Place du Territoire Français des Afars et des Issas entre Afrique et Asie." Revue Française d'Etudes Politiques Africaines, 124, April 1976, 67-90.
   A summary of attitudes of the governments of the Horn of Africa.

639 THOMPSON, VIRGINIA and ADLOFF, RICHARD. Djibouti and the Horn of Africa. London: Oxford University. Stanford: Stanford University, 1968. 246 p., bibl., ill., index, map.
   A chapter on external relations with Ethiopia, Somalia, Aden, South Africa, and Egypt is included.

640 THOMPSON, W. SCOTT and ZARTMAN, I. WILLIAM. "The Development of Norms in the African System." In: El-Ayouty, Yassin, (ed.). The Organization of African Unity after Ten Years: Comparative Perspective. New York, 1975, 3-46.
   OAU deliberations are an appropriate source for the study of the emergence of norms of behavior in African relations.

641 THURSTON, RAYMOND. "Détente in the Horn." Africa Report, 14, 2, February 1969, 6-13.

642 TOMLIN, B. W. and BUHLMAN, M. A. "Relative Status and Foreign Policy: Status Partitioning and the Analysis of Relations in Black Africa." Journal of Conflict Resolution, 21, 2, June 1977, 187-216, bibl., tabls.
   A revision and extension of status-field theory is applied to the analysis of foreign policy interactions among 32 black African states in the middle 1960s. The findings support the utility of the partioning procedure and confirm the relative importance of power in relations between states.

643 TOUVAL, SAADIA. "Africa's Frontiers: Reactions to a Colonial Legacy." International Affairs, 42, 4, October 1966, 641-54.
   African borders are no more artifical tham borders anywhere else. So far there have been remarkably few conflicts over borders, but this may change in the future.

645 _____. The Boundary Politics of Independent Africa. Cambridge: Harvard University, 1972. 334 p., index.
   Widespread belief that Africa would suffer serious boundary conflicts after independence have been unfounded. "This book attempts to explain why."

646 _____. "The Organization of African Unity and African Borders." International Organization, 21, 1, Winter 1967, 102-27.
   African acceptance of colonial borders and the role of the OAU in this acceptance are examined. The OAU role in resolving several border disputes is described.

647 _____. Somali Nationalism: International Politics and the Drive for Unity in the Horn of Africa. Cambridge: Harvard University, 1963. 214 p., index.
   "This is a study of one of the more complex and little-known problems of contemporary Africa - the Somali claims for national self-determination and unification, and their effect upon regional and international politics."

648 TOUVAL, S. "The Sources of Status Quo and Irredentist Policies." In: Widstrand, Carl G., (ed.). African Boundary Problems. New York, 1969, 101-18.
Why have some states adopted irredentist policies but others have advocated adherence to the status quo?

649 TRIKI, MAHMOUD. "La Crisi del Sahara Occidentale." Affari Esteri, 37, January 1978, 118-34.
"The Crisis of the Western Sahara." An analysis of the dispute over the territory of the ex-Spanish Sahara between Algeria, Morocco, Mauritania, and elements of the territory's population.

650 UNITED NATIONS. Economic and Social Council. Economic Commission for Africa. Report of the Conference on the Legal, Economic and Social Aspects of African Refugee Problems. Addis Ababa, 1969. 224 p.

651 VERDUN, L. G. "Les Relations Sénégal-Gambie." Revue Juridique et Politique, 20, 3, July-September 1966, 477-88.
The agreement of February 1965 between Senegal and Gambia is the basis of their relations with each other.

652 WATERS, ROBERT. "Inter-African Boundary Disputes: A List and a Map." In: Widstrand, Carl G., (ed.). African Boundary Problems. New York, 1969, 183-5.

653 WELADJI, C. "The Cameroon-Nigeria Border." Abbia, 27/28, June 1974, 157-72; 29/30, 1975, 163-95, map; 31/32/33, February 1978, 173-93.
Mainly documents with some interpretation.

654 WHITEMAN, K. "Guinea in West African Politics." World Today, 27, 8, August 1971, 350-8.
Guinea's position in West African politics has been uncompromising. It has been difficult to maintain cooperative relationships with neighboring states.

655 WIDSTRAND, CARL G., (ed.). African Boundary Problems. Uppsala: Scandanavian Institute of African Studies. New York: Africana, 1969. 202 p., map, tabls.
This volume contains ten essays and two appendices. Boundary disputes, economic effects of boundaries, and the OAU role in boundary conflicts are among the topics covered.

656 ———. "Some African Boundary Problems: A Discussion." In: Widstrand, Carl G., (ed.). African Boundary Problems. New York, 1969, 168-82.
A summary of the comments made by participants in a seminar on African boundary problems.

657 WILD, PATRICIA, B. "The Organization of African Unity and the Algerian-Moroccan Border Conflict." In: Tharp, Paul A., Jr., (ed.). Regional International Organizations. New York, 1971, 182-99.
   This essay is in a section titled "Rule Adjudication" in a book that applies systems theory to regional organizations.

658 _____. "The Organization of African Unity and the Algerian-Moroccan Border Conflict. A Study of New Machinery for Peace-Keeping and for the Peaceful Settlement of Disputes among African States." International Organization, 20, 1, Winter 1966, 18-36.
   The Algerian-Moroccan border dispute was the first test of the OAU's dispute settlement procedures. The OAU met with considerable success.

659 WILLIAMS, D. "How Deep the Split in West Africa?" Foreign Affairs, 40, 1, October 1961, 118-27.
   The Casablanca-Monrovia split in African international relations is not permanent. The main obstacle to African unity is the anglophone-francophone division.

660 WINE, JAMES. "The U. S. State Department's Attitude toward the Problem of Refugees." In: Brooks, Hugh C. and El-Ayouty, Y., (eds.). Refugees South of the Sahara. Westport, 1970, 201-8.
   A statement of US policy by a State Department officer. Some program description is included.

661 WORONOFF, JON. "Différends Frontaliers en Afrique." Revue Française d'Etudes Politiques Africaines, 80, August 1972, 58-78.
   Case studies of African border conflicts (Algeria-Morocco and Somalia-Ethiopia/Kenya) indicate that only partial solutions have been reached. Border problems may still harass Africa unless general solutions to the basic problems are now agreed upon.

662 _____. "L'Organisation de l'Unité Africaine et le Problème des Réfugiés." Revue Française d'Etudes Politiques Africaines, 93, September 1973, 86-97.
   A variety of organizations and agreements within the OAU assist in the solution of Africa's refugee problems.

663 YAKEMTCHOUK, R. "A Propos de Quelques Cas de Reconnaissance d'Etat et de Gouvernement en Afrique." Revue Belge de Droit International, 6, 2, 1970, 504-26.
   Recognition of states and governments has posed important political problems for African regimes. The author presents a typology of such problems and analyzes several cases as examples.

664 YAKEMTCHOUK, R. "Les Frontières Africaines." Revue Générale de Droit International Public, 74, 1, January-March 1970, 27-68.
International law has played an important role in the solution of boundary disputes in Africa.

665 ZARTMAN, I. WILLIAM. "Decision-making among African Governments on Inter-African Affairs." Journal of Development Studies, 2, 2, January 1966, 98-119. Also in: Doro, M. E. and Stultz, N. M., (eds.). Governing in Black Africa. Englewood Cliffs, N. J., 1970, 129-43.

666 _____. "The Foreign and Military Politics of African Boundary Problems." In: Widstrand, Carl G., (ed.) African Boundary Problems. New York, 1969, 79-100.
"Any African state can have boundary problems if it wants." Discusses the underlying factors that lead to boundary conflicts.

667 _____. "The Politics of Boundaries in North and West Africa." Journal of Modern African Studies, 3, 2, August 1965, 155-74.
A general survey of boundary problems with detailed analysis of five cases and the manner in which several other cases were settled.

Sub-continental Regionalism

668 ABANGWU, GEORGE C. "Systems Approach to Regional Integration in West Africa." Journal of Common Market Studies, 13, 1/2, 1975, 116-33.
Enough technical-economic studies on West African integration have been conducted, but there is not sufficient "human input" in the integration movement.

669 _____. "Towards West African Unity." Africa Contemporary Record, 1974/75, c190-3.
The author develops a new typology for classifying regional organizations and applies it to the West African case.

670 ABDEL-SALAM, OSMAN H. "The Evolution of African Monetary Institutions." Journal of Modern African Studies, 8, 3, October 1970, 339-62.
There are six major monetary groups on the continent, four of which are directly linked to European states. Currency characteristics and the structure of the monetary and banking institutions remain tied to the former colonial administration.

671 ADAMS, JOHN. "External Linkages of National Economies in West Africa." African Urban Notes, Fall 1972, 97-116, diagrs., tabls.
The author examines supranational linkages in West Africa in an attempt to analyze "a general enthusiasm for greater West African trade and economic cooperation."

672 ADEDEJI, ADEBAYO. "Prospects of Regional Economic Cooperation in West Africa." Journal of Modern African Studies, 8, 2, July 1970, 213-31.
"The purpose of this article is to evaluate the experience of West Africa in bringing about some form of regional co-operation during the past decade."

673 ADERIBIGBE, A. B. "West African Integration, an Historical Perspective." Nigerian Journal of Economic and Social Studies, 5, March 1963, 9-14.

674 AJOMO, M. A. "Regional Economic Organizations: The African Experience." International and Comparative Law Quarterly, 25, 1, January 1976, 58-101.
Author describes and analyzes numerous difficulties for integration in Africa, but argues that a major positive factor is the growing recognition of the need for integration.

675 AKINTAN, S. A. The Law of International Economic Institutions in Africa. Leyden: A. W. Sijthoff, 1977. 222 p., index.
After a general introduction there are chapters analyzing the ECA, ADB, EAC, EA Development Bank, UDEAC, ECOWAS, and a final chapter on Air Afrique, WAMU, and other specialized organizations.

676 AKIWUMI, A. M. "La Solución de Conflictos en los Procesos de Integración Económica de Africa, con Especial Referencia a la Comunidad de Africa Oriental." Derecho de la Integración, 10, April 1972, 77-95.

677 ALUKO, SAMUEL A. "Regional Economic Development in Africa." In: Arkhurst, F. S., (ed.). Arms and African Development. New York, 1972, 119-27.
The conditions for and benefits from regional economic integration are examined and several African organizations (EAEC, South African Customs Union, et al.) are described.

678 AMACHREE, IGOLIMA T. D. "Problems of Integration in West Africa." The American Journal of Economics and Sociology, 32, 2, April 1973, 216-9.

679 ANGLIN, DOUGLAS G. "Zambian Disengagement from Southern Africa and Integration with East Africa, 1964-1972: A Transaction Analysis." In: Shaw, T. M. and Heard, K. A., (eds.). Cooperation and Conflict in Southern Africa. Washington, D. C., 1977, 228-89, tabls.
"This paper attempts (1) to assess the scope of Zambia's disengagement from the Southern African system and of its integration with East Africa, (2) to compare the extent of change in different functional issue areas, and (3) to distinguish between behavioural patterns at the governmental and societal levels."

680 ARNOLD, GUY and WEISS, RUTH. Strategic Highways of Africa. New York: St. Martin's, 1977. 178 p., bibl., index, maps.
Major sections on Southern, Central, and West Africa.

681 ASFOUR, EDMOND Y. "International Cooperation in Africa and the Establishment of the Economic Commission for Africa." Civilizations, 10, 2, 1960, 181-9.

682 ASKEROV, A. "OCAM." International Affairs, 1, January 1975, 143-5.

683 ASSEFA, MEHRETU. Regional Integration for Economic Development of Greater East Africa. Kampala: Uganda Publishing House, 1973. 150 p., bibl., ill., maps.
Mathematical models of economic integration.

684 BALANDIER, G. "Remarques sur les Regroupements Politiques Africains." Revue Française de Science Politiques, 10, 4, December 1960, 841-9.
As Africa becomes independent, regional integration or cooperation may be an answer to many problems, but there are many difficulties with such plans.

685 BALLANCE, FRANK. Zambia and the East African Community. Syracuse, N. Y.: Program of Eastern African Studies, Syracuse University, 1971. 139 p., bibl.
Analysis of Zambian attempts to join East Africa and to withdraw from Southern Africa.

686 BANFIELD, JANE. "Federation in East Africa." International Journal, 18, 2, Spring 1963, 181-93.
A history of federation in East Africa and analysis of the EACSO.

687 _____. "The Structure and Administration of the East African Common Services Organization." In: Leys, Colin and Robson, P., (eds.). Federation in East Africa. Nairobi, 1965, 30-40.
A constitutional analysis with recommendations for strengthening EACSO.

688 Banque Centrale des Etats de l'Afrique de l'Ouest. Une Communauté Economique Ouest Africaine? Paris: BCEAO, 1968. 200 p., index.
This bibliography contains 984 annotated references organized by subject matter. An author index and list of relevant organizations are included.

689 BARBOUR, KENNETH M. "Industrialization in West Africa: The Need for Sub-Regional Groupings within an Integrated Economic Community." Journal of Modern African Studies, 10, 3, October 1972, 357-82, ill., map, tabls.
The OAU and the ECA must reflect in regional planning the interests being shown in African governments and universities.

690 BELSHAW, D. G. R. "Agricultural Production and Trade in the East African Common Market." In: Leys, Colin and Robson, P., (eds.). Federation in East Africa. Nairobi, 1965, 83 - 101, tabls.
Agriculture poses special problems for the EACM.

691 BENNOUNA, M. "Le Maghreb entre le Mythe et la Réalité." Intégration, 1, 1974, 9-46.
Regional cooperation in North Africa has been more talked about than acted upon.

692 BERRON, HENRI. "L'Association Air Afrique - UTA et les Transports Aériens dans les Pays d'Afrique Noire Francophone." Les Cahiers d'Outre-Mer, 29, 114, April/June 1976, 113-6, maps.
"A frail institution whose good running depends on a good relationship between the associated countries."

693 BIFUKO, BAHARANYI. "Etude des Obstacles Politiques et Economiques aux Regroupements Régionaux en Afrique: Cas Particulier de la Triparitite Burundi-Rwanda-Zaïre." Cahiers Zaïrois d'Etudes Politiques et Sociales, 3, October 1974, 155-70.
Analysis of problems faced in attempts at regional integration.

694 BIRCH, A. H. "Opportunities and Problems of Federation." In: Leys, Colin and Robson, P., (eds.). Federation in East Africa. Nairobi, 1965, 6-29.
A general analysis of the federal principle with numerous references to African cases.

695 BLAGOJEVIĆ, N. "Integracioni Pokreti u Africi." Medunarodni Problemi, 13, 1, January-March 1961, 67-89.
Analysis of various influences for and against regional integration in Africa.

696 BONZON, SUZANNE. "Les Dahoméens en Afrique de l'Ouest." Revue Française de Science Politique, 17, 4, August 1967, 718-26.

697 BORELLA, F. "L'Union des Etats de l'Afrique Centrale." Annuaire Français de Droit International, 14, 1968, 167-77.
An analysis of the economic organization formed in 1968 by Zaire, Chad, and the Central African Republic.

698 BORNSTEIN, RONALD. "The Organization of Senegal River States." Journal of Modern African Studies, 10, 2, July 1972, 267-83.
A pessimistic forecast for this regional organization of Guinea, Mali, Mauritania, and Senegal.

699 BRETT, EDWARD A. "Closer Union in East Africa." In: Austin, D. and Weiler, H., (eds.). Inter-State Relations in Africa. Freiburg, 1965, 51-67.
   The author briefly outlines the historical background and notes the major problems to be solved.

700 BUGNICOURT, J. "Illusions et Réalités de la Région et du Développement Régional en Afrique." Tiers-Monde, 19, 73, January-March 1978, 109-38, bibl.

701 CAMARA, S. S. "La Guinée et la Coopération Économique en Afrique de l'Ouest." Cultures et Développement, 8, 3, 1976, 517-32.
   Guinea remains largely isolated economically and politically from its neighbors. Several causes for this are presented.

702 CARRINGTON, C. E. "Frontiers in Africa." International Affairs, 36, 4, October 1960, 424-39.
   There are no struggling nations in Africa, only political parties. Railway lines will be the basic factor in remaking the map of Africa.

703 CERVENKA, Z. "Africa as a Case of Intraregional Contradictions." International Social Science Journal, 28, 4, 1976, 736-53.

704 CHIME, C. Integration and Politics among African States: Limitations and Horizons of Mid-term Theorizing. Uppsala: Scandinavian Institute of African Studies, 1977. 437 p., bibl., maps, tabls.

705 COLLINS, J. D. "The Clandestine Movement of Groundnuts across the Niger-Nigeria Boundary." Canadian Journal of African Studies, 10, 2, 1976, 259-78, tabls.
   An indication of the inadequacy of African foreign trade data.

706 COLLONTON, CAROL A. "Political Integration in East Africa." Journal of International and Comparative Studies, 5, 3, Fall 1972, 1-26.

707 "La Communauté Economique Est-Africaine." Revue de Défense Nationale, 25, January 1969, 118-24.

708 CONSTANTIN, FRANÇOIS. "L'Intégration Régionale en Afrique Noire. Esquisse sur Dix Années de Recherches." Revue Française de Science Politique, 22, 5, October 1972, 1074-110.
   The wide variety of attempts at regional integration in Africa makes the continent a highly suitable research location.

709 CONSTANTIN, F. "Lagos et les Regroupements en Afrique de l'Ouest." <u>Revue Française d'Etudes Politiques Africaines</u>, 57, September 1970, 12-4.

710 _____. "Régionalisme International et Pouvoirs Africains." <u>Revue Française de Science Politique</u>, <u>26</u>, 1, February 1976, 70-102, tabl.
A discussion of the theories and the realities of regional integration in Africa.

711 COX, RICHARD. <u>Pan-Africanism in Practice - An East African Study: PAFMECSA 1958-1964</u>. London: Oxford University, 1964. 95 p., index, map.

712 CROWDER, MICHAEL. "Colonial Rule in West Africa: Factor for Division or Unity?" <u>Civilisations</u>, <u>14</u>, 3, 1964, 167-78. Also <u>in</u>: Doro, M. <u>and</u> Stultz, N., (eds.). <u>Governing in Black Africa</u>. Englewood Cliffs, N. J., 1970, 299-310.

713 CURRY, ROBERT L., JR. "A Note on West African Economic Cooperation." <u>Journal of Modern African Studies</u>, <u>11</u>, 1, March 1973, 136-8.

714 CURTI GIALDINO, C. <u>and</u> PALMIERI, M. G. "Valorizzazione Internazionale del Bacino del Senegal." <u>Comunita Internazionale</u>, <u>29</u>, 1/2, 1974, 63-80.
Description of the various organizations that have been formed by the states of the Senegal River basin for development and economic cooperation.

715 DECRAENE, P. "Indépendance et Regroupements Politiques en Afrique au Sud du Sahara." <u>Revue Française de Science Politique</u>, <u>10</u>, 4, December 1960, 850-79.
The importance of Pan-Africanist thought and Kwame Nkrumah in African international relations are very great but the author describes numerous obstacles to political integration, regional or continental.

716 DELANCEY, MARK W. "Early Attempts at West African Unity: Ghana, Guinea and Mali." <u>Indian Journal of Politics</u>, <u>7</u>, 1, June 1973, 47-55.
An analysis based on the concepts of Karl Deutsch.

717 _____. "The Ghana-Guinea-Mali Union - A Bibliographic Essay." <u>African Studies Bulletin</u>, <u>9</u>, 2, September 1966, 35-51.

718 DIAGNE, PATHÉ. <u>Pour l'Unité Ouest-Africaine: Micro-états et Intégration Economique</u>. Paris: Anthropos, 1972. 370 p., bibl., tabls.
This volume is divided into three sections - the economic effects of colonial rule, the economy in the first years of independence, and the need and strategies for economic integration.

719 DIENG, DIAKHA. "From UAM to OCAM." African Forum, 1, Fall 1965, 29-35.

720 DOBOSLEWICZ, Z. "Traditional Trade and Population Migration as Integrating Factors in West Africa." Africana Bulletin, 14, 1971, 145-59, map.
Is there a historical basis for integration in West Africa?

721 DOIMI Di DELUPIS, INGRID. The East African Community and Common Market. Stockholm: Norstedt, 1969. London: Longman, 1970. 184 p., bibl.

722 DONAHEY, ROXANNE. "East African Unity: Political Differences vs. Economic Interests, 1920-1972." Journal of International and Comparative Studies, 5, 3, Fall 1972, 74-86.

723 DRESANG, DENNIS L. and SHARKANSKY, IRA. "Public Corporations in Single-country and Regional Settings: Kenya and the East African Community." International Organization, 27, 3, Summer 1973, 303-28, tabls.
What contributions do public corporations make in the regional integration process?

724 DU BOIS, VICTOR D. "The Search for Unity in French-Speaking Black Africa." American Universities Field Staff Reports, West Africa, 8, 3-6, 1965, 1-24, 1-17, 1-12, 1-26.
Part I. "The Founding of OCAM." Part II. "New Bonds between Ex-French and Ex-Belgian Colonies." Part III. "Mauritania's Disengagement from Black Africa." Part IV. "Relations between the 'Moderate' and the 'Revolutionary' States: The Case of Guinea."

725 "Economic Integration in East Africa: The Treaty for East-African Co-operation." Columbia Journal of Transitional Law, 7, 2, Fall 1968, 302-32.

726 Economist Intelligence Unit and Société d'Etudes pour le Développement Economique et Social. A Study on Possibilities of Economic Cooperation between Ghana, the Ivory Coast, Upper Volta, Niger, Dahomey and Togo. London: African Development Bank, 1970. 4 vols.

727 EKUE, A. K. "L'Organisation Commune Africaine et Mauricienne." Revue Française d'Etudes Politiques Africaines, 130, October 1976, 52-65.
Consideration of the future of OCAM after the withdrawal of Gabon.

728 ELKAN, WALTER and NULTY, LESLIE. The Economic Links between Kenya, Uganda, and Tanzania. Nairobi: Institute for Development Studies, University of Nairobi, 1972. 24 p., bibl. [Discussion Paper, 143].

729 ESSIEN, E. and EKUSSAH, M. "Spotlight on ECOWAS." Africa, an International Business, Economic and Political Monthly, 52, December 1975, 45-69, ill., tabls.

730 "European Economic Community - African and Malagasy States: Convention of Association and Annexed Documents." International Legal Materials, 9, 1, January 1970, 484-506.

731 EWING, A. F. "Prospects for Economic Integration in Africa." Journal of Modern African Studies, 5, 1, May 1967, 53-67.
    The need is great and the prospects are fair.

732 FAJANA, OLUFEMI. "Nigeria's Inter-African Economic Relations: Trends, Problems, and Prospects." In: Akinyemi, A. B., (ed.). Nigeria and the World. Ibadan, 1978, 17-31, tabls.
    "Recent developments augur well for the future economic relations between Nigeria and other African countries."

733 FANIRAN, A. "Drainage Basin Development and Political Boundaries in Africa." Nigerian Journal of Economic and Social Studies, 16, 3, November 1974, 445-59, bibl., figs., maps, tabls.
    Political limitations on river-basin development.

734 FEKETE, J. "Economic Groupings of the African Countries." In: Fekete, J. The Necessity of Economic Groupings and Their Main Features. Budapest, 1974, 21-31, tabl.

735 FESSARD DE FOUCAULT, B. "Crise ou Fin de l'OCAM?" Défense Nationale, November 1973, 75-87.
    An explanation of the crisis within OCAM.

736 _____. "Vers un Réaménagement des Relations entre les États Riverains du Fleuve Sénégal." Revue de Défense Nationale, 28, February 1971, 244-57.
    Senegal, Guinea, Mali, and Mauritania join together in the Organization of Senegal River States.

737 FRANCK, THOMAS M. East African Unity through Law. New Haven, Conn.: Yale University, 1964. 184 p., index.
    "An examination of its legal-constitutional techniques. Laws and institutions have played a particularly important role, perhaps too important, in promoting the unity of East Africa."

738 FREY, H. K. "Regionale Zusammenarbeit und Nationaler Aufbau im Widerstreit." Schweizer Monatschefte, 51, 7, October 1971, 478-85.
    Kenya and Rhodesia as competing centers of economic integration.

739 GABRIEL, J. M. "Regionale Integration in Afrika: Ein Neues Modell." Annuaire Suisse de Science Politique, 14, 1974, 43-56.
Functional integration theories do not apply to Africa, as shown by the UDEAC and East African experiences. Rather, we shall see the development of core states (Kenya, Ivory Coast, Nigeria, Cameroon, etc.) and dependent peripheral states.

740 GAM, P. "Les Causes de l'Eclatement de la Fédération du Mali." Revue Juridique et Politique, 20, 3, July-September 1966, 411-70.
The Federation of Mali was an attempt at political union between the now separate states of Mali and Senegal.

741 GAUTRON, JEAN-CLAUDE. "La Communauté Economique de l'Afrique de l'Ouest: Antécédents et Perspectives." Annuaire Français de Droit International, 21, 1975, 197-215.
The problems and prospects of the CEAO.

742 _____. "Les Métamorphoses d'un Groupement sous Régional: L'Organisation des Etats Riverains du Sénégal." Année Africaine 1970, 1970, 143-59.
The institutions, purposes, and problems of the Organization of Senegal River States.

743 _____. "Les Organisations Régionales Africaines." Année Africaine 1973, 1973, 224-84 and 1974, 1974, 163-82.
Chronological descriptions of the activities of the OAU and other regional organizations in Africa.

744 _____. "Le Régionalisme Africain et le Modèle Interaméricain." Annales Africaines 1966, 1966, 49-87.
Regional organizations would promote stability in Africa.

745 GHAI, DHARAM. "Territorial Distribution of the Benefits and Costs of the East African Common Market." In: Leys, Colin and Robson, P., (eds.). Federation in East Africa. Nairobi: 1965, 72-82, tabls.
How much of East Africa's economic growth is a result of the common market? How is this resultant growth distributed among the members?

746 GHAI, YASH. "Some Legal Aspects of an East African Federation." In: Leys, Colin and Robson, P., (eds.). Federation in East Africa. Nairobi, 1965, 172-82.
What type of constitution and legal system are necessary for an East African federation? Also see the esaay by S. A. de Smith.

747 GITELSON, S. A. "Can the United Nations Be an Effective Catalyst for Regional Integration? The Case of the East African Community." *Journal of Developing Areas*, 8, 1, October 1973, 65-82.
   The neofunctionalist approach has failed.

748 GLADDEN, E. N. "The East African Common Services Organization." *Parliamentary Affairs*, 16, 4, Autumn 1963, 428-39.
   Description of the institutions and operations of the EACSO. A very positive view.

749 GREEN, REGINALD H. and KRISHNA, K. G. Y. *Economic Co-operation in Africa: Retrospect and Prospect*. Nairobi: Oxford University for University College, 1967. 160 p.
   The major portion of the book is a summary of then-current thought on economic integration. Abstracts of several papers presented at a Nairobi conference in 1965 are included.

750 HADZI-VASILEVA, JOKICA. "Inter-African Economic Co-operation." *In*: Widstrand, Carl G., (ed.). *African Boundary Problems*. New York, 1969, 161-4.
   A note on the advantages for African states of economic cooperation.

751 HAMANI, D. "La Coopération Régionale en Afrique. La Communauté Economique de l'Afrique de l'Ouest." *Chronique de Politique Étrangère*, 27, 2, March 1974, 167-79.
   ECOWAS may serve as a major development in African regionalism. But various changes in state behavior are necessary before success will be achieved.

752 HAZELWOOD, ARTHUR, (ed.). *African Integration and Disintegration: Political and Economic Case Studies*. London: Oxford University, 1967. 414 p., bibl., figs., index, maps, tabls.
   Includes ten essays on economic and political unions. See references under Hazelwood, Robson, Julienne, and Hoskyns.

753 _____. "The Co-ordination of Transport Policy." *In*: Leys, Colin and Robson, P., (eds.). *Federation in East Africa*. Nairobi, 1965, 111-23.
   Investment in transport, particularly roads and railways, and the inter-territorial implications of progress in transport are considered.

754 _____. "Economic Integration in East Africa." *In*: Hazelwood, Arthur, (ed.). *African Integration and Disintegration*. London, 1967, 69-114, map, tabls.
   Although signs of deterioration were evident in East African integration, the author concludes this essay with an optimistic note.

755 HAZELWOOD, A. Economic Integration: The East African Experience. London: Heinemann, 1975. 180 p., bibl., index, tabls.
A history and theoretical analysis of the EAC.

756 _____. "Problems of Integration among African States." In: Hazelwood, Arthur, (ed.). African Integration and Disintegration. London, 1967, 3-25, map, tabl.
"The difficulties of establishing and maintaining systems of integration between independent states" are analyzed.

757 HEDRICH, MANFRED and VON DER ROPP, KLAUS. "Chancen Regionaler Integration in Westafrika." Aussenpolitik, 29, 1, 1978, 84-97. In English Edition: 29, 1, 1978, 87-101, tabls.
A comparison of the East African Community and CEAO.

758 HELLEINER, GERALD K. "East African Community Approaching the E. E. C." Africa Report, 13, 4, April 1968, 37-42. See also: "The East African Community and the E. E. C." In: Tandon, Y., (ed.). Readings in African International Relations, Vol. I. Nairobi, 1972, 358-67.
Sets out the main areas of interest even though written before the Arusha Agreement.

759 _____. "Nigeria and the African Common Market." Nigerian Journal of Economic and Social Studies, 4, 3, March 1963, 283-98.

760 HIPPOLYTE, M. "Regroupements en Afrique Francophone de Nouakchott à Niamey: L'Itinéraire de l'OCAM." Revue Française d'Etudes Politiques Africaines, 34, October 1968, 34-53. Also in English in: Africa Quarterly, 8, 4, January-March 1969, 343-57.
The francophone states have had a series of organizations - OAMCE, UAM, UAMCE, and OCAM. The author presents a brief review of the first three and longer analysis of the fourth.

761 HODDER-WILLIAMS, RICHARD. "Changing Perspectives in East Africa." World Today, 34, 5, May 1978, 166-74.
The events involved in breaking apart the EAC.

762 HOWELL, JOHN B. East African Community: Subject Guide to Official Publications. Washington, D. C.: Library of Congress, 1976. 272 p., index.
Contains 1812 references each with a brief annotation.

763 HUGHES, A. J. East Africa: The Search for Unity - Kenya, Tanganyika, Uganda, and Zanzibar. Harmondsworth: Penguin, 1963. 278 p. Rev. ed., 1969. 270 p., index, maps.
Domestic politics in each member state and the organizations that join these states in the EAC are discussed.

764 HUMMER, WALDEMAR and HINTERLEITNER, REINHOLD. "Überregionale, Regionale und Subregionale Kooperation und Integration auf dem Afrikanischen Kontinent." Jahrbuch des Öffentlichen Rechts der Gegenwart, 26, 1977, 631-81.
Parallels can be drawn between attempts at economic integration and efforts at national construction in Africa.

765 IGUE, O. J. "Un Aspect des Echanges entre le Dahomey et le Nigeria: Le Commerce du Cacao." Bulletin IFAN, B, 38, 3, July 1976, 636-69, maps, tabls.

766 _____. "Evolution du Commerce Clandestin entre le Dahomey et le Nigeria depuis la Guerre du 'Biafra'." Canadian Journal of African Studies, 10, 2, 1976, 235-57, map, tabls.
Impact of the war on illicit trade: the direction of benefits was altered in favor of Dahomey.

767 IJERE, M. O. "Problems of Post-independence Integration Movements in Africa." Pan-African Journal, 5, 3, Fall 1972, 330-46.

768 JARMACHE, ELIE. "Les Organisations Régionales Africaines." Année Africaine 1975, 1975, 116-35.
A review of regional organization developments.

769 JOHNS, D. H. "East African Unity - Problems and Prospect." World Today, 19, 12, December 1963, 533-40.
A general survey of the chances of federation and/or the success of economic integration in East Africa.

770 _____. "Einigungsbestrebungen in Ostafrika: Das Projekt einer Ostafrikanischen Föderation." Europa-Archiv, 18, 17, September 1963, 633-40.
A review of steps leading to the formation of an East African federation.

771 JOHNSON, CAROL A. "Political and Regional Groupings in Africa." Internationl Organization, 16, 2, Spring 1962, 426-48, bibl.
Summarizes the significance and results of several early attempts at regional organizations.

772 JORGE, E. F. "La Ideologia Panafricanista y Sus Bases de Sustentacíon." Foro Internacional, 7, 27, January-March 1967, 211-32.
"Pan-African Ideology and Its Bases." Description of the major elements of Pan-Africanism.

773 JULIENNE, ROLAND. "The Experience of Integration in French-speaking Africa." In: Hazelwood, Arthur, (ed.). African Integration and Disintegration. London, 1967, 339-53, map.
Although African unity is not feasible, French-speaking Africa may be a small enough unit to successfully integrate.

774 KAGWE, SOLOMAN. "Seminar on Regional Co-operation in Africa: Prospects and Problems in Monrovia." Journal of Modern African Studies, 8, 1, April 1970, 140-8.

775 KAPPELER, DIETRICH. "Causes et Conséquences de la Désintégration de la Communauté Est-Africaine." Politique Etrangère, 43, 4, 1978, 319-30.
The EAC broke apart for economic and political reasons.

776 KAPPELER, FRANZ. "Markwirtschaft und Planwirtschaft in West-Afrika." Schweizer Monatshefte, 49, 6, September 1969, 538-9.

777 KHALIL, K. H. "The Conventional Theory of Economic Integration and the African Conditions." L'Egypte Contemporaine, 344, April 1971, 169-88.

778 KIANO, J. GIKENYO. "The Foreign Policy of East African Countries." In: Black, J. E. and Thompson, K. W., (eds.). Foreign Policies in a World of Change. New York, 1963. 407-23, bibl., map.
A survey of policy stands by a high-ranking official of the government of Kenya.

779 ———. "From PAFMECA to PAFMECSA and Beyond." African Forum, 1, Fall 1965, 36-49.
Description of attempts at African unity in East Africa.

780 KLOMAN, E. H., JR. "African Unification Movements." International Organization, 16, 2, Spring 1962, 387-404, tabl. Also in: Padelford, N. J. and Emerson, R., (eds.). Africa and World Order. New York, 1963, 119-35, tabl.
Nkrumah's drive for African unity has spurred reaction from leaders who oppose his lead.

781 KOFI, T. A. "The Need for and Principles of a Pan-African Economic Ideology." Civilisations, 26, 3/4, 1976, 205-31.

782 KUMAR, ASHOK and OSAGIE, EGHOSA. "Problems of the Economic Community of West African States." In: Akinyemi, A. B., (ed.). Nigeria and the World. Ibadan, 1978, 45-56, tabls.
"This paper will consider some of the practical difficulties that are likely to face the ECOWAS countries in the organisation's initial take-off period."

783 KURTS, D. M. "Political Integration in Africa: The Mali Federation." Journal of Modern African Studies, 8, 3, October 1970, 405-24.
An application of the Haas/Schmitter analytic framework.

784 LAMPUÉ, P. "Les Groupements d'Etats Africains." Revue Juridique et Politique, 18, 1, January-March 1964, 21-57
Examples of various types of integration and cooperation.

785 LANGHAMMER, R. J. "Die Wirtschaftsgemeinschaft Westafrikanischer Staaten (ECOWAS): Ein Neuer Integrationsversuch." Europa-Archiv, 31, 5, March 1976, 163-8.
   The origins of ECOWAS are more political than economic. There are at least three cause for doubt that the organization will succeed: variations in national economic structures, low level of cooperative possibilities, and bilateral conflicts between several pairs of its members.

787 LATREMOLIERE, JACQUES. "Vers l'Eclatement Politique et Economique de l'Afrique Orientale." Afrique Contemporaine, 15, 83, January/February 1976, 1-8.
   A review of attitudes on unification in each member state of the EAC.

788 LATTRE, J. M. DE. "Organisation Africaine et Malgache de Cooperation Economique." Politique Etrangére, 25, 6, 1960, 584-604.
   The ex-French colonies grouped together for purposes of economic cooperation between themselves and between Africa and Europe.

789 LECHINI, GLADYS T. "La Communidad de Africa Oriental." Revista de Política Internacional, 158, July/August 1978, 117-46.

790 LE VINE, VICTOR T. "The Politics of Partition in Africa: The Cameroons and the Myth of Unification." Journal of International Affairs, 18, 2, 1964, 198-210.
   What are the lessons of Cameroon integration for Pan-Africanism?

791 LEYMARIE, P. "Les Dix Ans de la Communauté Est Africaine: Une Expérience Originale à la Recherche d'un Second Souffle." Revue Française d'Etudes Politiques Africaines, 133, January 1977, 37-55.

792 LEYS, COLIN. "Recent Relations between the States of East Africa." International Journal, 20, 4, Autumn 1965, 510-23.

793 _____ and ROBSON, PETER, (eds.). Federation in East Africa: Opportunities and Problems. Nairobi: Oxford University, 1965. 244 p., index, tabls.
   Twelve essays presented at a conference conducted in Nairobi in 1963 and several appendices are included in this volume. See references under Arkadie, Banfield, Belshaw, Birch, D. Ghai, Y. Ghai, Hazelwood, Mboya, Newman, Nye, Robson, and de Smith.

794 LIEBENOW, J. GUS. "The Quest for East African Unity: 'One Step Forward, Two Steps Backward.'" In: DeLancey, M. W., (ed.). Aspects of International Relations in Africa. Bloomington, Ind., 1979, 126-58.
"An analysis of the problems of functional integration on a regional basis in East Africa."

795 LIGOF, M. and DEVERNOIS, G. "L'Union Africaine et Malgache. Une Annee d'Existence." Revue Juridique et Politique d'Outre-Mer, 16, 3, July-September 1962, 317-38.
The UAM and the various agreements it had signed with France are described.

796 LINIGER-GOUMAZ, MAX. "Transsaharien et Transafricain: Essai Bibliographie." Genève-Afrique, 7, 1, 1968, 70-85.

797 MABOGUNJE, AKIN L. Regional Mobility and Resource Development in West Africa. Montreal: McGill-Queen's University, 1972. 154 p., bibl., index, maps, tabls.
"The author discusses the social and economic aspects of regional mobility in West Africa. . . . Regional mobility . . . represents a uniquely West African approach to the area's social, economic, and political problems."

798 MALECELA, J. S. "What Next for the East African Community? The Case for Integration." African Review, 2, 1, June 1972, 211-7.
Progress in integration will require free mobility of labor and a common policy on wages and social insurance.

799 MANOUAN, A. "L'Evolution du Conseil de l'Entente." Penant, 746, October-December 1974, 447-97; 747, January-March 1975, 19-38; 748, April-June 1975, 211-36; and 749, July-September 1975, 309-45.
The Conseil is a regional organization of ex-French colonies in West Africa (Ivory Coast, Dahomey-Benin, Upper Volta, Niger, and Togo).

800 MARIÑAS OTERO, LUIS. "La Communidad del Africa Oriental." Revista de Política Internacional, 141, September/October 1975, 113-48.
A very positive view of East African integration. The author suggests that federation may result from current economic interactions.

801 _____. "La Communidad Económica de los Países de los Grandes Lagos y Otros Intentos de Cooperación Multinacional en Aquella Región Africana." Revista de Política Internacional, 149, January/February 1977, 89-98.
"The Economic Community of the Great Lakes States and Other Attempts at Multinational Cooperation in this Region of Africa."

802 MARIÑAS OTERO, L. "Las Conferencias de Estados del Africa Central y Oriental." Revista de Política Internacional, 136, November/December 1974, 149-59.
   The author describes informal conferences of 16 Central and East African states and the success of this forum in acting as an international pressure group.

803 _____. "Las Conferencias Presidenciales Franco-Africanas." Revista de Política Internacional, 148, November/December 1976, 59-73.

804 _____. "El Consejo de la Entente." Revista de Política Internacional, 142, November/December 1975, 175-85.
   The Conseil de l'Entente is one of the oldest of African international organizations.

805 _____. "Cooperación Economica en el Sahara." Revista de Política Internacional, 156, March/April 1978, 231-6.

806 _____. "La Dificil Ruta de la Integración Africana." Revista de Política Internacional, 146, July-August 1976, 27-51.
   Analysis of major impediments to African unity.

807 _____. "El Fracaso de la Comunidad del Africa Oriental." Revista de Política Internacional, 154, November/December 1977, 159-83.
   The East African Community broke apart for economic reasons, not political factors.

808 _____. "Los Intentos de Integracion del Africa Occidental." Revista de Política Internacional, 132, March/April 1974, 73-92.
   In spite of difficulties caused by tribalism, colonial boundaries, trade links with Europe, and monetary divisions, ECOWAS and the UDAO are attempts at economic integration in West Africa.

809 _____. "La OCAM: Evolution de una Organización Africana de Integración." Revista de Política Internacional, 138, March/April 1975, 165-79.
   Description of the structures of OCAM, analysis of its functions, and explanation of the withdrawals of Zaire, Chad, Congo, and Cameroon.

810 _____. "La Organización para el Desarrollo de Rio Senegal." Revista de Política Internacional, 147, September/October 1976, 141-6.
   The Senegal River has irrigation, transport, and hydroelectric potentials that might be developed on a regional basis. The author reviews the history of such regional organizations.

811 MARIÑAS OTERO, L. "El Proceso de Integración entre Gambia y Senegal." Revista de Política Internacional, 144, March-April 1976, 273-82.
   The integration of Gambia and Senegal is going forward in spite of obstacles.

812 _____. "La Unión del Río Mano como Ejemplo de Integración Económica Africana." Revista de Política Internacional, 152, July/August 1977, 207-14.
   The Rio Mano Union between Liberia and Sierra Leone is the first serious attempt at economic integration in West Africa.

813 MARTIN, JANE. A Bibliography on African Regionalism. Boston: African Studies Center, Boston University, 1969. 121 p.

814 MAZRUI, ALI A. "Tanzania versus East Africa: A Case of Unwitting Federal Sabotage." Journal of Commonwealth Political Studies, 3, 3, November 1965, 209-25.
   The author describes several policies and actions of the Tanzanian government which have been destructive of the concept of an East African federation.

815 MBOYA, T. J. "East African Labour Policy and Federation." In: Leys, Colin and Robson, P., (eds.). Federation in East Africa. Nairobi, 1965, 102-10.
   After a strong reaffirmation of the desire of the peoples and leaders of East Africa to form a federation, the author discusses the role of labor and labor policy in forming such a federation.

816 MEAD, DONALD C. "Economic Cooperation in East Africa." Journal of Modern African Studies, 7, 2, July 1969, 277-87.
   Despite the euphoria surrounding the 1967 Treaty for East African Co-operation, this author points out numerous conflicts and problems that were left unresolved. The distribution of benefits continues to be a difficult question.

817 MENÉNDEZ, J. "Al Sur y Al Este del Congo: Impaciencia o Temor." Politika Internacional, 50/51, July-October 1960, 125-60.
   There is a need for greater cooperation among African states in order to bring about economic stability. The UN should encourage such a development.

818 MOES, JOHN E. "Foreign Exchange Policy and Economic Union in Central Africa." Economic Development and Cultural Change, 14, 4, July 1966, 471-83.
   A case study of the application of exchange controls in the underdeveloped states of Rwanda and Burundi.

819 MORTIMER, DELORES; RAO, GITA; and HOWELL, SANDRA ANN. Economic and Regional Integration in Africa. Washington: African Bibliographic Center, 1973. 49 p. [Current Reading List Series, 10, 6].
Contains references to items published, 1969-1972.

820 MUGOMBA, AGRIPPAH T. "Regional Organizations and African Underdevelopment: The Collapse of the East African Community." Journal of Modern African Studies, 16, 2, June 1978, 261-72.
The cardinal principles of unity and solidarity in purpose have been eroded due to ideological differences.

821 MUSHKAT, M. "Problems of Political and Organizational Unity in Africa." African Studies Review, 13, 2, September 1970, 265-90. Also in: Roland, Joan G., (ed.). Africa: The Heritage and the Challenge. Greenwich, Conn., 1974, 401-15.
The author depicts the various arguments concerning the importance and feasibility of various forms of African unity and concludes that the numerous organizations and structures that exist are the basis for the development of an important international subsystem.

822 MUTHARIKA, B. W. T. "Multinational Corporations in Regional Integration: The African Experience." African Review, 5, 4, 1975, 365-90.
The greater the strength of MNCs in an African state, the less the chances of effective participation of that state in regional integration.

823 _____. Toward Multinational Economic Cooperation in Africa. New York: Praeger, 1972. 434 p., bibl., map, tabls.
Economic and political aspects are discussed "in more simplified terms understandable to the politician and the layman." Past experiences and proposals for the future are included.

824 MYTELKA, LYNNE K. "Foreign Aid and Regional Integration: The UDEAC Case." Journal of Common Market Studies, 12, 2, December 1973, 138-58.
A survey of the literature indicates several variants of the hypothesis that foreign aid can be used to assist regional integration. UDEAC as a case study shows that the opposite is true. Aid has promoted disintegration in UDEAC.

825 _____. "A Genealogy of Francophone West and Equatorial African Regional Organizations." Journal of Modern African Studies, 12, 2, 1974, 297-320, bibl., tabl.
This review and description of the various international organizations in francophone Africa includes an extensive bibliography.

826 MYTELKA, L. K. "UDEAC: Problems of a Common Market." Africa Report, 15, 7, October 1970, 14-7.

827 NDONGKO, W. A. "Trade and Development Aspects of the Central African Customs and Economic Union." Cultures et Developpement, 7, 2, 1975, 337-55, tabls.
   Examines inter-state trade in UDEAC in order to learn the extent to which traditional customs theory helps in analyzing the potential for and the development of such trade in economic and customs unions.

828 NEAL, D. F. "Economic Community of West African States: Its Evolution, Significance, and Prospects." Liberian Economic and Management Review, 3, 2, 1975/1976, 1-30.

829 NEWMAN, PETER. "The Economics of Integration in East Africa." In: Leys, Colin and Robson, P., (eds.). Federation in East Africa. Nairobi, 1965, 56-71.
   Recommendations for changes in the EACM in order to strengthen that organization.

830 NGUODI, NGOM. La Réussite de l'Intégration Economique en Afrique. Paris: Présence Africaine, 1971. 141 p., bibl.
   Industrialization is necessary for African development, and industrialization will be the basis for the integration of Africa. Industrialization requires integration.

831 NIXSON, F. I. Economic Integration and Industrial Location: An East African Case Study. London: Longman, 1973. 181 p., bibl., index, map, tabls.
   "An attempt to comprehend in its totality the dynamic interaction between industrial location and economic integration in East Africa. Inevitably, this involves a consideration of economic, political and historical forces."

832 NOWZAD, BAHRAM. "Economic Integration in Central and West Africa." In: Tharp, Paul A., Jr., (ed.). Regional International Organizations. New York, 1971, 201-29, tabls.
   Background and potential for integration and brief notes on organizations existing in 1971.

833 NSIBAMBI, A. R. "Political Commitment and Economic Integration: East Africa's Experience." African Review, 2, 1, June 1972, 189-210. Also in: Mazrui, A. A. and Patel, H., (eds.). Africa in World Affairs. New York, 1973, 201-18.
   The East African example indicates that Africa needs full economic and political federation.

834 NYE, JOSEPH S. "The Extent and Viability of East African Co-operation." In: Leys, Colin and Robson, P., (eds.). Federation in East Africa. Nairobi, 1965, 41-55.

835 NYE, J. S. Pan Africanism and East African Integration. Cambridge: Harvard University, 1965. 307 p., bibl., index, maps, tabls.
"An attempt to evaluate the impact of a system of ideas that East Africans call Pan-Africanism upon a particular case of integration and attempted unification."

836 O'CONNER, A. M. "A Wider East African Union? Some Geographic Aspects." Journal of Modern African Studies, 6, 4, December 1968, 485-93, map, tabl.
"Geographical realities place distinct limits on the viable membership of economic unions."

837 OFOEGBU, MAZI RAY. "The Chad Basin Commission." In: Akinyemi, A. B., (ed.). Nigeria and the World. Ibadan, 1978, 82-95, tabls.
It is in Nigeria's interest to revitalize and make successful the Commission.

838 _____. "Functional Co-operation in West Africa." Odu: A Journal of West African Studies, n.s. 6, October 1971, 21-53.

839 _____. "Functional Co-operation in West Africa: Ghana and Its Neighbours." Insight and Opinion: Quarterly for Current African Thinking, 6, 1, 1971, 22-36.
"A study of the political and legal foundations for the joint development of river and lake basins in West Africa."

840 OGUNDANA, BABAFEMI. "Seaport Development: Multi-National Cooperation in West Africa." Journal of Modern African Studies, 12, 3, 1974, 395-407, map.
"This article examines the prospects and problems of co-operation on a regional basis, with particular reference to West Africa, in a specific economic field - namely, seaport development."

841 OLOFIN, SAM. "ECOWAS and the Lomé Convention: An Experiment in Complementary or Conflicting Customs Union Arrangements." Journal of Common Market Studies, 16, 1, September 1977, 53-72.

842 Organisation Commune Africaine, Malgache et Mauricienne. Coopération Interafricaine au Sein de l'OCAM. Yaoundé: OCAM, 1972. 203 p.
History, activities, special organizations within OCAM, and the future.

843 OSAGIE, EGHOSA. "Monetary Disintegration and Integration in West Africa." Nigerian Journal of International Studies, 1, 1, July 1975, 19-27.

844 OUKO, R. J. "The East African Community." International Review of Administrative Sciences, 35, 1, 1969, 47-51.

845 PANTER-BRICK, KEITH. "The Union Africaine et Malgache." In: Austin, D. and Weiler, H., (eds.). Interstate Relations in Africa. Freiburg, 1965, 68-84.
   The formation, structure and functioning, and future of the UAM are examined.

846 PEUREUX, G. "La Création de l'Union Africaine et Malgache et les Conférences des Chefs d'Etat d'Expression Française." Revue Juridique et Politique d'Outre-Mer, 15, 4, October-December 1961, 541-56.
   The author traces the events leading to the formation of an organization of ex-French colonies in Africa and outlines some of the results of the first meetings of the international organization that was formed.

847 PIQUEMAL, M. "Les Problèmes des Unions d'Etats en Afrique Noire." Revue Juridique et Politique d'Outre-Mer, 16, 1, January-March 1962, 21-58.
   A comparison of the Union of African States (Ghana, Guinea, and Mali) and l'Union Africaine et Malgache, with emphasis upon the latter.

848 PLESSZ, NICHOLAS G. Problems and Prospects of Economic Integration in West Africa. Montreal: McGill University, 1968. 91 p., index, map, tabls.
   Foreign trade, international organization, and population factors are discussed. The author concludes that "economic integration is likely to proceed at a very slow pace, if at all, in the foreseeable future."

849 PROCTOR, J. H. "The Effort to Federate East Africa: A Postmortem." Political Quarterly, 37, 1, January-March 1966, 46-69.
   Major problems leading to the failure of the move to federate include external influences (especially from Ghana) and the internal growth of separate nationalisms.

850 _____. "The Gambia's Relationship with Senegal: The Search for Partnership." Journal of Commonwealth Political Studies, 5, 2, July 1967, 143-60.
   Analysis of relations between the two states and the various treaties of cooperation signed by them.

851 RAMAMURTHI, T. G. "The Dynamics of Regional Integration in West Africa." India Quarterly, 26, 3, July-September 1970, 249-57.
   The numerous experiments in regional integration in West Africa fail because only a very small elite is involved in the integration process.

852 RAMCHANDANI, R. R. "East African Community Relations." India Quarterly, 31, 1, January-March 1975, 74-81.
   Causes of strain in the East African system.

853 RAMOLEJE, A. M. R. and SANDERS, A. J. G. M. "The Structural Pattern of African Regionalism." Comparative and International Law Journal of Southern Africa, 4, 2, July 1971, 155-92; 4, 3, November 1971, 293-323; 5, 1, March 1972, 21-55; 5, 2, July 1972, 171-88; 5, 3, November 1972, 299-338; and 6, 1, March 1973, 82-105.
   The OAU; OCAM; and regional cooperation in East, North, and South Africa are described.

854 RANA, R. S. "OCAM: An Experiment in Regional Cooperation." Africa Quarterly, 8, 2, July-September 1968, 158-65.
   The rapid success of OCAM shows clearly that economic integration must preceed political integration.

855 RAZAFIMBAHINY, J. A. "L'O. A. M. C. E." Revue Juridique et Politique d'Outre-Mer, 17, 2, April-June 1963, 177-93.
   A description of the structures of the OAMCE.

856 RENNIGER, JOHN P. Multinational Cooperation for Development in West Africa. Elmsford, N. Y.: Pergamon, 1978. 162 p.
   Emphasis on ECOWAS and its potential.

857 ROBINSON, PEARL T. "The Political Context of Regional Development in the West African Sahel." Journal of Modern African Studies, 16, 4, 1978, 579-95.
   A description and analysis of the Inter-state Committee for Drought Control in the Sahel (CILSS).

858 ROBSON, PETER. Economic Integration in Africa. Evanston: Northwestern University. London: Allen and Unwin, 1968. 320 p., index, tabls.
   This general survey includes case studies of the EAC and UDEAC.

859 _____. "Economic Integration in Equatorial Africa." In: Hazelwood, A., (ed.). African Integration and Disintegration. London, 1967, 27-69, map, tabls.
   Congo (B), Gabon, Central African Republic, Chad, and Cameroon are Equatorial Africa. Essay is mainly a discussion of UDEAC.

860 _____. "Economic Integration in Southern Africa." Journal of Modern African Studies, 5, 4, December 1967, 469-90, tabls.

861 _____. "Federal Finance." In: Leys, Colin and Robson, Peter, (eds.). Federation in East Africa. Nairobi, 1965, 124-45.
   General problems of finance in federations are applied in particular to the possible East African federation.

862 ROBSON, P. "The Problem of Senegambia." Journal of Modern African Studies, 3, 3, October 1965, 393-408.
The differences in economic organization in Gambia and Senegal preclude economic integration but a free trade area may be worthwhile.

863 _____. "Problems of Integration between Senegal and Gambia." In: Hazelwood, A., (ed.). African Integration and Disintegration. London, 1967, 115-28, figs., map.
A revised version of the author's article in the Journal of Modern African Studies.

864 RODNEY, WALTER. "The Entente States of West Africa." International Journal, 21, 1, Winter 1965/1966, 78-92.
Analysis of the general conditions of the members of the Entente - Dahomey, Ivory Coast, Niger, and Upper Volta. Domestic and external factors are described.

865 ROMANO, A. "Il Movimento verso l'Unità nelle Relazione Tragli Stati Africani." Rivista di Studi Politici Internazionali, 31, 3, July-September 1964, 395-455.
Study of attempts at African unity indicate that political factors are more important than economic or social in African interrelations.

866 RONEN, DOV. "Alternative Patterns of Integration in African States." Journal of Modern African Studies, 14, 4, December 1976, 577-96.

867 ROPP, K. VON DER. "Ansätze zu Regionaler Integration in Schwarzafrika." Europa-Archiv, 26, 12, June 1971, 429-36.
"The Beginnings of Regional Integration in Black Africa." Compares moves toward integration in francophone and anglophone Africa. The EAC, OCAM, and the Entente as case studies.

868 _____. "Chancen der Integration in Zentralafrika." Aussenpolitik, 23, 5, May 1972, 286-94.
"Chances of Integration in Central Africa." UDEAC has many serious problems but it does continue to survive.

869 _____. "Westafrikanische Wirtschaftsgemeinschaft." Aussenpolitik, 24, 4, 1973, 468-76.
"The West African Economic Community." The CEAO groups together Ivory Coast, Senegal, Mauritania, Mali, Niger, and Upper Volta in a customs union with a variety of economic agreements and structures.

870 ROSBERG, C. G., JR. and SEGAL, A. "An East African Federation." International Conciliation, 543, May 1963, 1-72.
An assessment of the factors favoring and opposing integration, the possible forms of a federation, and the geographical extent of such a federation.

871 ROTBERG, R. I. "The Federation Movement in British East and Central Africa, 1889-1953." Journal of Commonwealth Political Studies, 2, 2, May 1964, 141-60.
This essay provides a historical background on the integration of East Africa and Central Africa (Rhodesia, Malawi, and Zambia).

872 ROTHCHILD, DONALD S. "African Federations and the Diplomacy of Decolonization." Journal of Developing Areas, 4, 4, July 1970, 509-24.
Argues that colonially-constructed federations in Africa were doomed to failure because of the lack of African participation in their founding and organization.

873 _____. "From Hegemony to Bargaining in East African Relations." Journal of African Studies, 1, 4, Winter 1974, 390-416.
The influence of environmental circumstances upon integration outcomes is examined in colonial and post-colonial periods.

874 _____. Politics of Integration: An East African Documentry. Nairobi: East African Publishing House, 1968. 343 p., bibl., map, tabls.
This collection contains a number of primary and secondary resources for the study of the East African federation process.

875 SCHMIDT, G. "Integrationspolitik 1977 in Afrika." International Afrikaforum, 13, 4, 1977, 349-53.

876 SEGAL, AARON. "The Integration of Developing Countries: Some Thoughts on East Africa and Central America." Journal of Common Market Studies, 5, 3, 1967, 252-82.
First considers the respective arguments for and against political and economic integration and their relevance to the developing countries, and then analyzes comparatively the integration efforts in Central America and East Africa.

877 SEIDMAN, A. W. "Problems and Possibilities for East African Economic Integration." In: Mazrui, A. A. and Patel, H. H., (eds.). Africa in World Affairs. New York, 1973, 219-34.
Potential gains from unity and obstacles to economic integration in East Africa.

878 SHARKANSKY, I. and DRESANG, D. L. "International Assistance: Its Variety, Coordination, and Impact among Public Corporations in Kenya and the East African Community." International Organization, 28, 2, Spring 1974, 207-31.
Kenya and the EAC allow public corporations to deal directly with governments and businesses.

879 SHAW, T. M. "Regional Cooperation and Conflict in Africa." International Journal, 30, 4, Autumn 1975, 671-88.

880 SIMMONS, A. "Economic Cooperation in West Africa." Western Political Quarterly, 25, 2, June 1972, 295-304.
Attempts at economic unity in West Africa are analyzed in an effort to find solutions to the problems faced by such organizations.

881 SINCLAIR, M. R. The Strategic Significance of Tanzania. Pretoria: Institute for Strategic Studies, University of Pretoria, 1979(?). 99 p., bibl., figs., maps, tables.
"A realistic and dynamic evaluation of the strategic significance of Tanzania to the other actors of the international political system."

882 SIRCAR, PARBATI. "Toward a Greater East African Community." Africa Quarterly, 16, 3, January 1977, 14-27.

883 SMITH, JOHN G. Regional Economic Cooperation and Integration in Africa: Some Bibliographical References. Montreal: Centre for Developing-Area Studies, McGill University, 1973. [Bibliographical Series, 2].

884 SMITH, S. A. DE. "Integration of Legal Systems." In: Leys, Colin and Robson, P., (eds.). Federation in East Africa. Nairobi, 1965, 158-71.
The author raises questions about the type of legal system and the constitutional form that should be adopted for an East African federation. Also see the essay by Yash Ghai.

885 SOHN, LOUIS B., (ed.). Basic Documents of African Regional Organizations. Dobbs Ferry, N. Y.: Oceana Publications for the Inter-American Institute of International Legal Studies, 1971-1973. 1854 p.
This four volume publication includes an introductory essay, a bibliography, and a set of documents for each regional organization in Africa. Volume I includes the OAU, The ADB, and French-speaking Africa. Volume II completes French-speaking Africa. Volume III includes West and East Africa and the Maghreb. And, Volume IV is concerned with association with the EEC.

886 SOUTHALL, ROGER J. "The Federal University and the Politics of Federation in East Africa." East Africa Journal, 9, 11, November 1972, 38-43.
The University of East Africa was one of several functional ties between the states of EACSO. President Obote of Uganda is depicted as a key factor in the break-up and nationalization of the University in 1970.

887 SOUTHALL, R. J. Federalism and Higher Education in East Africa. Nairobi: East African Publishing, 1975. 160 p., bibl.
One of the institutions of the East African Community was a university. This and other aspects of education are examined.

888 TAMUNO, O. G. Cooperation for Development: A Bibliography on Interstate Relations in Economic, Technical, and Cultural Fields in Africa, 1950-1968. Ibadan: Nigerian Institute of Social and Economic Research, 1969. 113 p.

889 TANDON, YASHPAL. "The Transit Problems of Uganda within the East African Community." In: Cervenka, Z., (ed.). Landlocked Countries of Africa. Uppsala, 1973, 79-97, figs., map.
Legal status, political-economic costs, and the threat of being cut-off are examined.

890 _____ and MAZRUI, ALI A. "The East African Community as a Subregional Grouping." In: El-Ayouty, Y. and Brooks, H. C., (eds.). Africa and International Organization. Hague, 1974, 182-205, tabls.
"We shall be dealing mainly with the operations of the East African Common Market."

891 TAWFĪQ, SABRĪ. "Al-Sūmāl fī al-Jāmi'ah al-'Arabīyah." Al-'Ahrām al-Iqtisādī, 445, 1 March 1974, 18-20.
The Somali Republic and its membership in the Arab League.

892 TEVOEDJRE, ALBERT. Pan-Africanism in Action: An Account of the U. A. M. Cambridge, Mass.: Harvard University, 1965. 88 p., tabls.
The author served as Secretary-General of the UAM and as Minister of Information in the government of Dahomey.

893 THOMPSON, VIRGINIA M. West Africa's Council of the Entente. Ithaca, N. Y.: Cornell University, 1972. 313 p., bibl., ill., index, map.
A study of the domestic politics and relationships between the members of the Council of the Entente.

894 "Tribe and Nation in East Africa: Separatism and Regionalism." Round Table, 207, June 1962, 252-8.
Regionalism, Pan-Africanism, negritude, and the difficulties of integrating the various states of Africa.

895 UDOKANG, OKON. "Nigeria and ECOWAS: Economic and Political Implications of Regional Integration." In: Akinyemi, A. B., (ed.). Nigeria and the World. Ibadan, 1978, 57-81.
"The problems and prospects of the newly formed Economic Community of West African States."

896 VINAY, BERNARD. L'Afrique Commerce avec l'Afrique. Paris: Presses Universitaires de France, 1968. 213 p.
   An examination of intra-African trade, legitimate and clandestine, and trends in its development. The role of monetary institutions is considered and the effects of customs unions and other existing regional organizations are examined.

897 _____. "Coopération Intra-Africaine et Intégration: L'Expérience de l'U. D. E. A. C." Penant, 733, July-September 1971, 313-31.
   The recent withdrawal of Chad from UDEAC exemplifies the difficulties of regional integration.

898 WALKER, FRANKLIN V. "Regional Economic Integration in Africa." In: Uppal, J. S. and Salkever, Louis R., (eds.). Africa: Problems in Economic Development. New York, 1972, 340-53, tabl.
   A brief survey of the merits and demerits of economic integration with examples from African experiences.

899 WALLERSTEIN, I. M. "Larger Unities: Pan Africanism and Regional Federations." In: McEwan, P. and Sutcliffe, R., (eds.). Modern Africa. New York, 1965, 217-28.

900 WELCH, CLAUDE E. Dream of Unity: Pan-Africanism and Political Unification in West Africa. Ithaca, N. Y.: Cornell University, 1966. 396 p., bibl., index, maps, tabls.
   After a brief introduction to interstate relations in West Africa, the author analyzes unification in the Ewe case, Cameroon, Senegambia, and the attempted unity of Ghana, Guinea, and Mali. A list of international organizations is included.

901 "West African States: Establishment of West Africa Rice Development Association." International Legal Materials, 10, 3, May 1971, 648-68.

902 WODIE, FRANCIS. Les Institutions Internationales Régionales en Afrique Occidentale et Centrale. Paris: Pichon et Durand-Auzias, 1970. 274 p., bibl.
   Economic, technical, and administrative regional organizations such as UDEAC, the Niger River Commission, Air-Afrique, and the Conseil de l'Entente are included.

903 YADI, M. "Promotion du Développement Industriel Equilibré des Pays-Membres de l'UDEAC et de la CAE." Études Internationales, 6, 1, March 1975, 66-102.
   A comparative analysis of the problem of equal distribution of gains in schemes for regional integration in underdeveloped areas. UDEAC and the EAC are the cases studied.

904 YANNOPOULOS, T. and MARTIN, D. "Domination et Composition en Afrique: Le Conseil de l'Entente et la Communauté Est-Africaine Face à Eux-Mêmes et Face Aux 'Grands'." Revue Algérienne des Sciences Juridiques, Politiques et Economiques, 9, 1, March 1972, 129-51, tabls.
The Conseil and the EAC are facing numerous internal problems which prevent the organizations from reaching their full potential as bargaining agents in the conflict between the Third World and the imperialistic states.

905 YANSANE, AGUIBOU Y. "The State of Economic Integration in North West Africa South of the Sahara: The Emergence of the Economic Community of West African States (ECOWAS)." African Studies Review, 20, 2, September 1977, 63-87.
Analysis of previous attempts at economic integration in West Africa provides lessons for making ECOWAS a success.

906 _____. "The State of Economic Integration in Subsaharan North West Africa." Ufahamu, 8, 2, 1978, 88-130, tabls.

907 _____. "West African Economic Integration: Is ECOWAS the Answer?" Africa Today, 24, 3, July-September 1977, 43-59.

908 YOLOYE, E. A. "Educational Exchange between West African Countries." West African Journal of Education, 14, 3, 1970, 172-6.
A brief note on an important form of international exchange and cooperation in Africa.

909 ZARTMAN, I. WILLIAM. "The Sahara - Bridge or Barrier." International Conciliation, 541, January 1963, 1-62.
The Saharan states as a focus for international cooperation.

The OAU, Pan-Africanism, and African Unity

910 ABDALLAH, R. "L'Unité Africaine: Des Premiers Congrès Panafricains à la Fin de la Première Décennie de l'OUA." Revue Juridique, 1, 1974, 157-90.
   A history of the development of Pan-African thought and its culmination in the OAU.

911 ABROUS, A. "OAU and Arms Sales to South Africa." Africa Quarterly, 11, 1, April-June 1971, 2-10.
   The OAU has called for restrictions on arms sales to South Africa for many years.

912 ADDONA, A. F. The Organization of African Unity. Cleveland: World Publishing, 1969. 224 p., bibl., ill., index.
   An elementary introduction, probably written for young readers.

913 ADEBO, S. O. "Reflections on the Future of International Organization in Africa." In: El-Ayouty, Y. and Brooks, H. C., (eds.). Africa and International Organization. Hague, 1974, 235-7.
   Author is former Executive Director of UNITAR.

914 AGUDA, O. "Pan-Arabism versus Pan-Africanism: A Dilemma for African Unity." Quarterly Journal of Administration, 7, 3, April 1973, 357-71.
   Despite a tendency to go separate ways, the tendency to work together is predominant.

915 AGYEMAN, OPOKU. "The Osagyefo, The Mwalimu, and Pan-Africanism: A Study in the Growth of a Dynamic Concept." Journal of Modern African Studies, 13, 4, December 1975, 653-75.
   The unification movement owes its failure as much to conflicts between its supporters as to its opponents. Nkrumah and Nyerere are presented as examples of a militant and a gradualist.

916 AJALA, ADEKUNLE. Pan-Africanism: Evolution, Progress and Prospects. New York: St. Martins, 1973. 442 p., bibl., index, maps, tabls.
   After a brief introduction to the history of the Pan-African movements, the majority of the book is an examination of African unity after 1960. Boundary disputes, the Mali Federation, UDEAC, EAEC, and the Nigerian Civil War are among the many subjects included.

917 AKARAOGUN, OLU. "Towards Pan-Africanism: African States and the Politics of Economic Freedom." Black World, 21, 12, October 1972, 79-87.
   Economic dependence remains for Africa and the technological assistance of black Americans is needed to aid in the drive for independence.

918 AKE, C. "Pan-Africanism and African Governments." Review of Politics, 27, 4, October 1965, 532-42.
   Pan-Africanism helps to focus African governments on common problems and solutions. As a set of ideas and a program of action, it is an important determinant of African governmental behavior.

919 AKINDELE, RAFIU. "Organization of African Unity and International Law." Nigerian Journal of International Studies, 1, 1, July 1975, 7-18.

920 AKINYEMI, A. BOLAJI. "The O. A. U. and the Concept of Non-interference in Internal Affairs of Member States." British Yearbook of International Law, 46, 1972/1973, 393-400.
   The OAU opposes the deprivation of human rights in white-dominated Southern Africa, yet members of the OAU trample on human rights in their own states. The Organization should consider resolutions on such domestic situations.

921 _____. "The Organization of African Unity - Perception of Neo-Colonialism." Africa Quarterly, 14, 1/2, 1974, 32-52.
   African regimes are neo-colonial products and so the OAU has been strongly opposed to neo-colonialism.

922 _____. "Organization of African Unity: The Practice of Recognition of Governments." Indian Journal of Political Science, 36, 1, January-March 1975, 63-79.
   The OAU has not solved the problem of defining a policy on the recognition of governments that come to power by a coup d'état. Ideological factors play a heavy role in each decision.

923 AKUCHU, GEMUH E. "Peaceful Settlement of Disputes: Unsolved Problem for the OAU (A Case Study of the Nigeria-Biafra Conflict)." Africa Today, 24, 4, October-December 1977, 39-58.
   Factors of ineffective performance of the OAU.

924 ALIMOV, Y. "OAU: Ten Years of Existence." International Affairs, 7, July 1973, 59-64.
Emphasizes the importance of the OAU as a collective effort against neo-colonialism and imperialism.

925 ALUKO, O. "The OAU Liberation Committee after a Decade: An Appraisal." Quarterly Journal of Administration, 8, 1, October, 1973, 59-68.
Problems of the Liberation Committee - its administration, financial difficulties, and external pressure.

926 ANDEMICAEL, BERHANYKUN. The OAU and the UN: Relations between the Organization of African Unity and the United Nations. New York: Africana Publishing, 1976. 331 p., bibl., figs., index. [UNITAR Regional Study, No. 2].
Two major sections are devoted to UN-OAU relations in the peace and security field and in the economic and social field. UNHCR, UNCTAD, UNIDO, UNICEF, UNDP, and the ECA. receive special attention.

927 _____. "OAU Collaboration with the United Nations in Economic and Social Development." In: El-Ayouty, Yassin, (ed.). The Organization of African Unity after Ten Years. New York, 1975, 213-36.
This essay is based on the research reported by the same author in his UNITAR publication.

928 _____. Peaceful Settlement among African States: Roles of the United Nations and the Organization of African Unity. New York: UNITAR, 1972. 68 p., bibl.
A brief version of his later volume published by Africana. Author analyzes cases of internal and interstate conflicts.

929 ANDRIAN, CHARLES F. "The Pan-African Movement: The Search for Organization and Community." Phylon, 23, 1, Spring 1962, 5-17.

930 ANSPRENGER, FRANZ. Die Befreiungspolitik der Organisation für Afrikanische Einheit, 1963 bis 1975. München: C. Kaiser, 1975. 232 p., bibl., tabl.

931 ARMAH, KWESI. Africa's Golden Road. London: Heinemann, 1965. 292 p., index.
A call for Pan-African unity with chapters on non-alignment, the Commonwealth, the UN, and a government for Africa.

932 AUSTIN, DENNIS. "Pan-Africanism, 1957-1963." In: Austin, Dennis and Weiler, Hans N., (eds.). Inter-state Relations in Africa. Freiburg, 1965, 1-29, tabl.
Between 1957 and 1963 Pan-Africanism had five major themes - territorial integrity, African Personality, closer union, economic co-operation, and anti-colonialism.

933 AUSTIN, D. and NAGEL, RONALD. "The Organization of African Unity." World Today, 22, 12, December 1966, 520-9.
Brief discussion of the failure of the OAU to live up to its original ideals and the difficulties it faces in the future.

934 AYAGA, ODEYO O. "OAU and Pan-Africanism in the 1970s: Some Proposals." Black World, 20, 10, August 1971, 41-2.
Amendments should be made to the Charter of the OAU.

935 AZIKIWE, NNAMDI. "The Future of Pan-Africanism." Présence Africaine, 12, 1, 1962, 7-29.

936 _____. "The Realities of African Unity." African Forum, 1, Summer, 1965, 7-22.

937 BADAL, R. K. "The Rise and Fall of Separatism in Southern Sudan." African Affairs, 301, October 1976, 463-74.
Analysis of the 1972 Addis Ababa accords for the settlement of the civil war in the Sudan.

938 BEDJAOUI, M. "Le Règlement Pacifique des Différends Africains." Annuaire Français de Droit International, 18, 1972, 85-99.
Analyzes the set of rules developed by African states to negotiate their conflicts. The role of the OAU is stressed.

939 BENNETT, GEORGE. "Pan-Africanism." International Journal, 18, 1, Winter 1962/1963, 91-6.

940 BINAISA, G. L. "Organization of African Unity and Decolonization: Present and Future Trends." Annals of the American Academy of Political and Social Science, 432, July 1977, 52-69.
A review of the Pan-African movement and its stands on the decolonization of Africa.

941 BISSELL, RICHARD E. "African Unity Twelve Years Later." Current History, 68, 405, May 1975, 193-6.
The OAU has been a partial success.

942 BLAGOJEVIĆ, N. "Aktivnost i Neki Problemi Organizacije Africkog Jedinstva." Medunarodni Problemi, 18, 1, January-March 1966, 77-89.
A brief summary of the first two years of the OAU.

943 _____. "Konferencija Sefova Nezavisnih Africkih Drzava." Medunarodni Problemi, 15, 2, April-June 1963, 67-85.
Tendencies to union and disunion in Africa and the formation of the OAU.

944 BLOCH, H. S. "Regional Development Financing." International Organization, 22, 1, Winter 1968, 182-203.
A profile and description of the financial resources of the ADB are included in this analysis of the role of regional development banks.

945 BORELLA, F. "Evolution Récente de l'Organisation de l'Unité Africaine." Annuaire Français de Droit International, 20, 1974, 215-25.
The successes of the OAU are much greater than those of OCAM. The OAU may achieve complete decolonization and the transformation of relations with the ex-colonial powers.

946 _____. "Le Système Juridique de l'OAU." Annuaire Français de Droit International, 17, 1971, 233-53.
The structures and norms of the OAU juridicial system.

947 BOTCHWAY, FRANCIS A. "A Reconsideration of the Pan-Africanism Reality." Pan-African Journal, 3, 1, Winter 1970, 34-48.
"My purpose in this article is to attempt to suggest a model for the study of inter-state integration in Africa."

948 BOUTROS-GHALI, B. "The Addis Ababa Charter." International Conciliation, 546, January 1964, 1-62.
The Addis Ababa Conference and the agreements made there in 1963 are analyzed in an effort to delineate their legal implications.

949 _____. "The Afro-Asian Movement: A Survey of Sources and Development." Revue Egyptienne de Droit International, 26, 1970, 18-57.

950 _____. Les Difficultés Institutionnelles du Panafricanisme. Genève: Institut Universitaire des Hautes Études Internationales, 1971. 57 p. [Conferences, 9].
The first section is a description of the concept Pan-Africanism. The second analyzes those institutions of the OAU most relevant to fostering African unity.

951 _____. "The League of Arab States and the Organization of African Unity." In: El-Ayouty, Yassin, (ed.). The Organization of African Unity after Ten Years. New York, 1975, 47-61.
A comparative study of the roles of the two organizations in decolonization, the settling of international disputes, and economic cooperation.

952 CASTAGNO, A. A. "The Somali-Kenyan Controversy: Implications for the Future." Journal of Modern African Studies, 2, 2, July 1964, 165-88, map.
Historical background and analysis of OAU efforts to resolve the conflict.

953 CERVENKA, ZDENEK. "Major Policy Shifts in the Organization of African Unity, 1963-1973." In: Ingham, K., (ed.). Foreign Relations of African States. London, 1974, 323-44.

954 _____. "The OAU and the Confrontation of Independent Africa with the White South." In: Shaw, T. M. and Heard, K. A., (eds.). Cooperation and Conflict in Southern Africa. Washington, D. C., 1977, 429-53.
   Changes in the balance of power in Southern Africa in 1974 came about "as a result of the success of two arms of OAU policy: its commitment to armed struggle in Southern Africa in the absence of meaningful negotiations, and its diplomatic role in the UN and the Third World."

955 _____. "The OAU and the Nigerian Civil War." In: El-Ayouty, Yassin, (ed.). The Organization of African Unity after Ten Years. New York, 1975, 152-73.
   The author considers the impact of the war on the OAU and on the relationships of its members.

956 _____. "L'OAU contre Vents et Marées." Revue Française d'Etudes Politiques Africaines, 93, September 1973, 61-79.
   Although not perfect, the OAU has been a positive factor in the solution of African economic and political problems.

957 _____. "OAU og Betydningen av den Nye Arabisk-Afrikanske Solidaritet." Internasjonal Politikk, 1, January-March 1974, 5-17.
   Similar to his article in Instant Research on Peace and Violence.

958 _____. The Organization of African Unity and Its Charter. New York: Praeger. London: C. Hurst, 1969. 253 p., index.
   The OAU from 1963 until 1969. Analysis of the Charter; peaceful settlement of disputes; relationships between the OAU and the UN, the ICJ, and regional organizations; and the Rhodesian and Biafran crises.

959 _____. "The Organization of African Unity in the Seventies." Verfassung und Recht in Übersee, 5, 1, 1972, 29-39.
   Will the OAU succeed in ending racial domination in Rhodesia and South Africa?

960 _____. "The Role of the OAU in the Peaceful Settlement of Disputes." In: El-Ayouty, Y. and Brooks, H. C., (eds.). Africa and International Organization. Hague, 1974, 48-68.
   Peaceful settlement of disputes has been supported by African states through the OAU (and prior to it) and in the UN.

961 _____. "The Settlement of Disputes among Members of the Organization of African Unity." Verfassung und Recht in Übersee, 7, 2, 1974, 117-38.

962 CERVENKA, Z. The Unfinished Quest for Unity: Africa and the OAU. New York: Africana, 1977. 251 p., index, map.
After a description of the structure of the OAU, the author examines the organization's role in the settlement of disputes, the Congo and Nigerian crises, the liberation of Southern Africa, and economic cooperation.

963 CHAPAL, P. "Le Rôle de l'Organisation de l'Unité Africaine dans le Règlement des Litiges entre Etats Africains." Revue Algérienne des Sciences Juridiques, Politiques et Economiques, 8, 4, December 1971, 875-910.
The OAU has devised official procedures for conflict resolution, but these are rarely used; rather, the OAU has played an important, but informal, role in solving intra-African disputes.

964 CHIKWE, J. "Africa: Unity and Differentiation." World Marxist Review, January 1972, 84-90.
Unity and disunity in African international politics.

965 CHRISMAN, ROBERT and HARE, NATHAN, (eds.). Pan-Africanism. Indianapolis: Bobbs-Merrill, 1974. 318 p., ill., tabls.
A collection of articles previously published in the journal, The Black Scholar.

966 COLA ALBERICH, JULIO. "La Explosiva Inestabilidad Africana." Revue Política Internacional, 77, January/February 1965, 57-75.
Largely an attack on the ideas of Nkrumah.

967 _____. "Las Islas Canarias y los Acuerdos de la OUA." Revista de Política Internacional, 156, March/April 1978, 49-66.

968 COLLIARD, C.-A. "Les Organisations Africaines." In: Colliard, C.-A. Institutions des Relations Internationales, 6th. ed. Paris, 1974, 551-63, map.

969 CONTRERAS GRANGUILLHOME, J. "La Necesidad de una Unificacion Económica Africana y la Interferencia de las Grandes Potencias." Revista Mexicana de Ciencia Política, 63, January-March 1971, 47-63.
African foreign trade is controlled by the ex-colonial powers. This prevents the formation of an African common market.

970 COWAN, L. GRAY. The Dilemmas of African Independence. New York: Walker, 1964. 162 p., bibl., figs., index, maps, tabls.
A major section is titled, "The Drive toward African Unity."

971 COX-GEORGE, N. A. "The Political Economy of African Unity."
Présence Africaine, 3, 31, 1960, 9-24.
Outlines the economic basis for African unity. Asserts that solving economic problems will clear the path for political unification.

972 DAVIS, LENWOOD G. "Pan-Africanism: An Extensive Bibliography." Genève-Afrique, 12, 1, 1973, 103-20.

973 DECRAENE PHILIPPE. Le Panafricanisme. 5th. Ed. Paris: Universitaires de France, 1976. 128 p., maps.
A history of the Pan-African movement with chapters on regional organizations, the OAU, and the role of the great powers.

974 _____. "Panafricanisme et Grandes Puissances." Civilisations, 13, 4, 1963, 445-62.
Author considers the attitudes of the Islamic states, the Communist states, and the USA toward Pan-Africanism.

975 _____. "Pan-Africanisme et Grandes Puissances." Politique Étrangère, 24, 4, December 1959, 408-21.
Author compares the attitudes of Great Britain and France toward Pan-Africanism.

976 DEGAN, V. D. "Commission of Mediation, Conciliation and Arbitration of the OAU." Revue Egyptienne de Droit International, 20, 1964, 53-80.
A description of the formal structure of the Commission and its responsibilities with comparisons between it and other pacific settlement institutions.

977 DIAITÉ, I. "L'OUA, l'ONU et le Règlement Pacifique des Conflits Interafricains." Annales Africaines, 1975, 9-34.
The OAU has been given sole responsibility for the settlement of inter-African disputes, but there are dangers in this and a UN role is recommended.

978 DIARA, AGADEM L. Islam and Pan-Africanism. Detroit: Agascha, 1973. 95 p., bibl.

979 DIOP, CHEIKH ANTA. Black Africa: The Economic and Cultural Basis for a Federated State. Westport, Conn.: Lawrence Hill, 1978. 88 p.
The author discusses a variety of topics concerned with culture, energy production, and industrialization. He presents "14 Steps to African Unity."

980 DUGARD, C. J. R. "The Organization of African Unity and Colonialism." International and Comparative Law Quarterly, 16, January 1967, 157-90.
Self-defense as a justification for the use of force.

981 DUIGNAN, PETER. "Pan-Africanism: A Bibliographic Essay." African Forum, 1, 1, Summer 1965, 105-7.

982 EFRAT, EDGAR S. "Pan-Africanism: Problems and Prospects." Bulletin of African Studies in Canada, 2, November 1964, 11-24.

983 EKPO, SMART A. "Eritrea: The O. A. U. and the Secession Issue." Africa Report, 20, 6, November/December 1975, 33-6, ill.
   The issue has caused a split in the OAU, but the organization may be forced to become involved in the settlement of the dispute.

984 EL-AYOUTY, YASSIN. "The OAU and the Arab-Israeli Conflict: A Case of Mediation that Failed." In: El-Ayouty, Yassin, (ed.). The Organization of African Unity after Ten Years. New York, 1975, 189-212.
   Africa's decision to support the Arab states is based on "Africa's capacity to make its diplomatic decisions in the light of its own long-range interests," not on Arab oil money.

985 _____. "O. A. U. Mediation in the Arab-Israeli Conflict." Genève-Afrique, 14, 1, 1975, 5-29.
   Analyzes the 1971 effort by African presidents to mediate the conflict between Egypt and Israel and the contribution of that failure to the transformation of the system of international relations between Africa and the Middle East.

986 _____, (ed.). The Organization of African Unity after Ten Years. New York: Praeger, 1975. 262 p., index, tabls.
   Contains twelve essays, six of which are case studies of specific OAU activities. Included are essays on the interactions between the OAU and other international organization, the OAU roles in conflict resolution and in Southern Africa, and the collective defense potential of the OAU.

987 ELIAS, T. O. "The Charter of the Organization of African Unity." American Journal of International Law, 59, 2, April 1965, 243-67.
   Description of the Charter and institutions of the OAU.

988 EL-KHAWAS, MOHAMED A. "Southern Africa: A Challenge to the OAU." Africa Today, 24, 3, July-September 1977, 25-41, tabls.
   Examines the broad range of assistance the OAU has given to various liberation movements and analyzes the role of the front-line states.

989 EMERSON, RUPERT. "Pan-Africanism." International Organization, 16, 2, Spring 1962, 275-90. Also in: Padelford, N. I. and Emerson, R., (eds.). Africa and World Order. New York, 1963, 7-22. Also in: Doro, M. and Stultz, N., (eds.). Governing in Black Africa. Englewood, Cliffs, N. J., 1970, 310-24. Also in: Markovitz, I. L., (ed.). African Politics and Society. New York, 1970, 444-58.
 A survey of the origins, meanings, and future of Pan-Africanism.

990 ESEDEBE, P. O. "Origins and Meaning of Pan-Africanism." Présence Africaine, 73, First Quarter 1970, 109-27.
 Attempts to find a working definition of Pan-Africanism.

991 _____. "What Is Pan-Africanism?" Journal of African Studies, 4, 2, Summer 1977, 167-87.
 Explores the development of Pan-Africanism, as both an idea and a movement, and urges writers on the subject to concentrate more on personalities and societies committed to its ideals.

992 ETINGER, YAKOV Y. African Solidarity and Neo-Colonialism. Moscow: Novosti, 1970 (?). 70 p., bibl.
 This translation of Afrikanskoe Edinstvo i Neokolonializm is a Soviet view of the OAU.

993 _____. Afrikanskaia Solidarnost i Proiski Neokolonializma. Moscow: Znanie, 1967. 29 p.
 Brief notes on the OAU.

994 _____. Mezhgosudarstvennye Organizatsii Stran Azii i Afriki. Moscow: Nauka, 1976. 191 p., bibl.
 A study of the political and economic aspects of international organizations in Africa and Asia.

995 EZE, O. C. "OAU Faces Rhodesia." African Review, 5, 1, 1975, 43-62, tabls.
 The OAU must provide economic and military assistance to African nationalist movements in Rhodesia and facilitate the consolidation of the new governments in Angola and Mozambique.

996 FOUQUET, DAVID. "African Unity -- In Europe." Africa Report, 20, 4, July-August 1974, 7-9.

997 FREDLAND, RICHARD A. "The OAU after Ten Years: Can it Survive?" African Affairs, 72, 288, July 1973, 309-16.
 Ten negative factors cloud the future of the OAU.

998 GAM, P. "L' O. U. A., l'Organisation de l'Unité Africaine." Revue Juridique et Politique, 20, 2, April-June 1966, 295-334.
 The place of the OAU in the movement toward African unity.

999 GEISS, IMANUEL. The Pan-African Movement: A History of Pan-Africanism in America, Europe and Africa. London: Methuen. New York: Africana, 1974. 575 p., bibl., maps;
Originally published as Panafrikanismus in 1968. This is mainly a history of Pan-Africanism up to the end of World War II.

1000 _____. "Pan-Africanism." Journal of Contemporary History, 4, 1, January 1969, 187-200.
The professed goals on political independence and unity will be reached only by a strong will to radical modernization, which stands no romantic and dangerous nonsense about preserving traditional African society.

1001 GITELSON, SUSAN AURELIA. "The OAU Mission and the Middle East Conflict." International Organization, 27, 3, Summer 1973, 413-9.
Can small states help resolve major conflicts in which the super powers are interested? An analysis of African intervention in the Middle East Crisis.

1002 GORDENKER, LEON. "The Organization of African Unity and the United Nations: Can they Live Together?" In: Mazrui, Ali A. and Patel, H. H., (eds.). Africa in World Affairs. New York, 1973, 105-19.
This is largely an analysis of the purposes and structures of the OAU and its relationships with the UN. "The tasks and resources of the two organizations are examined in the light of their own aims."

1003 GREEN, R. H. and SEIDMAN, A. Unity or Poverty? The Economics of Pan-Africanism. Harmondsworth: Penguin, 1968. 363 p., figs., index, tabls.
"The fragmented economies of a patchwork of states set clear limits to the prospect of economic growth, and the wealth of the whole continent is being distorted and divided by the dependence of its various political parts on foreign assistance."

1004 GUITON, R. J. "Der Beginn der Zusammenarbeit unter den Afrikanischen Staaten." Europa-Archiv, 17, 8, April 1962, 273-80.
A brief historical sketch of the Pan-African movement, with emphasis on the late 1950s and early 1960s.

1005 GUPTA, A. "The Rhodesian Crisis and the Organisation of African Unity." International Studies, 9, 1, July 1967, 55-64.
What role has the OAU played in settling the Rhodesian crisis?

1006 HAWLEY, EDWARD A. "The People's Republic of Angola Joins the OAU." Africa Today, 23, 1, January-March 1976, 5-6.

1007 HELLEN, J. A. "Independence or Colonial Determinism? The Africa Case." International Affairs, 44, 4, October 1968, 691-708.
The case for economic unity first, then political unity.

1008 HIPPOLYTE-MANIGAT, M. "Le Groupe de l'Organisation de l'Unité Africaine à l'OUA." Revue Française d'Etudes Politiques Africaines, 104, August 1974, 61-91.
Although the African states have made an appearance of Third World leadership, it has been only an appearance.

1009 HOOVER, ALICE. "Pan-Africanism: A Selective Bibliography." Current Bibliography on African Affairs, 4, 1, January 1971, 10-24.
Lists books and articles in English. Sections on the OAU, black Americans and Africa, Pan-Africanism, the "Fatherland," and cultural Pan-Africanism.

1010 HOSKYNS, CATHERINE. "Pan-Africanism and Integration." In: Hazelwood, A., (ed.). African Integration and Disintegration. London, 1967, 354-93.
"The purpose of the final chapter is to take a continental view and to see to what extent pan-African ideas and organisations have been able, between 1957 and 1966, to promote any degree of integration."

1011 _____. "Trends and Developments in the Organization of African Unity." Year Book of World Affairs, 21, 1967, 164-78.

1012 IQBAL, M. H. "The Organization of African Unity, 1969-1973." Pakistan Horizon, 26, 4, 1973, 50-60.
The OAU serves as a forum for the discussion of African problems.

1013 JOHNSON, CAROL A. and RUSSELL, SARA S. "Selected Bibliography: Africa and International Organization." International Organization, 16, 2, Spring 1962, 449-64.
This lengthy bibliography is in a special edition of this journal devoted to studies of African international organizations.

1014 JONAH, JAMES O. C. "The U. N. and the O. A. U.: Roles in the Maintenance of International Peace and Security in Africa." In: El-Ayouty, Y. and Brooks, H. C., (eds.). Africa and International Organization. Hague, 1974, 127-51.
"This paper will focus on the political areas of the UN-OAU relationship."

1015 JOUVE, E. "L'OUA et la Libération de l'Afrique." Annuaire du Tiers Monde, 1, 1975, 149-69.
Notes on the roles - political, economic, military, and cultural - of the OAU in the complete liberation of Africa.

1016 KALE, NDIVAKOFELE. "Crisis in African Leadership: O. A. U.'s Secretary-General and the Lonrho Agreement." Pan-Africanist, 5, September 1974, 12-25.
The author is a Cameroon political scientist. Agreements between the OAU and Lonrho led to a crisis and the dismissal of the OAU Secretary-General.

1017 KAMAL, HUMAYUN A. "Organization of African Unity." Pakistan Horizon, 26, 1, 1973, 36-47.

1018 KAMANU, ONYEONORO S. "Secession and the Right of Self-Determination: An O. A. U. Dilemma." Journal of Modern African Studies, 12, 1, 1974, 355-76.
Secession movements present a powerful political problem for the OAU, which has as a basic provision of its existence the safe-guarding of the territorial integrity of its members.

1019 KAPUNGU, LEONARD T. "The OAU's Support for the Liberation of Southern Africa." In: El-Ayouty, Yassin, (ed.). The Organization of African Unity after Ten Years. New York, 1975, 135-51.
In addition to providing moral and material support to liberation movements, the OAU also serves to unify such movements. The possibility of direct OAU intervention is considered.

1020 KARAOSMANOGLU, ALI L. "Conflits Internes et Règlement Pacifique des Différends dans le Cadre des Organisations Internationales." Turkish Yearbook of International Relations, 15, 1975, 37-62.
The OAU is one of four international organizations examined in this study of the role of international organization in the peaceful settlement of disputes.

1021 KAPPELER, DIETRICH. "Territorialkonflikte in Afrika und die Grundsätze der Organisation der Afrikanischen Einheit." Europa-Archiv, 33, 17, 10 September 1978, 561-72.

1022 KENIG, MARIA MAGALENA. "Les Problèmes du Caractère Juridique de l'Organisation de l'Unité Africaine." Africana Bulletin, 25, 1976, 7-16.

1023 KHALID, MANSOUR. "The Southern Sudan Settlement and its African Implications." In: El-Ayouty, Yassin, (ed.). The Organization of African Unity after Ten Years. New York, 1975, 174-88.
This is mainly a discussion of the internal conflict with only brief comments on the OAU role in its resolution.

1024 KLOMAN, E. H., JR. "New Directions in the Drive toward African Unity." Orbis, 6, 4, Winter 1963, 575-92.

1025 KNOTHE, TOMASZ. "The OAU as a Regional Organization of Collective Security." Studies in Developing Countries, 5, 1974, 62-91.

1026 KUCZYNSKI, B. "African-Arab Co-operation and the Evolution in the Position of the Organization of African Unity on the Middle East Conflict." Studies on Developing Countries, 9, 1978, 29-47.

1027 LECLERC, C. "Addis Abéba." Revue Juridique et Politique d'Outre-Mer, 17, 2, April-June 1963, 220-34.
The significance of the establishment of the OAU.

1028 LEGUM, COLIN. "Changing Ideas of Pan-Africanism." African Forum, 1, 2, Fall 1965, 50-61.

1029 _____. "The Organisation of African Unity: Success or Failure?" International Affairs, 51, 2, April 1975, 208-19.
The OAU serves as a mediator, conciliator, and arbitrator. Like the UN, its power is limited.

1030 _____. Pan-Africanism: A Short Political Guide. New York: Praeger, 1962. 296 p., maps. [Revised edition, 1965. 326 p.].
One of the most outstanding analyses of the development of Pan-African thought.

1031 _____. "Pan-Africanism and Communism." In: Hamrell, Sven and Widstrand, Carl G., (eds.). The Soviet Bloc, China and Africa. Uppsala, 1964, 9-29.
There has been a long-term confrontation between communism and Pan-Africanism.

1032 _____. "Possible Political Unities: The Thrust of Labor Movements in Pan-Africanism." In: Davis, John A. and Baker, James K., (eds.). Southern Africa in Transition. New York, 1966, 383-422.
A discussion of PAFMECA, AAPC, PAFMECSA, and EACSO.

1033 _____. "The Specialised Commissions of the Organisation of African Unity." Journal of Modern African Studies, 2, December 1964, 587-90.
Lists the various commissions and indicates the functions and roles of each.

1034 LOEWE, H. "Interafrikanische Zusammenschlüsse bis zur Organisation Afrikanischen Einheit 1963." Zeitschrift für Ausländisches Öffentlisches Recht und Volkerrecht, 24, 1, February 1964, 122-55.
A brief review of Pan-African meetings before the formation of the OAU is followed by a description of the capabilities and role of the OAU.

1035 MAKONNEN, RAS. Pan-Africanism from Within. Nairobi: Oxford University, 1973. 293 p., ill., index.
Makonnen has been an important figure in the Pan-African movement. In this volume he has focused attention on the cluster of attitudes essential to his own brand of Pan-Africanism.

1036 MANIGAT, M. "L'Organisation de l'Unité Africaine." Revue Française de Science Politique, 21, 2, April 1971, 382-401.
Since the inception of the OAU, African international conflicts have been settled by that body rather than by the UN. Why has this been so? How does the OAU resolve conflicts?

1037 MARCUM, JOHN A. "Pan-Africanism or Fragmentation?" In: Lewis, William H., (ed.). New Forces in Africa. Washington, 1962, 25-41.
The trend to further proliferation of states in Africa could be stopped. However, there are important conflicts between proponents of monolithic and plural unity in the Pan-African movement.

1038 MARIÑAS OTERO, L. "El Banco Africano de Desorrollo." Revista de Política Internacional, 133, May/June 1974, 159-70.
The African Development Bank was created in 1964 by the ECA. It has been successful.

1039 MARKAKIS, J. "The Organisation of African Unity: A Progress Report." Journal of Modern African Studies, 4, 2, October 1966, 135-53.
Description and analysis of the various structures of the OAU.

1040 MARTIN, D. and YANNOPOULOS, T. "L'Unité Africaine Face au Pouvoir Blanc." L'Univers Politique, 3, 1970, 276-95.
A discussion and defense of President Boigny's decision to accept the dialogue policy of South Africa.

1041 MATHEWS, K. "The Organization of African Unity." India Quarterly, 33, 3, July-September 1977, 308-24.
After 15 years of life, the OAU is still not capable of effective action. It has some gains, but on the whole it has not been of value and it is in need of revitalization.

1042 MAYALL, JAMES. "African Unity and the Organisation for African Unity: The Place of a Political Myth in African Diplomacy." Yearbook of World Affairs, 27, 1973, 110-33.
The major value of the OAU is not the policy it formulates. It is the values which are promoted throughout Africa by its existence.

1043 _____. "The OAU and the African Crisis." Optima, 27, 2, 1977, 82-95, ill.

1044 MAZRUI, ALI A. "Africa and the Black Diaspora: The Future in Historical Perspective." International Journal, 30, 3, Summer 1975, 569-86.
    Pan-Africanism exists at five levels - Sub-Saharan, Trans-Saharan, Trans-Atlantic, West-Hemispheric, and Global.

1045 _____. "African Diplomatic Thought and Supranationality." In: Mazrui, Ali. A. and Patel, H. H., (eds.). Africa in World Affairs. New York, 1973, 121-33.
    Africanism is a concept based upon "our continent" not "our race."

1046 _____. "On the Concept of 'We Are All Africans'." American Political Science Review, 57, 1, March 1963, 88-97. Also in: Mazrui, A. A. Towards a Pax Africana. London, 1967, 42-56.
    A discussion of the underlying unity of African peoples.

1047 MBOYA, TOM. "Conflicting Demands of Regionalism: Pan-Africanism and the Commonwealth." In: Tandon, Yashpal, (ed.). Readings in African International Relations. Vol. I. Nairobi, 1972, 271-7.
    Speech by the late Kenya Minister at Makerere University College, August 1964.

1048 McCAIN, J. A. "Ideology in Africa: Some Perceptual Types." African Studies Review, 18, 1, April 1975, 61-87.
    The author analyzes the meanings of African socialism to Africans and Africanists through the use of interviews and quantitative analysis of the data. Three variants of African socialism appear to exist - pragmatic, scientific, and internationalist. Julius Nyerere is seen by the respondents as a common element in all three variants.

1049 McKAY, VERNON. "Progress toward African Unity." Current History, 56, 333, May 1969, 257-62.
    The author concludes that there is little chance for an African federation or other forms of unitary government.

1050 McKEON, N. "The African States and the Organization of African Unity." International Affairs, 42, 3, July 1966, 390-409.
    The original concept of the OAU in 1963 was too ambitious and incapable of realization. This has led to disillusionment financial crisis, and poor attendance at meetings.

1051 McWILLIAM, MICHAEL D. "Pan-Africanism: Some Economic Considerations." In: Austin, D. and Weiler, H. N., (eds.). Inter-State Relations in Africa. Freiburg, 1965, 85-105.
    Analysis of the basic structure of the African economy, independence, the political characteristics of the new states, and institutional arrangements to put Pan-Africanism into effect.

1052 MEYERS, B. DAVID. "An Analysis of the OAU's Effectiveness at Regional Collective Defense." In: El-Ayouty, Yassin, (ed.). The Organization of African Unity after Ten Years. New York, 1975, 118-34.

"To examine the collective defense activities undertaken by the OAU. The responses of the OAU, and of its individual members, to cases of extraregional aggression will then be used to test claims concerning collective defense that have been made by advocates of regionalism."

1053 _____. "Intraregional Conflict Management by the Organization of African Unity." International Organization, 28, 3, Summer 1974, 345-74.

Several hypotheses concerning the role of regional organizations in conflict resolution are examined. In general, the OAU has not been effective in this respect.

1054 _____. "The OAU's Administrative Secretary-General." International Organization, 30, 3, Summer 1976, 521-31.

A framework for the analysis of the role of secretary-general is established and applied to the OAU, an example of a weak secretary-general.

1055 _____. "The Organization of African Unity: An Annotated Bibliography." Africana Journal, 5, 4, Winter 1974, 308-21.

Covers the scholarly research literature dealing directly with the organization and its activities. Includes books, dissertations, articles, and reviews from professional journals. English sources only.

1056 MINI, MOLEFI. "Class Struggle and African Unity: Ten Years of the OAU." African Communist, 54, 1973, 16-55.

1057 MITTELMAN, JAMES H. "The Development of Post-Colonial African Regionalism and the Formation of the Organization of African Unity." Kroniek van Afrika, 2, 1971/1972, 83-105.

1058 MORJANE, KAMEL. "L'Organisation de l'Unité Africaine et le Règlement Pacifique des Différends Interafricains." Revue Egyptienne de Droit International, 31, 1975, 17-74.

1059 MUSHKAT, M. "Zur Evolution und Rolle der Einheitsbestrebungen auf dem Afrikanischen Kontinent." Politische Studien, 200, November/December 1971, 571-87 and 201, January/February 1972, 25-41.

"The Evolution and Role of Efforts at Unification on the Continent of Africa." The first half of this essay is a historical sketch of Pan-Africanism. The second half describes current international organizations on the continent.

1060   NAGEL, RONALD K. "Documentary Resources of the Organization of African Unity on Its First Quintade." Current Bibliography on African Affairs, 3, 11/12, November/December 1970, 19-24.
        Twenty-six references are described.

1061   NELKIN, DOROTHY. "Pan-African Trade-Union Organization." In: Butler, J. and Castagno, A. A., (eds.). Transition in African Politics. New York, 1967, 39-66, tabls.
        "The innumerable attempts that have been undertaken to create a single Pan-African trade-union body remain thwarted."

1062   NEUBERGER, BENYAMIN. "Pan-Africanism: A Comparative Analysis." Canadian Review of Studies in Nationalism, 4, 1, Fall 1976, 100-27.

1063   NKRUMAH, KWAME. Africa Must Unite. New York: Praeger, 1963. London: Heinemann, 1963. New York: International, 1970. 229 p., index.
        Nkrumah's major statement on African unity.

1064   NWORAH, D. "The Integration of the Commission for Technical Co-operation in Africa with the Organization of African Unity: The Process of the Merger and the Problems of Institutional Rivalry and Complementarity." African Review, 6, 1, 1976, 55-67.
        Various problems of integrating otherwise duplicative organs of the OAU and the CCTA.

1065   _____. "The Organisation of African Unity and the International Labour Organisation, 1963-1973: A Decade of International Co-operation." Africa Quarterly, 16, 4, 1977, 27-34.

1066   NYANG, SULAYMAN SHEIH. "Islam and Panafricanism." L'Afrique et l'Asie Modernes, 101, 1974, 42-50.

1067   NYERERE, JULIUS. "Black Unity and Human Freedom." Africa Report, 20, 5, September/October 1974, 1-6.

1068   _____. "The Nature and Requirements of African Unity." African Forum, 1, Summer 1965, 38-52.

1069   _____. "A United States of Africa." Journal of Modern African Studies, 1, March 1963, 1-6.

1070   OTERO, LUIS MARIÑAS. "VI Congreso Panafricano (Dar es Salaam, 1974)." Revista de Política Internacional, 137, January/February 1975, 77-98.

1071   PADELFORD, NORMAN J. "The Organization of African Unity." International Organization, 18, 3, Summer 1964, 521-42.
        An introduction to the young OAU.

1072 PANOFSKY, HANS E. "Pan-Africanism: A Bibliographic Note on Organizations." African Forum, 1, 2, Fall 1965, 62-4.

1073 PERSON, YVES. "Les Contradictions du Nationalisme Etatique en Afrique Noire." In: Ingham, K., (ed.). Foreign Relations of African States. London, 1974, 239-57.
   The growing conflict between nationalism and supranationalism and the OAU's problems with the issue of Biafra are considered.

1074 _____. "L'OUA ou une Décennie d'Epreuves pour l'Unité." Revue Française d'Etudes Politiques Africaines, 93, September 1973, 29-60.
   Although the OAU has not proven capable of effective action, it is important as a symbol, a reminder, of the Pan-African ideal.

1075 PINDIĆ, D. "Organizacija Ujedinjenih Nacija Organizacija Africkog Jedinstva." Jugoslevenska Revija za Medunarodno Prava, 12, 1, May-August 1965, 258-71.
   "The United Nations and the Organization of African Unity." A brief review of the activities of the OAU indicates its basic harmony with the objectives and principles of the UN.

1076 POLHEMUS, JAMES H. "Organisation of African Unity and the Changing African International System." Nigerian Journal of International Studies, 1, 1, July 1975, 41-55.

1077 _____. "The Provisional Secretariat of the O. A. U., 1963-1964." Journal of Modern African Studies, 12, 2, 1974, 287-95.
   The creation of an international secretariat is the innovation that transforms a series of conferences into an organization. The author looks at the operation, effectiveness, and importance of the provisional secretariat and finds that it set many important patterns which were later followed by the General Secretariat.

1078 POVOLNY, MOJMIR. "Africa in Search of Unity: Model and Reality." Background, 9, 4, February 1966, 297-318, bibl.
   Applies an expanded systems (input-output) model to the study of attempts at interstate integration. Makes a tentative appraisal, and suggests tasks for further research along the same lines.

1079 PROVIZER, NORMAN W. "Themes, Images and Modes: A Note on Pan-Africanism." Southern Quarterly, 16, 1, October 1977, 59-66.
   Dignity, freedom, unity and redistribution are the major areas of concern in Pan-Africanism.

1080 QUÉNEUDEC, J. P. "La Commission de Médiation, de Conciliation et d'Arbitrage de l'O. U. A." Annales Africaines, 1966, 9-58.
Purposes and structures of the OAU's Commission of Mediation, Conciliation, and Arbitration.

1081 RAINERO, R. "L'Edificazione dell'Africa Nuova e la Conferenza di Addis Abeba." Comunità Internazionale, 18, 4, October 1963, 551-69.
"The Construction of the New Africa and the Addis Ababa Conference." A review of the conferences and meetings which preceded the Addis conference and which were steps in the construction of African unity.

1082 _____. "Il Problema dell'Unità Africana: Sviluppi e Prospettive." Comunità Internazionale, 20, 4, October 1965, 549-62.
"The Problem of African Unity: Developments and Prospects." A review of the growth of African unity and the obstacles facing such unity from the perspective of activities at the OAU.

1083 RIVKIN, ARNOLD. "The Organization of African Unity." Current History, 48, 248, April 1965, 193-200, and 240-2.
Posits that the OAU is an organization for establishing unity and not one of unity. It provides a structure and a framework within which to discuss and define problems among African states and to reconcile differences and resolve disputes.

1084 ROSSI, G. "The OAU: Results of a Decade." International Journal of Politics, 4, 4, Winter 1974/1975, 15-34.
Draws a rough balance sheet of the OAU's activities with respect to major continental problems since 1963.

1085 RUBIO GARCÍA, L. "El Arduo Camino de la Organización Africana." Revista de Política Internacional, 112, November/December 1970, 85-100.
"The Difficult Way to African Organization." Author exposes problems for African unity and for the OAU.

1086 _____. "El Discurrir del Panafricanismo en un Mundo de Estados Africanos Independientes." Revista de Política Internacional, 119, January/February 1972, 47-70.

1087 _____. "Ideología y Realidades en la Dinámica de la OUA." Revista de Política Internacional, 123, September/October 1972, 117-27; 124, November/December 1972, 57-89; 126 March/April 1973, 149-75.
"Ideology and Realities in the Dynamics of the OAU." These essays present a review of the activities of the OAU, its purposes and goals, and its strengths and weaknesses.

1088 SADIQALI, SHANTI. "OAU and Conflict Situations in Southern Africa." IDSA Journal, 9, 4, April-June 1977, 376-403.
African unity, the OAU and their roles in the liberation of Southern Africa.

1089 SAENZ, P. "The Organization of African Unity in the Subordinate African Regional System." African Studies Review, 13, 2, September 1970, 203-25.
The OAU is the institutional framework for the African subordinate regional system.

1090 _____. "A Summary of Pan-African Achievements." Rocky Mountain Social Science Journal, 5, 2, October 1968, 86-96.
A description of the achievements of the OAU in promoting continental cooperation.

1091 SALIH, GALOBAWI M. "The Role of the OAU in Public Administration and Management." In: El-Ayouty, Yassin, (ed.). The Organization of African Unity after Ten Years. New York, 1975, 237-50, fig.
What steps can the OAU take to improve public administration in its member states? ECA, CAFRAD, AAPAM, and the AOAS are also discussed.

1092 SANGER, CLYDE. "Toward Unity in Africa." Foreign Affairs, 42, 2, January 1964, 269-81.
A general review of the progress of African unity and its future after the founding of the OAU.

1093 SCIANÒ, FEDERICO. "Africa is Crisis." Affari Esteri, 36, October 1977, 730-47.
In spite of numerous crises in Africa the OAU has become an important factor in settling disputes, in the liberation of Southern Africa, and in forming an African stand on international issues.

1094 SEGAL, A. "La Integración en África: Problemas y Perspectivas." Revista de la Integración, 5, November 1969, 152-90.
Obstacles to integration are real but not insurmontable.

1095 SELASSIE, HAILE, I. "Towards African Unity." Journal of Modern African Studies, 1, 3, September 1963, 281-91. Also in: Roland, Joan G., (ed.). Africa: The Heritage and the Challenge. Greenwich, Conn., 1974, 388-400.
A statement by one of Africa's elder statesmen.

1096 SENGHOR, LEOPOLD S. "Le Dixième Anniversaire de l'OUA et l'Eurafrique. Vortag, Gehalten vom Staatspräsidenten der Republik Senegal." Osterreichische Zeitschrift für Aussenpolitik, 13, 5, 1973, 323-7.

1097 SHEPPERSON, GEORGE. "Pan-Africanism and 'Pan-Africanism': Some Historical Notes." Phylon, 23, 4, Winter 1962, 346-58.

Alleging that the term "Pan-Africanism" has been bandied about in recent years with disturbing inaccuracy, the author traces the origins of "Pan-Africanism" and "pan-Africanism" as separate developments.

1098 SINGLETON, S. "Les Etats Africains et le Congo. Recherche de Définition des Règles du Jeu." Cahiers Economique et Sociaux (Kinshasha), 8, 3, September 1970, 417-33.

A major problem for African states is the regulation of internal conflict and the avoidance of external activity therein. The role of the OAU may be significant in this.

1099 SMITH, GEORGE BUNDY. "Pan-Africanism and Nkrumah's Foreign Policy." Pan-African Journal, 1, Fall 1968, 195-200.

Maintains that while Nkrumah was for many years an important advocate of Pan-Africanism, he was also for a time the chief architect of Ghana's foreign policy, and that on most occasions he acted in the national interest when the two were in conflict.

1100 STAAL, GILLES DE. "Le Panafricanisme et le Contenu des Indépendances Nationales." Revue Française d'Etudes Politiques Africaines, 62, February 1971, 70-80.

1101 TANDON, YASHPAL. "Die Organisation der Afrikanischen Einheit in einer Phase der Konsolidierung." Europa-Archiv, 26, 23, December 1971, 835-42.

"The Organization of African Unity in a Phase of Consolidation." Three major problems faced the OAU in 1971 - a policy on dialogue, the admittance of a new delegation from Uganda, and Tanzania's demand for a new policy on OAU membership.

1102 _____. "The Organization of African Unity." Round Table, 246, April 1972, 221-30.

Introductory and descriptive.

1103 _____. "The Organization of African Unity and the Liberation of Southern Africa." In: Potholm, C. P. and Dale, R., (eds.). Southern Africa in Perspective. New York, 1972, 245-61, figs.

"This chapter examines the conflict from the perspective of the OAU. It will describe the OAU's response to the challenge posed by Southern Africa and will evaluate the OAU's strategy of liberation."

1104 _____. "The Organization of African Unity and the Principle of Universality of Membership." African Review, 1, 4, April 1972, 52-60.

Tanzania's views on OAU membership criteria.

1105 TANDON, Y. "The Organization of African Unity as an Instrument and Forum of Protest." In: Rotberg, R. and Mazrui, A. A., (eds.). Protest and Power in Black Africa. New York, 1970, 1153-84.
"The Organization is a very weak instrument of protest against the outside world . . . . It is more potent as a forum of protest within Africa itself."

1106 _____. "South Africa and the O. A. U.: The Dialogue on the Dialogue Issue." Instant Research on Peace and Violence, 2, 1972, 54-66.
Dialogue vs. armed confrontation with South Africa, a major issue settled this year at the OAU.

1107 TELLI, DIALLO. "The Organization of African Unity in Historical Perspective." African Forum, 1, 2, Fall 1965, 7-28.

1108 THOMAS, TONY and ALLEN, ROBERT. Two Views on Pan-Africanism. New York: Pathfinder, 1972. 21 p.

1109 THOMPSON, VINCENT BAKPETU. Africa and Unity: The Evolution of Pan-Africanism. London: International University Booksellers, 1969. New York: Humanities, 1969. 412 p., bibl., ill., index.
Origins and modern history of the Pan-African movement. A wide range of topics are examined, including brief studies of the major personalities involved in the modern period. An extensive set of documents is also included.

1110 THOMPSON, W. S. and BISSELL, R. E. "Development of the African Subsystem: Legitimacy and Authority in the OAU." Polity, 5, 3, Spring 1973, 335-61.
Authors argue that the OAU has had partial success in creating legitimacy but almost none in creating authority. Definitions are based upon Huntington's work.

1111 _____ and _____. "Legitimacy and Authority in the OAU." African Studies Review, 15, 1, April 1972, 17-42.
"Our purpose in this essay is to study the development of the African system by examining the Organization of African Unity." Definitions are based upon Ernst Haas' work.

1112 TIEWUL, S. AZADON. "Relations Between the United Nations and the Organization of African Unity in Settlement of Secessionist Conflicts." Harvard International Law Journal, 16, 1, Winter 1975, 259-302.

1113 TIMMLER, M. "Zehn Jahre nach der Gründung der OAE." Aussenpolitik, 24, 3, 1973, 320-32.
A positive summary of the accomplishments of the OAU during its first ten years of existence.

1114   TOLEN, AARON. "The Addis Ababa Conference." Présence Africaine, 21, 1, 1964, 33-47.

1115   TOWA, MARCIEN. Léopold Sédar Senghor: Négritude ou Servitude? Yaoundé: Editions CLE, 1971. 117 p.
       A severe criticism of Senghor's view of négritude.

1116   TWITCHETT, KENNETH J. "African Modernization and International Institutions." Orbis, 14, 4, Winter 1971, 868-90.
       Most African diplomatic activity occurs within the UN, the OAU, and the EEC.

1117   UKPABI, S. C. "The OAU and the Problems of African Unity." Africa Quarterly, 15, 4, 1976, 25-55.

1118   UKU, SKYNE R. The Pan-African Movement and the Nigerian Civil War. New York: Vantage, 1978. 106 p.

1119   UTETE, C. MUNHAMU BOTISO. "Crisis in the Organization of African Unity." Black World, 20, 12, October 1971, 23-31.
       The failure to end colonialism in Southern Africa and the debate over "Dialogue" may lead to the break-up of the OAU.

1120   VAILLANT, JANET G. "Dilemmas for Anti-Western Patriotism: Slavophilism and Négritude." Journal of Modern African Studies, 12, 3, September 1974, 377-94.

1121   VUKADINOVIĆ, R. "Organizacija Africkog Jedinstva." Politicka Misao, 3, 1/2, January-August 1966, 87-103.
       "The Organization of African Unity." The OAU has had only limited success in solving African economic and political problems. There are numerous reasons for this.

1122   WALLERSTEIN, IMMANUEL. Africa, the Politics of Unity: An Analysis of a Contemporary Social Movement. New York: Random House, 1967. 274 p., index, map.
       "An interpretation of the major political developments in Africa between 1957 and 1965 from the perspective of a major social movement on the continent, the movement toward African unity."

1123   _____. "The Early Years of the OAU. The Search for Organizational Pre-eminence." International Organization, 20, 4, Autumn 1966, 774-87.
       The OAU has attempted to create a central role for itself in African international relations.

1124   _____. "The Role of the Organization of African Unity in Contemporary African Politics." In: El-Ayouty, Y. and Brooks, H. C., (eds.). Africa and International Organization. Hague, 1974, 18-28.
       The OAU as an independent variable.

1125 WARD, B. "Problèmes Africains." Politique Etrangère, 27, 2, 1962, 140-50.
African unity is a powerful and important force in African international relations, but there are many problems to be solved upon which the African states have not yet reached agreement.

1126 WATERS, ALAN R. "A Behavioral Model of Pan African Disintegration." African Studies Review, 13, 3, December 1970, 415-33.
The failure of Pan-Africanism is linked to several factors.

1127 WELCH, CLAUDE E., JR. "The OAU and International Recognition: Lessons from Uganda." In: El-Ayouty, Yassin, (ed.). The Organization of African Unity after Ten Years. New York, 1975, 103-17.
Recognition of the Amin regime became a focus of conflict between radicals and moderates in the OAU. An analysis of the coup is included.

1128 WHITEMAN, KAYE. "The OAU and the Nigerian Issue." World Today, 24, November 1968, 449-53.
Sees differences and similarities between the way the Nigerian civil war affected the politics of independent Africa, and the way the Congo crisis did in the first half of the decade of the 1960s.

1129 WILD, PATRICIA BERKO. "Radicals and Moderates in the O. A. U.: Origins of Conflict and Bases for Coexistence." In: Tharp, Paul A., Jr., (ed.). Regional International Organizations. New York, 1971, 36-50.
An illustration of interest articulation and aggregation in international organization.

1130 WILLIAMS, OLU. "The Test of Pan-Africanism: The Invasion of Guinea and Aftermath." Mazungumzo, 2, 2, Winter 1972, 27-51.

1131 WOLFERS, MICHAEL. Politics in the Organization of African Unity. London: Methuen, 1976. 229 p., index, map.
A study of the OAU from 1963 to 1973, based largely on unpublished material.

1132 WORONOFF, JON. "African Unity: The First Ten Years." Africa Report, 18, 3, May/June 1973, 27-8.
A review of the successes of the OAU.

1133 _____. "Interview: Nzo Ekangaki, OAU Secretary-General." Africa Report, 17, 8, September/October 1972, 21-2.

1134 WORONOFF, J. "The OAU and Sub-Saharan Regional Bodies." In: El-Ayouty, Yassin, (ed.). The Organization of African Unity after Ten Years. New York, 1975, 62-78.
   Unlike the UN system, the international organizations of Africa often compete, conflict, and interact in a seemingly unorganized way.

1135 _____. Organizing African Unity. Metuchen, N.J.: Scarecrow, 1970. 708 p., bibl., index, maps.
   An analysis of the Pan-African movement with major emphasis on the OAU, its purposes and practices.

1136 WÜNSCHE, RENATE and STÖBER, HORST. "The Organization of African Unity and the Anti-imperialist Struggle of the Present Time." In: Büttner, T. and Brehme, G., (eds.). African Studies: Afrika Studien. Berlin, 1973.

1137 YAKEMTCHOUK, ROMAIN. "The OAU and International Law." In: El-Ayouty, Yassin, (ed.). The Organization of African Unity after Ten Years. New York, 1975, 79-102.
   The author attempts "to draw up a more or less valid balance of an international institution and the place it holds in relation to the positive law of nations."

1138 _____. "L'Organisation de l'Unité Africaine." Cultures et Développement, 6, 1, 1974, 23-59.
   This analysis of the OAU provides insight into the major elements of African international relations and the developing body of law and custom by which relations between African states are regulated.

1139 YTURRIAGA, J. A. DE. "L'Organisation de l'Unité Africaine et les Nations Unies." Revue Générale de Droit International Public, 69, 2, April-June, 1965, 370-94.
   The OAU is not incompatible with the UN and there is a certain amount of agreement in their goals and cooperation in their activities.

1140 ZAJACZOWSKI, A. "Le Ideologie dell'Africa Occidentale Contemporanea." Rivista di Sociologia, 4, 10, May-August 1966, 111-28.
   Pan-Africanism and négritude are compared.

The UN and International Law

1141 "Africa's Impact on the United Nations." Bulletin of the Africa Institute of South Africa, September 1972, 315-27, tabl.
Examines the impact of the Afro-Asian-Communist alliance at the UN.

1142 AHMAD, S. H. "The Fourth Committee and the Cases of Some Disputed Territories under Chapter XI." Indian Journal of Politics, 3, 1/2, January-December 1969, 143-61.
The Spanish Sahara and Ifni conflict is used as a case study.

1143 AKINSANYA, A. "The Nigerian Territorial Waters Decrees of 1967 and of 1971 in International Law." Revue Égyptienne de Droit Internacional, 32, 1976, 141-54.

1144 AKPAN, MOSES E. African Goals and Diplomatic Strategies in the United Nations: An In-depth Analysis of African Diplomacy. North Quincy, Mass.: Christopher, 1976. 165 p., bibl., index, maps, tabls.
The South West African, Rhodesian, and South African problems and African actions on these at the UN are the main topic. The UN and multilateral economic assistance are also discussed.

1145 AMACHREE, GODFREY K. J. "U. N. Civilian Operations in the Congo." In: Davis, J. A. and Baker, J. K., (eds.). Southern Africa in Transition. New York, 1966, 305-17, tabls.
A description of UN technical assistance activities in the Congo (Zaire) by a Nigerian who served as UN Under Secretary in charge of UN Civilian Operations in the Congo.

1146 AYAGA, O. O. "The U. N. Security Council's African Safari." International Studies, 12, 1, January-March 1973, 111-32, tabls.
The Security Council should recognize governments-in-exile for the white-ruled states of Africa.

1147 BARBIER, MAURICE. Le Comité de Décolonisation des Nations Unies. Paris: R. Pichon et Durand-Azias, 1974. 757 p., bibl., index, maps, tabl.
Description and analysis of the role of the Committee in a variety of case studies of decolonization, including Equatorial Guinea, the Portuguese territories, South West Africa, Kenya, Gambia, Rhodesia, Zanzibar, the B. L. S. states, and Spanish Sahara. There is a major section on the structure and functioning of the Committee and a very useful analytic bibliography.

1148 _____. "L'ONU et Le Tiers Monde." Annuaire du Tiers Monde, 1, 1975, 420-35.
The UN has served the Third World by its continuing pressure on South Africa and Rhodesia.

1149 BARRIE, G. N. "A Legal View of Transkeian Recognition and So-Called 'Statelessness'." Politikon, 3, 2, October 1976, 31-5.
Analysis of Transkei's independence in the light of international law.

1150 BENIPARRELL, C. DE. "Cooperacion Internacional en Africa." Política Internacional, 50/51, July-October 1960, 61-77.
Analyzes the role of international organizations in African development since the end of World War II. The UN, FAO, UNICEF, and others are examined.

1151 _____. "El Problema Politico de la Federacion de Rhodesia - Nyasalandia." Revue Política Internacional, 61, May/June 1962, 141-50.
Description and analysis of the effects of the 1962 General Assembly Resolution on Rhodesia. The resolution was proposed by Ghana.

1152 BENNOUNA, MOHAMED. "L'Affaire du 'Sahara Occidental' devant la Cour Internationale de Justice. Essai d'Analyse 'Structurale' de l'Avis Consultatif de 16 Octobre 1975." Revue Juridique, Politique et Economique du Maroc, 1, December 1976, 81-106.

1153 BEYERLIN, ULRICH. "Die Israelische Befreiungsaktion von Entebbe in Völkerrechtlicher Sicht." Zeitschrift für Ausländisches Öffentliches Recht und Völkerrecht, 37, 2, 1977, 213-43.
Analysis of the legal aspects of the Israeli raid on Entebbe in 1977.

1154 BIPOUN-WOUM, JOSEPH-MARIE. Le Droit International Africain: Problèmes Généraux - Règlement des Conflits. Paris: Pichon et Durand-Auzias, 1970. 327 p., bibl.
The OAU and inter-African conflicts are the major topics.

1155 BISHOP, A. S. and MUNRO, R. D. "The United Nations Regional Economic Commission and Environment Problems." International Organization, 26, 2, Spring 1972, 348-71.
   Contains some information on the ECA role in environmental problems.

1156 BLEICHER, S. A. "United Nations v. IBRD: A Dilemma of Functionalism." International Organization, 24, 1, Winter 1970, 31-47.
   Problems of the campaign to oust South Africa from international organizations.

1157 BLOM-COOPER, L. J. "Republic and Mandate." Modern Law Review, 24, 2, March 1961, 256-60.
   Considers the legal effects on the status of South West Africa when the Union of South Africa became the Republic of South Africa.

1158 BOEG, P. "The United Nations and the Decolonization of Africa." Genève-Afrique, 4, 2, 1965, 185-205.
   Direct UN influence was felt in the decolonization of the trust territories and Rhodesia, but in general the main role of the UN has been its influence on world public opinion.

1159 BOLLECKER, B. "L'Avis Consultatif du 21 Juin 1971 dans l'Affaire de la Namibie-Sud-Ouest Africain." Annuaire Français de Droit International, 17, 1971, 281-333.
   The action changed nothing in Namibia but it may have increased African respect for the ICJ.

1160 BOTHE, MICHAEL. "Völkerrechtliche Aspekte des Angola-Konflikts." Zeitschrift für Ausländisches Öffentliches Recht und Völkerrecht, 37, 3/4, 1977, 572-603.
   "The Angolan Conflict and International Law." Is this a domestic or an international conflict? Which body of law is applicable?

1161 BROOKS, ANGIE. "The U. N. Position - And a Projection for the Future." In: Davis, J. A. and Baker, J. K., (eds.). Southern Africa in Transition. New York, 1966, 59-71.
   Principally a description of the UN role in the South West Africa-Republic of South Africa question.

1162 CAHIN, GERARD and CARKACI, DENIZ. "Les Guerres de Libération Nationale et le Droit International." Annuaire du Tiers Monde, 2, 1976, 32-56.
   What is the status of wars of national liberation in international law?

1163 CARILLO SALCEDO, JUAN ANTONIO. "Libre Determinación de los Pueblos e Integridad Territorial de los Estados en el Dictamen del TIJ sobre el Sahara Occidental." Revista Española de Derecho Internacional, 29, 1, 1976, 33-49.
Analysis of the ICJ opinion on the Western Saharan question.

1164 CARROLL, FAYE. South West Africa and the United Nations. Lexington: University of Kentucky, 1967. 123 p., index.
"The purpose of this study is to provide a historical background to the dispute, reviewing the actions of the United Nations and of the South African government, evaluating the effectiveness of the UN in this matter, and considering the probable future consequences of the dispute."

1165 CARTER, GWENDOLEN M. "The Impact of the African States in the United Nations." In: Gardiner, R. K. A., (ed.). Africa and the World. Addis Ababa, 1970, 19-27.

1166 CASSESE, A. "L'Azione delle Nazione Unite contra Apartheid." Communita Internacional, 25, 3-5, July-October 1970, 619-45.
"The Action of the UN against Apartheid."

1167 CHAPPEZ, JEAN. "L'Avis Consultatif de la Cour Internationale de Justice du 16 Octobre 1975 dans l'Affaire du Sahara Occidental." Revue Générale de Droit International Public, 80, 4, October-December 1976, 1132-87.
The opinion rendered by the ICJ on the Western Saharan case was quite ambiguous and thus the various interested parties have interpreted it in their own ways.

1168 CHEMILLIER-GENDREAU, MONIQUE. "La Question du Sahara Occidental." Annuaire du Tiers Monde, 2, 1976, 270-80.
The ICJ has given an ambiguous opinion in the Western Saharan case, but international law is very unclear. So, the case must actually be settled by economic and political means.

1169 CHIMANGO, L. J. "The Relevance of Humanitarian International Law to the Liberation Struggles in Southern Africa - the Case of Moçambique in Retrospect." Comparative and International Law Journal of Southern Africa, 8, 3, November 1975, 287-317.

1170 CHOURAQUI, GILLES. "L'Afrique et le Droit de la Mer." Revue Juridique et Politique. Indépendance et Coopération, 31, 4, December 1977, 1129-39.
African states have played an important role in the development of the law of the sea.

1171 COHEN, A. "The New Africa and the United Nations." International Affairs, 36, 4, October 1960, 476-88.
Africa's impact on the UN and the work of the UN in Africa.

1172 CORET, A. "Problèmes de Participation à la Commission Economique pour l'Afrique." Penant, 696, April/May 1963, 182-97.
Limitations on membership in the ECA.

1173 _____. "Les Provinces Portugaises d'Outre-mer et l'ONU." Revue Juridique et Politique d'Outre-mer, 16, 2, April-June 1962, 173-221.
The role of the UN in Portuguese decolonization.

1174 DAYAL, RAJESHWAR. Mission for Hammarskjold: The Congo Crisis. London: Oxford University, 1976. 348 p, bibl., ill., index, map.
A description of UN activities in the Congo. The author was head of the Operation for nine months.

1175 DECALO, S. "Africa and the UN Anti-Zionism Resolution: Roots and Causes." Cultures et Développement, 8, 1, 1976, 89-117.
A review of African-Israeli relations and analysis of the great change that occurred in African attitudes during the mid-1970s. Arab oil money is seen as significant but not sufficient as an explanation of changes in policy.

1176 DERDA, J. "Sadrzina i Putevi Resenja Juznoafrickog Problema." Medunarodni Problemi, 16, 1, January-March 1964, 19-33.
The UN could take adequate action to resolve the South African problem.

1177 DOBELL, W. M. "United Nations: Sea Law, Peacekeeping, and Southern Africa." International Journal, 33, 2, Spring 1978, 415-23.

1178 DRAGHICI, M. "Le Processus de Décolonisation et Quelques Considérations Concernant le Statut Juridique de la Namibie." Revue Roumaine d'Etudes Internationales, 20/21, 1973, 157-74.

1179 DRISSI ALAMI, MOHAMED. "La Récupération du Sahara et le Droit Positif." Revue Juridique, Politique et Economique du Maroc, 2, June 1977, 9-35.
The ICJ and the Western Saharan question.

1180 DUCAT, M. "L'Affaire de la Namibie ou du Sud-Ouest Africain après l'Avis de la Cour Internationale de Justice." Penant, 737, June-September 1972, 301-31.

1181 DUGARD, J. "Namibia (South West Africa): The Court's Opinion, South Africa's Response, and Prospects for the Future." Columbia Journal of Transnational Law, 11, 1, Winter 1972, 14-49.
　　　The 1971 opinion of the ICJ on South Africa's presence in Namibia is not a model of perfection.

1182 _____. "The Revocation of the Mandate for South West Africa." American Journal of International Law, 62, 1, January 1968, 78-9.
　　　The legal status of General Assembly, ICJ, and Security Council actions on the mandate/trust status of South West Africa.

1183 _____. "South West Africa and the 'Terrorist Trial'." American Journal of International Law, 64, 1, January 1970, 19-41.
　　　International tutelage, such as the mandate system, should include judicial review by international courts of the administration of the territory.

1184 _____, (ed.). The South West Africa - Namibia Dispute: Documents and Scholarly Writings on the Controversy between South Africa and the United Nations. Berkeley: University of California, 1973. 585 p., bibl., index.
　　　Legal and political aspects are examined in depth. After the actions of the UN, the ICJ, and the Republic of South Africa the author concludes that a free plebiscite would be best in keeping with the stated goals and values of all parties concerned.

1185 EIDE, ASBJØRN. "International Law in Relation to Humanitarian Intervention in Biafra." Internasjonal Politikk, 3, 1969, 389-405.

1186 _____. "Prospects for U. N. Intervention in the Conflict in Nigeria-Biafra." Internasjonal Politikk, 3, 1969, 406-14.

1187 EL-AYOUTY, YASSIN, (ed.). Africa and International Organisation. Hague: Nijhoff, 1974. 250 p., fig., index, tabls.
　　　Fourteen essays under four topics are included. Group behavior in international organizations, international law and peaceful settlement; human rights, enforcement and security questions; integration and unity; and relational impacts and the future are the major topics. See references under Adebo, Brooks, Cervenka, El-Ayouty, Engo, Farah, Franck, Hovet, Jonah, Kapungu, Tandon, and Wallerstein.

1188 _____. "Africa's 'Burning Issues' and United Nations Action." Issue, 3, Fall 1972, 44-8.
　　　African states and the UN on issues of colonialism and apartheid.

1189 EL-AYOUTY, Y. "Legitimization of National Liberation: The
United Nations and Southern Africa." In: El-Ayouty, Y.,
(ed.). Africa and International Organization. Hague, 1974,
209-29, fig. Also in: Issue, 2, 4, 1972, 36-45.
An important function of the UN in the liberation struggle has been to serve as a source of legitimacy for liberation movements.

1190 _____. "The United Nations and Decolonization, 1960-70."
Journal of Modern African Studies, 8, 3, October 1970, 462-67.

1191 _____. The United Nations and Decolonization: The Role
of Afro-Asia. The Hague: Martinus Nijhoff, 1971. 286 p.,
bibl., index.
The Charter, constitutional evolution, and institutional
behavior are examined in the analysis of the role of the
Afro-Asian states in promoting decolonization through the UN.
Some information on the UN role in Namibia, Rhodesia/Zimbabwe,
and South Africa is contained.

1192 ELIAS, T. O. Africa and the Development of International
Law. Dobbs Ferry, N. Y.: Oceana Publications. Leiden:
Sitjhoff, 1972, 261 p.
The history of African international relations, Africa
in the UN, the ICJ and Africa (with stress on Namibia), the
OAU, human rights in Africa, and the legal aspects of foreign
investment are discussed.

1193 _____. "The Contribution of Asia and Africa to Contemporary International Law." Africa Quarterly, 16, 1, July
1976, 60-75.

1194 EL-KHAWAS, MOHAMED A. "Africa vs. South Africa: Quest for
Racial Equality." Current Bibliography on African Affairs,
4, 5, September/October 1971, 327-39.
Discusses the failure of the African states, in spite
of their unity of action and purpose, to bring about effective UN action against South Africa.

1195 _____. "The Afro-Asian Group in the United Nations."
Current Bibliography on African Affairs, 3, 11/12, November/
December 1970, 5-18, bibl.
A review essay with discussion of possible methodologies
for research.

1196 _____. "Mozambique and the United Nations." Issue, 2,
4, Winter 1972, 30-5.
UN action to end colonial rule by Portugal.

1197 EL-KHAWAS, M. "The Third World Stance on Apartheid: The United Nations Record." Journal of Modern African Studies, 9, 3, October 1971, 443-52.
   The Third World has been united in its action against apartheid, but this action has not caused much change in South Africa's policy.

1198 ELLIS, WILLIAM W. and SALZBERG, JOHN. "Africa and the United Nations - A Statistical Note." American Behavioral Scientist, 8, April 1965, 30-2.
   Relationship between UN voting and aid/trade.

1199 ENGERS, J. F. "The United Nations Travel and Identity Document for Namibians." American Journal of International Law, 65, 3, July 1971, 571-77.

1200 ENGO, PAUL BAMELA. "Peaceful Co-existence and Friendly Relations among States: The African Contribution to the Progressive Development of Principles of International Law." In: El-Ayouty, Y. and Brooks, H. C., (eds.). Africa and International Organization. Hague, 1974, 31-47.
   Author is Minister Plenipotentiary of the Foreign Service of Cameroon.

1201 EVANS, G. "The United Nations Economic Commission for Africa: The Tangier Meeting." World Today, 16, 4, April 1960, 176-80.
   Reviews the deliberation of the first two meetings of the ECA.

1202 EZE, O. C. "Prospects for International Protection of Human Rights in Africa." African Review, 4, 1, 1974, 79-90.
   The conditions in Africa are not appropriate for the construction of international machinery for the protection of human rights on the continent.

1203 FALK, RICHARD A. "The South West Africa Cases: An Appraisal." International Organization, 21, 1, Winter 1967, 1-23.
   The effects of the South West Africa decisions on Third World attitudes toward the ICJ.

1204 FARAH, A. A. "Southern Africa: A Challenge to the United Nations." Issue, 2, 2, Summer 1972, 14-24.
   African attempts to influence South African developments through the UN.

1205 FARER, TOM. Africa's Goals: The Options of International Law. New York: Center for International Studies, New York University, n.d. 24 p.
   The use of international law as a means to obtain African goals.

1206 FARINA, N. "Osservazioni sugli atta Internazionali di Riconoscimento della Guinea Bissau." Rivista Trimestrale di Diritto Pubblico, 3, 1975, 1521-26.
"Remarks on the International Instruments of Recognition of Guinea-Bissau."

1207 FLETCHER-COOKE, J. "Some Reflections on the International Trusteeship System, with Particular Reference to Its Impact on the Governments and Peoples of the Trust Territories." International Organization, 13, 3, Summer 1959, 422-30.
Tanganyika as a case study of the role of the Trusteeship system.

1208 FLORY, MAURICE. "L'Avis de la Cour Internationale de Justice sur le Sahara Occidental, 16 Octobre 1975." Annuaire Français de Droit International, 21, 1975, 253-77.
The opinion rendered by the ICJ on the Western Saharan case was quite ambiguous and each of the interested parties has its own interpretation of that opinion.

1209 FOELL, EARL W. "Africa's Vanishing Act at the UN." Africa Report, 14, 7, November 1969, 31-40.

1210 FOX, H. "The Settlement of Disputes by Peaceful Means and the Observance of International Law: African Attitudes." International Relations, 3, 6, October 1968, 389-410, 443.
African states wish to make changes in international law, but they do not reject international law.

1211 FRANCK, THOMAS M. "United Nations Law in Africa: The Congo Operation as a Case Study." Law and Contemporary Problems, 27, 4, Autumn 1962, 632-52.

1212 GAREAU, FREDERICK H. "The Impact of the United Nations upon Africa." Journal of Modern African Studies, 16, 4, 1978, 565-78.
"A general assessment of this impact with supporting data."

1213 GERGER, HALUK B. and GÜREL, SÜKRÜ S. "Oil Crisis and the Voting Patterns in UN General Assembly (1967-1975)." Turkish Yearbook of International Relations, 15, 1977, 63-80.
An analysis of voting patterns in the General Assembly is used as an indication of Third World attitudes on the Arab-Israeli debate.

1214 GITELSON, SUSAN A. "U. N. - Middle East Voting Pattern of the Black African States, 1967-1974." Middle East Review, Spring/Summer 1975, 33-7.

1215 GONZALEZ-SOUZA, LUIS. "La Orientación Política vs. el Tradicionalismo: ¿Un Cisma Irreparable en la Concepción del Derecho Internacional? El Caso de Namibia ante la Corte Internacional." Foro Internacional, 70, October-December 1977, 287-323.
 The ICJ and the problem of Namibia.

1216 GORDENKER, L. "Conor Cruise O'Brien and the Truth about the United Nations." International Organization, 23, 4, August 1969, 897-913.
 A review of O'Brien's writings about the UN, in particular its African operations.

1217 GORDON, EDWARD. "Old Orthodoxies Amid New Experiences: The South West Africa (Namibia) Liberation and the Uncertain Jurisprudence of the International Court of Justice." Denver Journal of International Law and Policy, 1, 1, Fall 1973, 65-92.

1218 GREEN, L. C. "De l'Influence des Nouveaux Etats sur le Droit International." Revue Générale de Droit International Public, 74, 1, January-March 1970, 78-106.
 Through the UN and the OAU the new states are influencing international law, especially with respect to problems of equality of states and self-determination.

1219 _____. "South-West Africa and the World Court." International Journal, 22, 1, Winter 1966/1967, 39-67.
 The 1966 judgement of the ICJ would not be repeated due to changes in Court membership.

1220 GREGG, R. W. "The United Nations Regional Economic Commissions and Integration in the Underdeveloped Regions." International Organization, 20, 2, Spring 1966, 208-32. Also in: Nye, J. S., (ed.). International Regionalism. Boston, 1966, 304-32.
 Analysis of the role of the regional economic commissions of the UN - including the ECA - in promoting economic integration.

1221 GROSS. E. A. "The South West Africa Case: What Happened?" Foreign Affairs, 45, 1, October 1966, 36-48.
 The ICJ decision in 1966 was inadequate and indicates that changes in the Statute of the Court are needed.

1222 GROSS, S. R. "The United Nations, Self-determination and the Namibia Opinions." Yale Law Journal, 82, 3, January 1973, 533-58.
 The General Assembly's role in self-determination and human rights as exemplified by the South West Africa situation.

1223 GRUHN, ISEBILL V. "The Commission for Technical Co-operation in Africa." Journal of Modern African Studies, 9, 3, October 1971, 459-69.
The CCTA existed from 1950 until 1965.

1224 _____. "International Development and Organization at the Crossroads: Choices Confronting Africa." Africa Today, 18, 2, April 1971, 56-65.

1225 _____. Regionalism Reconsidered: The Economic Commission for Africa. Boulder, Col.: Westview, 1979. 160 p., bibl., index.
Analysis and description of ECA structure and activities with some data on other regional organizations in the appendix.

1226 GYEKE-DAKO, K. Economic Sanctions under the United Nations. Tema: Ghana Publishing, 1974. 128 p., bibl.
See "Southern Rhodesia and Economic Sanctions," 50-9.

1227 HERZOG, CHAIM. "UN at Work: The Benin Affair." Foreign Policy, 29, Winter 1977-78, 140-59.
The UN debate over the 1977 invasion of Benin as an example of the irrelevance of UN debate and the waste of public money by the international organization.

1228 HEVENER, NATALIE K. "The 1971 South-West African Opinion: A New International Juridicial Philosophy." International and Comparative Law Quarterly, 24, October 1975, 791-810.

1229 HIDAYATULLAH, M. The South West Africa Case. New York: Asia Publishing, 1967. 144 p.
The book "is designed to bring before the ordinary reader an account of the dispute and an analysis of the judgements of the International Court of Justice."

1230 HIGGINS, R. "The Advisory Opinion on Namibia: Which United Nations Resolutions Are Binding under Article 25 of the Charter?" International and Comparative Law Quarterly, 21, 2, April 1972, 270-86.
Is the 1971 decision of the Court based on UN resolutions that are decisions bearing obligations on the members?

1231 _____. "The International Court and South-West Africa. The Implications of the Judgement." International Affairs, 42, 4, October 1966, 573-99.
The Court's 1966 decision was unfortunate, both for the future of the Court and of South West Africa.

1232 HOFFMAN, S. "In Search of a Thread: The United Nations in the Congo Labyrinth." International Organization, 16, 2, Spring 1962, 331-61. Also in: Padelford, N. and Emerson, R., (eds.). Africa and World Order. New York, 1963, 63-93.
What lessons for international organization are to be learned from the Congo operation?

1233 HOLDER, WILLIAM E. "1971 Advisory Opinion of the International Court of Justice on Namibia (South West Africa)." Federal Law Review, 5, 1973, 115-24.

1234 HOSKYNS, CATHERINE. "The African States and the United Nations, 1958-1964." International Affairs, 40, 3, July 1964, 466-80.
Africa and the UN; Pan-Africanism; Africa, India, and the USSR; and US policy in Africa.

1235 HOUSE, ARTHUR H. The UN in the Congo: The Political and Civilian Efforts. Washington, D. C.: University Press of America, 1978. 435 p.

1236 HOVET, THOMAS, JR. Africa in the United Nations. Evanston: Northwestern University, 1963. 336 p., bibl., figs., index, tabls.
An analysis of the African voting bloc at the UN.

1237 _____. "African Politics in the United Nations." In: Spiro, H., (ed.). Africa: The Primacy of Politics. New York, 1966, 116-49, fig.
"This paper attempts to analyze various aspects of the participation of Africa in the United Nations as possible avenues of insight into the nature of African politics on the world scene."

1238 _____. "Effect of the African Group of States on the Behavior of the United Nations." In: El-Ayouty, Y. and Brooks, H. C., (eds.). Africa and International Organization. Hague, 1974, 11-7.
"The theme of this chapter . . . will be to suggest illustrations in the UN which have been influenced directly or indirectly by the infusion of the African United Nations system."

1239 _____. "The Role of Africa in the United Nations." Annals of the American Academy of Political and Social Science, 354, July 1964, 122-34.
African states are operating in concert on issues directly relevant to Africa and on some non-African issues.

1240 IDOWU, E. O. A. Regulation and Control of the Seas: The United Nations Proposed Conference: An Opportunity or a Trap: An African View. Washington, D. C.: World Peace Through Law Center, 1973. 87 p.
The author considers the present laws of the sea as detrimental to African interests and makes several proposals for changes to be pursued at the 1974 Santiago Conference.

1241 IJALAYE, DAVID. "Was Biafra at Anytime a State in International Law?" American Journal of International Law, 65, 3, July 1971, 551-9.
It is "difficult to establish that Biafra attained statehood."

1242 IMISHUE, R. W. South-West Africa: An International Problem. London: Pall Mall, 1965. 80 p., bibl., map.
The mandate and trusteeship concepts and their application in South West Africa/Namibia. Appendices include relevant sections of the Charter and the Covenant and the Mandate for South West Africa.

1243 "International Court of Justice: U. N. Security Council Request for an Advisory Opinion on the Legal Consequences for States of the Continued Presence of South Africa in Namibia." International Legal Materials, 10, 4, July 1971, 677-829.

1244 JACOBSON, H. K. "ONUC's Civilian Operations: State-preserving and State-building." World Politics, 17, 1, October 1964, 75-107.
An evaluation of the effectiveness of the UN's non-military programs in the Congo.

1245 JACQUIER, B. "L'Autodétermination du Sahara Espagnol." Revue Générale de Droit International Public, 78, 3, 1974, 683-728.
The author seeks precedents in international experience for the settlement of the problem of Spanish Sahara.

1246 JINADU, ADELE L. "South West Africa: A Study in the 'Sacred Trust' Thesis." African Studies Review, 14, 3, December 1971, 369-87.
"The purpose of this paper is to analyze critically the concept of accountability to an international organization."

1247 JOHNSON, D. H. N. "The South-West Africa Case." International Relations, 3, 3, April 1967, 157-76.
Analysis of the 1966 judgement of the ICJ.

1248 KAREFA-SMART, JOHN. "Africa and the United Nations." International Organization, 19, 3, Summer 1965, 764-73.
Discussion of African views of the role of the UN.

1249 KAY, DAVID A. "The Impact of African States on the United Nations." International Organization, 23, 1, Winter 1969, 20-49.
The admission of many new African states to the UN has had many effects upon the UN political system.

1250 _____. The New Nations in the United Nations, 1960-1967. New York: Columbia University, 1970. 254 p.
Because of the addition of many new members, mostly African, we can expect the pattern of demands made upon the UN system to change.

1251 _____. "The Politics of Decolonization: The New Nations and the United Nations Political Process." International Organization, 21, 4, August 1967, 786-811.
Rhodesia as a case study of the role of the UN in decolonization.

1252 KOCHAN, RAN; GITELSON, SUSAN A.; and DUBEK, EPHRAIM. "Black African Voting Behavior in the UN on the Middle East Conflict." In: Curtis, M. and Gitelson, S. A., (eds.). Israel in the Third World. New Brunswick, N. J., 1976, 289-317, tabls. Also in: Jerusalem Journal of International Relations, 1, 2, 1975, 21-52.
How do small states relate to conflicts within regional subsystems external to their subsystem?

1253 KUNZMANN, K. H. "Die Bundnisfreiheit der Afro-Asiatischen Staaten. Eine Darstellung ihres Abstimmungsverhaltens in der Vollversammlung der Vereinten Nationen." Europa-Archiv, 16, 32, December 1961, 695-708.
The voting behavior of the Afro-Asian states in the General Assembly.

1254 LASS, H. D. "Die Vereinten Nationen und die Dekolonisation der Portugiesischen Überseegebiete." Politik und Zeitgeschichte, 40, 4, October 1975, 3-27.
After Portugal's entry into the UN in 1955 that organization became the scene of some of the efforts to end Portuguese colonial rule in Africa.

1255 LEFEVER, ERNEST W. Crisis in the Congo: A United Nations Force in Action. Washington, D. C.: The Broookings Institution, 1965. 215 p., bibl, index, maps.
An analysis of the UN role and the roles of the various foreign powers involved in the Congo crisis.

1256 LEFEVER, E. W. "The Limits of the United Nations Intervention in the Third World." Review of Politics, 30, 1, January 1968, 3-19.
UN intervention in South Africa and Rhodesia has little chance of success.

1257 _____. Uncertain Mandate: Politics of the United Nations Congo Operations. Baltimore: John Hopkins University, 1967. 254 p., bibl., fig., ill., index, map, tabls.
This is an analysis of the international politics of the ONUC. Chapters on the policies of the UN, USSR, Belgium, France, Britain, and Afro-Asian and small western powers are included.

1258 LEGUM, COLIN. "Economic Commission for Africa: Progress Report." World Today, 17, 7, July 1961, 299-307.
What is the ECA doing? Legum presents a positive assessment.

1259 _____. The United Nations and Southern Africa. Falmer, England: Institute for the Study of International Organisation, University of Sussex, 1970. 40 p., bibl.

1260 LEISS, AMELIA, C., (ed.). Apartheid and United Nations Collective Measures: An Analysis. New York: Carnegie Endowment for International Peace, 1965. 170 p., maps, tabls.
Analysis of the potential of various actions the UN might undertake.

1261 LETOUCHA, TADEUSZ. "Problem of Succession of African States in Respect of Post-colonial Boundaries in the Light of Practice." Studies in Developing Countries, 1, 1972, 131-55.

1262 LEWAN, K. M. "Rhodesien und die Zuständigkeitsgrenzen der Vereinten Nationen." Österreichische Zeitschrift für Aussenpolitik, 10, 1, 1970, 16-25.
It is doubtful that the UN had competence to act against Rhodesia, but the actions that have been taken have been tantamont to recognition of the existence of a sovereign state.

1263 LISSITZYN, O. J. "International Law and the Advisory Opinion on Namibia." Columbia Journal of Transnational Law, 11, 1, Winter 1972, 50-73.
The opinion was not well-written, although it should have significance for international law.

1264 LUARD, EVAN. "The Civil War in the Congo." In: Luard, E., (ed.). The International Regulation of Civil Wars. London, 1972, 108-24.
The UN's role in the Congolese Civil War prevented large scale intervention by the world powers.

1265 LUZZATO, L. "L'ONU e il Diritto all'Indipendenza dei Populi delle Colonie Portoghesi." Democrazia e Diritto, 13, 4, 1973, 145-84.
   UN actions have aided in the recognition of liberation movements as the governments of Guinea-Bissau and Mozambique.

1266 MAGEE, JAMES S. "ECA and the Paradox of African Cooperation." International Conciliation, 580, November 1970, 64 p.
   Author concludes that the ECA is of marginal concern to African states.

1267 _____. "What Role for ECA? - or Pan-Africanism Revisited." Journal of Modern African Studies, 9, 1, May 1971, 73-89.
   At its tenth anniversary, the ECA is in decline, and its adaptation of radical Nkrumahist policy is hastening its demise.

1268 MAIRATA LAVINA, J. "La Cuestión de Rhodesia del Sur, a la Luz de las Naciones Unidas." Revista de Política Internacional, 122, July/August 1972, 177-91.
   The Rhodesian problem is a source of hypocrisy for members of the UN who vote for economic sanctions but continue to trade with Rhodesia. Great Britain is especially culpable.

1269 MALHOTRA, RAM C. "Apartheid and the United Nations." Annals of the American Academy of Political and Social Sciences, 354, July 1964, 135-44.
   A summary of actions in the UN to oppose apartheid.

1270 MANNING, C. A. W. "The South West Africa Cases. A Personal Analysis." International Relations, 3, 2, October 1966, 98-110.

1271 MAZRUI, ALI A. "Africa and the United Nations: The Last Twenty-five Years." In: Reflections on the First Decade of Negro-African Independence. Paris: Présence Africaine, 1971, 67-80.
   An analysis of UN activities that benefit Africa especially in the areas of racial equality and economic development.

1272 _____. "The United Nations and Some African Political Attitudes." International Organization, 18, 3, Summer 1964, 449-520. Also in: Mazrui, A. A. On Heroes and Uhuru Worship. London, 1967, 183-208.
   The author shows relationships between several African political attitudes and the goals of the United Nations.

1273 McDOUGAL, M. S. and REISMAN, W. M. "Rhodesia and the United Nations: The Lawfulness of International Concern." American Journal of International Law, 62, 1, January 1968, 1-19.
Various cases against the legality of UN action in the Rhodesian case are put forward and then shown to be false.

1274 MEYERS, B. DAVID. "African Voting in the United Nations General Assembly." Journal of Modern African Studies, 4, 2, October 1968, 213-27.
The author identifies three African voting clusters and shows that, in general, African states were united on issues of self-determination, denuclearization, increased African representation, and apartheid.

1275 MITTELMAN, JAMES H. "Collective Decolonisation and the U. N. Committee of 24." Journal of Modern African Studies, 14, 1, March 1976, 41-64, tabl.
Investigates decolonisation by looking at attempts to implement General Assembly Resolution 1514 (XV) - the Declaration on the Granting of Independence to Colonial Countries and Peoples.

1276 MUSHKAT, M. "L'Afrique et les Problèmes du Droit des Gens." Verfassung und Recht in Übersee, 7, 1, 1974, 3-18.
African views of international law, including a legal justification for international economic aid.

1277 _____. "The Voice and Problems of Africa at the United Nations." Africa, 25, 4, December 1970, 387-412.
A study of blocs within the UN.

1278 MUTHARIKA, A. PETER. "Treaty Acceptance in the African States." Denver Journal of International Law and Policy, 3, 2, Fall 1973, 185-96.

1279 "Namibia: South Africa's Presence Found To Be Illegal; United Nation's Measures Declared Void." New York University Journal of International Law and Politics, 5, 1, Spring 1972, 117-38.

1280 NEUHOLD, H. "Völkerrechtliche Aspekte des Bürgerkrieges in Nigeria." Österreichische Zeitschrift für Aussenpolitik, 9, 2, 1969, 63-87.
"International Law Aspects of the Civil War in Nigeria." Although civil wars have become common, there are many questions in the international law of civil war that are unanswered. The author relates such question to the Nigerian case.

1281 NEWCOMBE, H.; ROSS, M.; and NEWCOMBE, A. G. "United Nations Voting Patterns." International Organization, 24, 1, Winter 1970, 100-21.
Factor analysis is used in this effort to portray dynamic aspects of bloc voting behavior. Over time, bipolarization is becoming more pronounced with the African and Asian states joining the Soviets and the Latin Americans joining the West.

1282 NICHOLAS, H. "United Nations Peace Forces and the Changing Globe." International Organization, 17, 2, Spring 1963, 321-37.
An examination of the significance of UN peace keeping forces based on the examples of Congo and Suez.

1283 NICOL, DAVIDSON. "Africa and the U. S. A. in the United Nations." Journal of Modern African Studies, 16, 3, September 1978, 365-95.

1284 _____. "The Attitudes of African States in the United Nations." Commonwealth, 16, 2, April 1972, 43-4.

1285 NISOT, J. "La Namibie et la Cour Internationale de Justice." Revue Générale de Droit International Public, 75, 4, October-December 1971, 933-43.
What are the consequences of the advisory opinion of 21 June 1971 of the ICJ for the members of the UN?

1286 NWORAH, DIKE. "The African Group in the United Nations, 1963-1966: The Absence of a Mechanism for Co-operation." International Studies, 14, 4, October-December 1975, 633-41.
Analysis and description of attempts to form a coherent Africa bloc at the UN.

1288 _____. "The United Nations in Africa, 1963-1973: Comparative Roles in Education, Food, and Finance." Genève-Afrique, 14, 2, 1975, 83-94.

1289 O'DONOVAN, P. "The Precedent of the Congo." International Affairs, 37, 2, April 1961, 181-8.
The precedent of intervention in the sovereign affairs of a state by the UN may be applicable in other instances of seemingly insoluble problems.

1290 OJO, MICHAEL A. "How the Trusteeship Colonies Became Independent." Africa Quarterly, 16, 4, 1977, 99-119.

1291 OJO, M. A. "UN and Freedom for Portugese Colonies." Africa Quarterly, 16, 1, July 1976, 5-28.

1292 PANTER-BRICK, S. K. "The Right to Self-Determination: Its Application to Nigeria." International Affairs, 44, 2, April 1968, 254-66.

1293 PECHOTA, VRATLISLAV. "The Right to Access to the Sea." In: Cervenka, Z., (ed.). Land-locked Countries of Africa. Uppsala, 1973, 37-43.
A review of the general status of all land-locked states. Also see the essay in this volume by A. P. Rubin.

1294 PERSAUD, MOTEE. "Namibia and the International Court of Justice." Current History, 68, 405, May 1975, 220-5.
The decisions and opinions of the ICJ, although bold, have had no affect. Peaceful change can only occur if the major powers and the UN take strong action.

1295 PETIT, GÉRARD. "Les Mouvements de Libération Nationale et le Droit." Annuaire du Tiers Monde, 2, 1976, 56-75.
What is the status of national liberation movements in international law?

1296 PEYROUX, E. "Les Etats Africains Face aux Questions Actuelles du Droit de la Mer." Revue Générale de Droit Internationale Public, 78, 3, 1974, 623-48.
African views on the law of the sea and the results of a 1972 conference in Yaoundé on this topic.

1297 PFEIFENBERGER, WERNER. "Die Sekretarisse-Generaal van die Verenigde Naises en Suid-Afrika." Politikon, 1, 1, June 1974, 64-85.
"Secretary-Generals of the UN and South Africa." A comparison of the roles of the four Secretary-Generals in the UN's dispute with South Africa.

1298 PIERSON-MATHY, P. "L'Action des Nations Unies contre l'Apartheid." Revue Belge de Droit International, 6, 1/2, 1970, 203-45.
A survey of UN action against apartheid and of restraints against such activities.

1299 PLENDER, RICHARD. "The Exodus of Asians from East and Central Africa: Some Comparative and International Law Aspects." American Journal of Comparative Law, 19, 2, Spring 1971, 287-324.

1300 POCKAEVA, M. V. "Ekonomiceskaja Komissija Organizacii Ob'- ediniennyh Nacij Stran Afrika." Sovetskoe Gosudarstvo i Pravo, 7, 1962, 116-22.
"The UN Economic Commission for African Countries."

1301 POLLOCK, A. J. "The South West Africa Cases and the Jurisprudence of International Law." <u>International Organization</u>, <u>23</u>, 4, Autumn 1969, 767-87.

Tensions between the concepts of sovereignty and human rights assist us in understanding the seeming contradictions between opinions and judgements of the ICJ and the UN and the lack of substantial UN action against South Africa.

1302 POMERANCE, M. "The Admission of Judges Ad Hoc in Advisory Proceedings: Some Reflections in the Light of the Namibia Case." <u>American Journal of International Law</u>, <u>67</u>, 3, July 1973, 446-64.

The main emphasis in this article in on the functioning of the ICJ, but some information on the Namibia case is included.

1303 RAMA RAO, T. S. "The Right of Self-Determination: Its Status and Role in International Law." <u>Internationales Recht und Diplomatie</u>, 1968, 19-27.

The author defines the concept of self-determination and traces the history of its development. Special attention is given to the question of the applicability of the concept with respect to South Africa and Rhodesia.

1304 RAO, P. S. "South West Africa Cases: Ethiopia v. South Africa; Liberia v. South Africa." <u>Africa Quarterly</u>, <u>6</u>, 3, October-December 1966, 236-53.

1305 RENNIGER, J. P. "After the Seventh Special (UN) General Assembly Session: Africa and the New Emerging World Order." <u>African Studies Review</u>, <u>19</u>, 2, September 1976, 35-48.

Policies of economic self-reliance and the establishment of a new world economic order are among the steps necessary for economic development in Africa. The actions taken at the Seventh Special Session are important in this respect.

1306 RIAD, F. A. M. United Nations Action in the Congo and Its Legal Basis." <u>Revue Egyptienne de Droit International</u>, <u>17</u>, 1961, 1-53.

The UN Operation in the Congo has so far been a failure due to the weakness of the mandate, the framework within which UNOC operates, and the inadequate financing by the UN.

1307 ROVINE, A. W. "The World Court Opinion on Namibia." <u>Columbia Journal of Transnational Law</u>, <u>11</u>, 2, Spring 1972, 203-39.

Negative aspects in the manner in which the ICJ handled the Namibia case.

1308 RUBIN, ALFRED P. "Land-locked Countries and Rights of Access to the Sea." <u>In</u>: Cervenka, Z., (<u>ed</u>.). <u>Land-locked Countries of Africa</u>. Uppsala, 1973, 44-62.

Also see the essay by V. Pechota in this volume.

1309 SAENZ, PAUL. "A Latin American-African Partnership." Journal of Inter-American Studies, 11, 2, April 1969, 317-27.
African and Latin American cooperation at the UN.

1310 SAGAY, ITSE. The Legal Aspects of the Namibian Dispute. Ile-Ife, Nigeria: University of Ife, 1975, 464 p.

1311 _____. "The Right of the United Nations to Bring Actions in Muncipal Courts in Order to Claim Title to Namibian Products Exported Abroad." American Journal of International Law, 66, 3, July 1972, 600-3.

1312 SALMON, J. J. A. "L'Accord O. N. U. - Congo (Léopoldville) du 27 Novembre 1961." Revue Générale de Droit International Public, 68, 1, January-March 1964, 60-109.
The agreement between the UN and the government of the Congo is not revolutionary in content, but it does contain some unusual characteristics.

1313 SAXENA, S. C. "Namibia and the United Nations." Indian Journal of Politics, 36, 3, July-September 1975, 274-96.
Consistent pressure through the UN has caused South Africa to change its Namibia policies.

1314 _____. "The Role of the U. N. Council for Namibia." Africa Quarterly, 17, 3, 1978, 5-31, tabl.

1315 SCHREIBER, M. "L'Année Internationale de la Lutte contre le Racisme et la Discrimination Raciale." Revue des Droits de l'Homme, 4, 2/3, July 1971, 311-40.
The campaign against apartheid occupies a major portion of this essay.

1316 SHEEHAN, JEFFEREY A. "The Entebbe Raid: The Principle of Self-Help in International Law as Justification for State Use of Armed Force." Fletcher Forum, 1, Fall 1976, 135-53.

1317 SHIHATA, I. F. I. "The Attitude of New States toward the International Court of Justice." International Organization, 19, 2, Spring 1965, 203-22.
African and Asian states have a positive view of international law as a protector of their interests.

1318 SLONIM, SOLOMON. South West Africa and the United Nations: An International Mandate in Dispute. Baltimore: John Hopkins University, 1973. 409 p., bibl., index, map.
International law aspects of the South West Africa case.

1319 SMOUTS, M. C. "Décolonisation et Sécession: Double Morale à l'ONU?" Revue Française de Science Politique, 22, 4, August 1972, 832-46.
UN practice in cases of secession such as Katanga and Biafra.

1320 SOUBEYROL, J. "Aspects de la Fonction Interprétative du Secrétaire Général de l'ONU lors de l'Affaire du Congo." Revue Générale de Droit International Public, 70, 3, July-September 1966, 565-631.
Analysis of the rules and practices regulating the behavior of the Secretary General's staff in the UN Operation in the Congo.

1321 _____. "Forum Prorogatum' et Cour Internationale de Justice: De la Procédure Contentieuse à la Procédure Consultative." Revue Générale de Droit International Public, 76, 4, 1972, 1098-104.
A case study of the 1971 ICJ opinion on the South West Africa issue.

1322 _____. "La Lutte contre l'Apartheid et le Colonialisme en Afrique." Année Africaine 1971, 1971, 167-272.
A summary of the campaign waged in international organizations against South African, Portuguese, and Rhodesian colonialism and racism.

1323 "South West Africa before the United Nations." World Today, 16, 8, August 1960, 334-45.
A review of South Africa's attempts to incorporate South West Africa and the Republic's refusals to agree to UN demands on this issue.

1324 SPENCER, J. H. "Africa at the United Nations: Some Observations." International Organization, 16, 2, Spring 1962, 375-86. Also in: McEwan, P. J. M. and Sutcliffe, R. B., (eds.). Modern Africa. New York, 1965, 393-405. Also in: Padelford, N. J. and Emerson, R., (eds.). Africa and World Order. New York, 1963, 107-18. Also in: Hanna, W., (ed.). Independent Black Africa. Chicago, 1964, 542-4.
African attitudes toward the UN are ambivalent, but there are strong hopes that the organization will further African interests.

1325 STEPHEN, M. "Natural Justice at the United Nations: The Rhodesian Case." American Journal of International Law, 67, 3, July 1973, 479-90.
The Security Council and the General Assembly must be bound by the rules of natural law; they have failed to be so bound in the Rhodesian case.

1326 STONE, J. "The International Court and World Crisis." International Conciliation, 536, January 1962, 1-64.
African views on international law are included.

1327 STREBEL, HELMUT. "Nochmals zur Geiselbefrieung in Entebbe." Zeitschrift für Ausländisches Öffentliches Recht und Völkerrecht, 37, 3/4, 1977, 691-710.
"Further Views on the Liberation of Hostages at Entebbe." Legal aspects of the Israeli action at Entebbe in 1977.

1328 SUKIJASOVIC, M. "Karakter Intervencije UN u Kongu." Medunarodni Problemi, 14, 2/3, April-September 1962, 65-87.
"The Nature of the UN Intervention in the Congo." A statement of the Third World view of the purposes of the UN Operation in the Congo.

1329 TRACHTMAN, JOEL. "The South-West Africa Cases and the Development of International Law." Millenium, 5, 3, Winter 1976/1977, 292-302.
Underlying the SWA cases is conflict between the positivist and the sociological schools of international law.

1330 TRISTRAM, UVEDALE. "Planning for Plenty in Africa: FAO's Indicative World Plan." Journal of Modern African Studies, 8, 3, October 1970, 468-70.

1331 TWITCHETT, K. J. "The Racial Issue at the United Nations: A Study of the African States' Reaction to the American-Belgian Congo Rescue Operation of November 1964." International Relations, 2, 12, October 1965, 830-46.
The first instance of black racism at the UN is seen in the reaction of the African states to the American-Belgian rescue operation in the Congo. Such racism is of benefit only to the PRC.

1332 UMOZURIKE, U. O. "International Law and Self-Determination in Namibia." Journal of Modern African Studies, 8, 4, December 1970, 585-603.
"Are the people of Namibia entitled to self-determination and how may they exercise that right? It will be necessary to refer back to the history of Namibia from the time of the mandate."

1333 _____. "The Namibia Cases 1950-1971." Africa Quarterly, 12, 1, April-June 1972, 41-58.
Several ICJ decisions make it clear that the people of Namibia have the right to self-determination. That right is violated through a variety of techniques.

1334 "The United Nations, Self-Determination and the Namibia Opinions." Yale Law Journal, 82, 3, January 1973, 533-58.

1335 U. N. SECRETARIAT. "Action Undertaken in Relation to Decolonization and the Elimination of 'Apartheid' by United Nations Bodies Primarily Connected with the Protection and Promotion of Human Rights." In: Stokke, O. and Widstrand, C., (eds.). The UN-OAU Conference on Southern Africa, Vol. II. Uppsala, 1973, 245-56.
Summarizes these actions by bodies such as the Third Committee of the General Assembly, the Economic and Social Council, the Commission on Human Rights, and the Sub-Commission on Prevention of Discrimination and Protection of Minorities.

1336 _____. "The United Nations Action on 'Apartheid ' in the Republic of South Africa." In: Stokke, O. and Widstrand, C., (eds.). The UN-OAU Conference on Southern Africa, Vol. II. Uppsala, 1973, 227-44.
Reviews the consideration of apartheid by the UN organs, indicates measures taken or recommended by them, and outlines a program of action as reflected in their resolutions.

1337 _____. "The United Nations and Decolonization: Principles, Objectives, Methods and Action." In: Stokke, O. and Widstrand, C., (eds.). The UN-OAU Conference on Southern Africa, Vol. II. Uppsala, 1973, 191-226.
Briefly discusses principles, objectives, and methods, and then lists in some detail relevant resolutions and other calls for action.

1338 VALLÉE, C. "L'Affaire du Sahara Occidental devant la Cour Internationale de Justice." Maghreb-Machrek, 71, January-March 1976, 47-55.
Description and analysis of the 1975 ICJ opinion on the Western Saharan question.

1339 VERNER, J. G. "Votación por Bloques en las Naciones Unidas: Comparación entre Africa y América Latina." Foro Internacional, 34, 9-2, October-December 1968, 121-41.
"Bloc Voting in the United Nations: A Comparison of Africa and Latin America." The main stress of this essay is on Latin American states' voting behavior in the General Assembly with some comparison to African behavior.

1340 VERZIJL, J. H. W. "The South West Africa Cases: Second Phase." International Relations, 3, 2, October 1966, 87-97.
A judgement on the juridical merits of the 1966 statement of the ICJ on the South West Africa Case.

1341 VILAKAZI, ABSOLOM L. "The Importance of International Organizations to African Diplomacy." Mawazo, 4, 3, 1975, 23-34.

1342 WAINHOUSE, DAVID W. Remnants of Empire: The United Nations and the End of Colonialism. New York: Harper and Row, 1964. 153 p., index, map.

Sections are included on Namibia and South Africa, Angola, and the British colonial empire.

1343 WALEFFE, F. "Pour une Coopération Internationale à l'Evolution Politique, Economique et Sociale du Congo Belge." Revue de l'Université de Bruxelles, 13, 1, October-December 1960, 47-56.

In a study completed before the independence of the Congo, the author argues that a major UN operation is needed to enable a successful transition from colony to independence for the Congo.

1344 WEIGERT, K. MAAS and RIGGS, R. E. "Africa and United Nations Elections: An Aggregate Data Analysis." International Organization, 23, 1, Winter 1969, 1-19.

Authors demonstrate a determinative association between UN office-holding and national capabilities, economic development, and UN participation.

1345 WEINSTEIN, W. "Africa's Approach to Human Rights and the United Nations." Issue, 6, 4, Winter 1976, 14-21.

Looks at the record of African states in the UN regarding human rights matters and then at the record of rights violations in Africa itself. Urges new OAU initiatives to remedy the present situation of gross violations and double-standards.

1346 WELENSKY, R. "The United Nations and Colonialism in Africa." Annals of the American Academy of Political and Social Sciences, 354, July 1964, 145-52.

The impact of Africa on the UN and of the UN on Africa.

1347 WELLINGTON, JOHN H. South-West Africa and Its Human Issues. Oxford: Clarendon, 1967. 461 p.

This volume contains history, geography, and political science. The mandate, UN relations, and ICJ decisions are discussed.

1348 WEST, R. L. "The United Nations and the Congo Financial Crisis: Lessons of the First Year." International Organization, 15, 4, Autumn 1961, 603-17.

Analysis of the lessons of ONUC.

1349 WISEBERG, LAURIE S. The Nigerian Civil War, 1967-1970: A Case Study in the Efficacy of International Law as a Regulator of Intrastate Violence. Santa Monica: Southern California Arms Control and Foreign Policy Seminar, 1972. 47 p., bibl.

1350 WODIE, F. "La Sécession du Biafra et le Droit International Public." Revue Générale de Droit International Public, 73, 4, October-December 1969, 1018-60.
The Nigerian Civil War or the attempted secession of Biafra was the focus of several debates over international law.

1351 WOHLGEMUTH, P. "The Portuguese Territories and the United Nations." International Conciliation, 545, November 1963, 1-68.
The UN and Portuguese views of the Portuguese overseas territories (such as Mozambique) were in distinct opposition. Even within the UN there was debate over what actions the organization could undertake.

1352 WOLF, FRANCIS. "L'Organisation Internationale du Travail et l'Afrique." La Comunita Internazionale, 26, 2, April 1971, 282-343.

1353 YAKEMTCHOUK, R. "L'Afrique en Droit International." Cahiers Economique et Sociaux (Kinshasha), 7, 4, December 1969, 383-410.
Boundaries, succession, recognition, and development-cooperation are the major areas of African concern in international law.

1354 YTURRIAGA, J. DE. "Les Relations entre les Nations Unies et les Organisations Régionales Africaines de Nature Technique." Revue Egyptienne de Droit International, 19, 1963, 33-80.
The OAMCE, the CCTA, and the ECA and their relations with the UN.

1355 ZEIGLER, J. "Les Nations Unies au Congo." Politico, 27, 4, December 1962, 826-35.
There have been five phases in the UN's policy in the Congo.

Southern Africa

1356 A. N. C. "The Liberation Struggle within South Africa and the International Community." In: Stokke, O. and Widstrand, C., (eds.). The UN-OAU Conference on Southern Africa, Vol. II. Uppsala, 1973, 183-90.
   Calls for mobilizing action in support of the oppressed and exploited masses of South Africa.

1357 ABRAHAMSE, L. G. "Current Trends in Southern Africa." South Africa International, 8, 3, January 1978, 137-47.
   Southern Africa and the EEC, strategic considerations, and economic factors.

1358 ADEBISI, B. O. "Alliance for Oppression: Pre-coup Portugal, Rhodesia and South Africa versus Blacks." Africa Quarterly, 16, 4, 1977, 5-26.

1359 "Africa 1971." Bulletin of the Africa Institute of South Africa, January/February 1972, 4-15.
   Both international communism and South Africa's policy of "dialogue" made substantial progress during 1971.

1360 AFRICA RESEARCH GROUP. Race to Power: The Struggle for Southern Africa. New York: Anchor, 1974. 341 p., ill., index, maps.
   South Africa's relations in Southern Africa and with the Western powers are examined.

1361 ANGLIN, DOUGLAS G. "The Politics of Transit Routes in Landlocked Southern Africa." In: Cervenka, Z., (ed.). Landlocked Countries of Africa. Uppsala, 1973, 98-133, tabls.
   Geographical, economic, political, and psychological factors influence a state's "landlockedness."

1362 ARKIN, MARCUS. "South Africa, Its Jews, and the Israel Connection." South Africa International, 8, 2, October 1977, 83-90.
Some discussion of the Jewish community in South Africa, but emphasis is on the Zionist sentiments of this community and ties between South Africa and Israel.

1363 ARNOLD, GUY. "Rhodesia: Increasing the Effectiveness of Sanctions." In: Stokke, O. and Widstrand, C., (eds.). The UN-OAU Conference on Southern Africa, Vol. II. Uppsala, 1973, 163-72.
Lists eight specific actions as having priority, including the strengthening of sanctions.

1364 _____. "Rhodesia under Pressure." Africa Report, 19, 4, July/August 1974, 16-20, ill., map.
Changes in Portugal have caused a dramatic shift in the political and military situation in Southern Africa, but the Rhodesian government does not seem to recognize this.

1365 BAINS, J. S. "Rhodesia and the United Nations." Journal of African and Asian Studies, 1, 1, Autumn 1968, 1-16.
Sanctions against Rhodesia will only succeed if extended to other white-dominated states of the area.

1366 BAKER, PAULINE H. "South Africa's Strategic Vulnerabilities: The 'Citadel Assumption' Reconsidered." African Studies Review, 20, 2, September 1977, 89-99.
The assumption of an impregnable white government in South Africa is open to question for several reasons.

1367 BANMEYER, D. G. A. "Indian Ocean Islands." South Africa International, 2, 4, April 1972, 202-8+.
Should South Africa become involved with the hundreds of islands off her east coast? And, if so, in what ways?

1368 BARBER, JAMES P. "Rhodesia and Interstate Relationships in Southern Africa." In: Potholm, C. P. and Dale, R., (eds.). Southern Africa in Perspective. New York, 1972, 129-33.
The author examines Rhodesia's role in the "Southern African bloc of states." He stresses the dynamic nature of the situation.

1369 _____. South Africa's Foreign Policy, 1945-1970. London: Oxford University, 1973. 325 p., bibl., ill., index, tabls.
A history of post-World War II relations with emphasis upon Britain and South Africa. Included are discussion of the UN, relations with African states, and Western policy.

1370 BARRATT, C. J. A. "Intra-Regional and International Relations." Bulletin of the Africa Institute of South Africa, November/December 1972, 420-41.

1371 BARRATT, JOHN. "The Department of Foreign Affairs." In: Worral, Denis, (ed.). South Africa: Government and Politics. Pretoria, 1971, 332-47.

1372 _____. "Détente in Southern Africa." World Today, 31, 3, March 1975, 120-30.
The Portuguese coup d'etat of 1974 has caused changes in the international relations of Southern Africa.

1373 _____. "South Africa's Outward Policy: From Isolation to Dialogue." In: Rhoodie, N., (ed.). South African Dialogue. Philadelphia, 1972, 543-61.
Analyzes the genesis of the outward policy, the impact of African politics on the outward movement, the Lusaka Manifesto, and domestic affairs as an international issue.

1374 _____. "Southern Africa: A South African View." Foreign Affairs, 55, 1, October 1976, 147-68.
Linkages between internal and external politics. External factors have impinged more heavily in recent years causing changes in attitude.

1375 _____. Southern Africa: Intraregional and International Relations. Johannesburg: South African Institute of International Affairs, 1973.

1376 _____. "Southern Africa: Intra-Regional and International Relations." In: Shaw, T. M. and Heard, K. A., (eds.). Cooperation and Conflict in Southern Africa. Washington, D. C., 1977, 44-80.
"It is intended to identify some of the major determinants in the foreign policies of the independent Black states of the African sub-continent which have a bearing on relations between the countries of the region."

1377 BARRETT, JOHN. "Dialogue in Africa: A New Approach." South Africa International, 2, 2, October 1971, 29-33.

1378 BATES, R. H. "A Simulation Study of a Crisis in Southern Africa." African Studies Review, 13, 2, September 1970, 253-64.
Zambia and the United States were used as the key parties in this simulation exercise.

1379 BISSELL, RICHARD E. Apartheid and International Organizations. Boulder, Col.: Westview, 1977. 231 p., bibl., index.
"Aside from analyzing the course of pressures on South Africa from international organizations, this study includes an attempt to determine the consequences of this dispute for South Africa, for international organizations, and, briefly, for the world at large."

1380 BISSELL, R. E. "How Strategic is South Africa?" In: Bissell, R. E. and Crocker, C. A., (eds.). South Africa into the 1980s. Boulder, Col., 1979, 209-32.

1381 _____ and CROCKER, CHESTER A., (eds.). South Africa into the 1980s. Boulder, Col.: Westview, 1979. 256 p., bibl., index.
Contains essays on South Africa's relations with its neighbors, military capabilities, Soviet-Cuban activities, etc.

1382 BOWMAN, LARRY W. Politics in Rhodesia: White Power in an African State. Cambridge, Mass.: Harvard University, 1973. 206 p., bibl., index, map, tabls.
This study of domestic politics included a chapter on the effects of international sanctions.

1383 _____. "The Problems of Labour Migration in Southern Africa." In: Cervenka, Z., (ed.). Land-locked Countries of Africa. Uppsala, 1973, 233-5.
A major form of international transaction in this region is labor migration.

1384 _____. "Southern Africa and the Indian Ocean." In: Cottrell, A. J. and Burrell, R. M., (eds.). The Indian Ocean: Its Political, Economic, and Military Importance. New York, 1972, 293-306.
Considers the effects of the decline of British hegemony and the emergence of great power competition. Also see the essay by P. Smit in this volume.

1385 _____. "The Subordinate State System of Southern Africa." International Studies Quarterly, 12, 3, September 1968, 231-61. Also in: McLellan, David S., et al, (eds.). The Theory and Practice of International Relations. Englewood Cliffs, 1974, 455-66. Also in: Shaw, T. M. and Heard, K. A., (eds.). Cooperation and Conflict in Southern Africa. Washington, D. C., 1977, 16-43.
An important essay in which the subsystem concept is refined and applied to Southern Africa.

1386 BOYD, J. BARRON. "Southern African Interactions: The Foreign Policy Perspective." In: Shaw, T. M. and Heard, K. A., (eds.). Cooperation and Conflict in Southern Africa. Washington, D. C., 1977, 98-117, figs., tabls.
The author utilizes event data analysis "to determine the structure and content of the foreign policy interactions which characterize Southern Africa."

1387 BOZIC, NEMANJA. "UN Economic Sanctions against Rhodesia." Medunarodni Problemi, 11, 3, 1969, 111-24.

1388 BREYTENBACH, W. J. Migratory Labour Arrangements in Southern Africa. Pretoria: Africa Institute of South Africa, 1972. 70 p., bibl., map.
One of the major ties of the Southern African regional system is labor migration.

1389 _____. South Africa's Involvement in Africa. Pretoria: Africa Institute of South Africa, 1978. 28 p., map, tabls.
Aid, trade, tourism, and cooperation are discussed. There is an interesting section titled "Requirements for Normalized Relations."

1390 BRIONNE, B. "Le Canal du Mozambique et le Sécurité de l'Afrique du Sud." Défense Nationale, February 1976, 125-41.
Developments internally in Mozambique and Madagascar affect their foreign policy and thus the security of South Africa.

1391 BULLIER, ANTIONE J. "L'Indépendance du Bophuthatswana et la Politique Ethnique de Prétoria." Revue Française d'Etudes Politiques Africaines, 145, January 1978, 98-126.
This new state will be refused international recognition as has been done to the Transkei.

1392 _____. "L'Indépendance du Transkei et la Politique de Développement Séparé." Revue Française d'Etudes Politiques Africaines, 129, September 1976, 28-49.
Argues that the UN and the OAU should recognize Transkei as a sovereign state.

1393 _____. "Pretoria et les Etats d'Afrique Noire." Revue Française d'Etudes Politiques Africaines, 10, 116, August 1975, 20-35.
Few of the black states of Africa can resist the dynamic political and economic detente policy of the Republic of South Africa.

1394 _____. "Les Relations entre l'Afrique du Sud et Israël." Revue Française d'Etudes Politiques Africaines, 119, November 1975, 54-65.
Matters of security and the common position of international isolation of these two states has led to the development of their relations.

1395 _____. "La République Sud-Africaine et l'Amérique Latine." Revue Française d'Etudes Politiques Africaines, 130, October 1976, 84-94.
The Republic of South Africa desires to be a major power in the southern hemisphere and is striving to develop strong relations with Latin American states.

1396 BUTTERWORTH, R. "The Future of South Africa." Yearbook of World Affairs, 31, 1977, 27-45.
Three variables will decide S. Africa's political future - the performance of the economy, white behavior, and the political development of black Africa. Progressive destabilization seems to be the most likely future.

1397 CARADON, HUGH M. F. Southern Africa in International Relations. London: Africa Bureau, 1970. 12 p.
South Africa and relations with Great Britain.

1398 CARTER, GWENDOLEN M. "Confrontation in Southern Africa." In: Paden,J. and Soja, E., (eds.). The African Experience, Vol. I. Evanston, 1970, 568-81.
A brief introduction to the international implications of the Southern Africa situation in a textbook designed as an introduction to African studies.

1399 _____ and O'MEARA, PATRICK, (eds.). Southern Africa in Crisis. Bloomington: Indiana University, 1977. 279 p., figs., index, maps, tabls.
The editors provide eight essays in an effort "to enable students and the general public to focus on the most significant features of . . . Southern Africa."

1400 CAVALLERO, T. and WEINSTEIN, WARREN. "Rhodesia: The United Nations and the Problems of Sanctions." Pan-African Journal, 5, 1, Spring 1972, 27-37.
Analysis of the effectiveness of sanctions in general, and against Rhodesia in particular. Finds international sanctions have been ineffective.

1401 CEFKIN, J. LEO. "African Development Strategies and the White-ruled States of Southern Africa." Africa Today, 20, 4, Fall 1973, 29-37.
The effects of white military and economic domination on the development plans of black-ruled neighbors in Southern Africa.

1402 _____. "The Rhodesian Question at the United Nations." International Organization, 22, 3, Summer 1968, 649-69.
More attention toward improving the effectiveness of sanctions is needed.

1403 CHALLENOR, HERCHELLE. "Towards South African Economic Hegemony in Africa." Review of Black Political Economy, 2, 3, Spring 1972, 56-80.

1404 CHARI, P. R. "The Military Balance." IDSA Journal, 9, 4, April-June 1977, 358-75.
Could South Africa become a nuclear garrison state? US, Soviet, Chinese, and regional nationalism are factors.

1405 CHENU, F. "Perspectives et Limites du Dialogue avec l'Afrique du Sud." Revue Française d'Etudes Politiques Africaines, 74, February 1972, 31-41.
Except in the Ivory Coast, South Africa's policy of dialogue has not been welcomed in Africa. The Southern African Common Market idea seems to be aimed at providing a protective wall between the Republic and the rest of Africa.

1406 CHETTLE, J. H. "U. S. Corporations and South Africa." South Africa International, 2, 1, July 1971, 47-54.
There are three options for a company operating within a foreign state: withdrawal, working within the law, or asserting a political and social attitude.

1407 CHHABRA, HARI SHARAN. "Southern Africa: The External Environment." IDSA Journal, 9, 4, April-June 1977, 349-57.
The roles of the frontline states (Zambia, Tanzania, Mozambique, Botswana, and Angola) in the Southern Africa conflicts.

1408 CHIROUX, R. "Le Tiers Monde et la Crise de l'Afrique Australe." Annuarie du Tiers Monde, 1, 1975, 170-87.
Southern Africa as a threat to world peace. UN, Third World, and Great Power activities and policies are critiqued.

1409 CLIFFE, LIONEL. "The Implications of the Tanzam Railway for the Liberation and Development of Southeastern Africa." In: Cervenka, Z., (ed.). Land-locked Countries of Africa. Uppsala, 1973, 293-9.
How will the railway change the balance of forces in the region? What effects will it have on industrialization and trade? How will Zambia's and Tanzania's dependency be altered?

1410 CLIFFORD-VAUGHAN, F. et al, (eds.). International Pressures and Political Change in South Africa. Cape Town: Oxford University, 1978. 109 p., bibl., maps.

1411 COCKRAM, GAIL-MARYSE. Vorster's Foreign Policy. Pretoria: Academica, 1970. 222 p., bibl., ill.

1412 COHEN, B. "Les Investissements Américains en Afrique Australe." Revue Française d'Etudes Politiques Africaines, 147, March 1978, 36-43.

1413 COLLINGS, FRANCIS D'A et al. "The Rand and the Monetary Systems of Botswana, Lesotho, and Swaziland." Journal of Modern African Studies, 16, 1, March 1978, 97-121.

1414 COLLINS, L. JOHN. "Activities of the International Defence and Aid Fund." In: Stokke, O. and Widstrand, C., (eds.). The UN-OAU Conference on Southern Africa, Vol II. Uppsala, 1973, 33-6.

1415 "The Commonwealth: A South African View. Three Schools of Thought." Round Table, 200, September 1960, 365-70.

1416 CONSTANTIN, FRANCOIS and COULON, CHRISTIAN. "La Diplomatie du Dialogue." Revue Francaise d'Etudes Politiques Africaines, 101, May 1974, 57-82.

1417 COULMAS, PETER. "Südafrika zur Stunde der Weltmächte." Merkur, 30, 11, November 1976, 1014-28.
"South Africa at the Time of the Big Powers." The last phase of the black drive for power in South Africa is underway and the US is changing to a policy more favorable to the black population.

1418 COURTNEY, WINIFRED and DAVIS, JENNIFER. Namibia: United States Corporate Involvement. New York: Africa Fund, 1972. 32 p., ill.
A major issue in the campaign to drive South Africa out of Namibia has been the role of US corporations in the trust territory.

1419 COWEN, D. V. "Towards a Common Market in Southern Africa: Economic Integration and Political Independence." Optima, 17, 2, 1967, 43-51, ill.
Political and economic discussions between South Africa and Lesotho, Botswana, and Swaziland, can - if imaginatively and generously conducted - lead to a situation that will be an object lesson for those parts of the world where people of different cultures and colors must co-exist.

1420 CROCKER, CHESTER A. "Current and Projected Military Balances in Southern Africa." In: Bissell, R. E. and Crocker, C. A., (eds.). South Africa into the 1980s. Boulder, Col., 1979, 71-106.

1421 CUMMINGS, RICHARD M. "The Rhodesian Unilateral Declaration of Independence and the Position of the International Community." New York University Journal of International Law and Politics, 6, 1, Spring 1973, 57-84.

1422 CURTIN, T. R. C. "Total Sanctions and Economic Development in Rhodesia." Journal of Commonwealth Political Studies, 7, 2, July 1969, 126-31.
Sanctions have a positive effect on economic development in Rhodesia.

1423 DALE, R. "The 'Glass Palace War' over the International Decolonization of South West Africa." International Organization, 29, 2, Spring 1975, 535-44.
An extended review of and commentary on two books by S. Slonim and J. Dugard on Namibia.

1424 DALE, R. "South Africa and the International Community." World Politics, 18, 2, January 1966, 277-313.
A review of three books edited or authored by Colin and Margaret Legum, A. C. Leiss, and R. Segal on the domestic and international ramifications of apartheid.

1425 DANNHAUSER, J. A. "Die Bepaling van die Seegrense van die Republiek van Sid-Afrika." Politikon, 3, 1, June 1976, 29-41.
"The Determination of the Sea Borders of the Republic of South Africa." South Africa's maritime boundaries with its neighbors are not yet delimited.

1426 DAY, JOHN. International Nationalism: The Extra-Territorial Relations of Southern Rhodesian African Nationalists. London: Routledge and Kegan Paul. New York: Humanities, 1967. 141 p., bibl.
The author analyzes growth of African nationalism and links of the nationalist movements and leaders with Britain, international organizations, and other external entities.

1427 _____. "The Rhodesian African Nationalists and the Commonwealth African States." Journal of Commonwealth Political Studies, 7, 2, July 1969, 132-44.
African states support nationalist policies in Rhodesia but not Rhodesian nationalists. The attitude of Zambia is discussed in detail.

1428 DE KIEWIET, C. W. "The Cape of Storms." South Africa International, 7, 3, January 1977, 125-48.
The US must counter Soviet moves in Southern Africa, but the RSA government must also alter its domestic policies. Suggestions are presented.

1429 _____. "The World and Pretoria." South Africa International, 1, 1, July 1970, 1-16.
South Africa is isolated and hated, but whites are optimistic and "confident that they can face their dangers and endure their unpopularity."

1430 DELANCEY, M. W. "The International Relations of Southern Africa: A Review of Recent Studies." Genève-Afrique, 15, 2, 1976, 82-135.
A general review of the literature in essay form with an extensive bibliographic list.

1431 DE ST. JORRE, JOHN. "South Africa: Up against the World." Foreign Policy, 28, Fall, 1977, 53-85.
South Africa views the USSR and the PRC as a threatening monolith. This is the key feature of South Africa's foreign policy. But new realities in Mozambique may cause changes in South African thinking.

1432 "Detente." Bulletin of the Africa Institute of South Africa, 13, 3, 1975, 75-83.

1433 DORO, MARION E. "A Bibliographical Essay on the November 1971 Rhodesian Settlement Proposals." Current Bibliography on African Affairs, 6, 4, Fall 1973, 411-30, bibl.

1434 DOXEY, MARGARET. "Alignments and Coalitions in Southern Africa." International Journal, 30, 3, Summer 1975, 518-35.
Regional relations are analyzed with emphasis on South Africa's African relations. Author concludes that the potential for increasing conflict is very great.

1435 _____. "International Sanctions: A Framework for Analysis with Special Reference to the United Nations and Southern Africa." International Organization, 26, 3, Summer 1972, 527-50.
An analytical framework is developed to test the viability of a sanctions policy. These criteria are then applied to UN Rhodesia sanctions. The UN fails to meet the stated criteria.

1436 _____. "The Rhodesian Sanctions Experiment." Yearbook of World Affairs, 25, 1971, 142-62.
Sanctions have not been as effective as expected. What were the grounds for such expectations?

1437 _____. "Sanctions Revisited." International Journal, 31, 1, Winter 1975/76, 53-78.

1438 DREIJMANIS, J. "The Rhodesian Question." Modern Age, 12, 4, Fall 1968, 371-8.
UN intervention is of questionable legality and sanctions are a failure. Britain should grant independence before majority rule is achieved.

1439 DUNCAN, P. "Towards a World Policy for South Africa." Foreign Affairs, 42, 1, October 1963, 38-48. Also in: Quigg, P., (ed.). Africa. New York, 1964, 250-64.
Controlled intervention with UN supervision is the best means to solve the South Africa problem, but this requires US leadership.

1440 DYKE, V. VAN. "Self-determination and Minority Rights." International Studies Quarterly, 13, 3, September 1969, 223-53.
Analysis of the UN General Assembly view of self-determination and comparison with the view of the government of the Republic of South Africa.

1441 EDMONDSON, LOCKSLEY. "The Challenges of Race: From Entrenched White Power to Rising Black Power." International Journal, 24, 4, Autumn 1969, 693-716.
Race in domestic and international politics, as symbolized by the United States and the Republic of South Africa.

1442 EGELAND, LEIF. South Africa's Role in Africa. Johannesburg: Jan Smuts House, 1967. 13 p.

1443 EISEMANN, PIERRE M. Les Sanctions contre la Rhodésie. Paris: A. Pedone, 1972. 154 p., bibl., maps, tabls.
Internal and external factors relevant to the success and failure of sanctions as well as the legal aspects of sanctions are examined.

1444 EL-KHAWAS, MOHAMED A. "South Africa and the Angolan Conflict". Africa Today, 24, 2, April-June 1977, 35-46.
Civil war in Angola led to intervention by South Africa to prevent the establishment of a radical regime. Opposition to South Africa's move led to increased involvement by other African states.

1445 _____. "U. S. Corporations in Southern Africa." Africa Today, 24, 2, April-June 1977, 65-9.
Review of White Wealth and Black Poverty: American Investment in Southern Africa, by Barbara Rogers, which documents the crucial role of US investment in supporting minority regimes.

1446 ERIKSEN, T. L. "Fra Rhodesia til Zimbabwe." Internasjonal Politikk, 1976, 659-88.
An analysis of recent events in Southern Africa and of the effects of Kissinger's diplomacy on those events.

1447 FARAH, ABDULRAHIM A. "South Africa's Apartheid Policy: An Assessment." In: El-Ayouty, Y. and Brooks, H. C., (eds.). Africa and International Organization. Hague, 1974, 71-102.
After introducing the reader to the meaning of apartheid, the author discusses various aspects of international relations, organization, and law relevant to the campaign to end apartheid.

1448 FARSŪN, SAMĪH. "Janūb Ifrīqīya wa-Isrā'īl: 'Ilāqah Khāsah." Shu'ūn Filastīnīyah, 47, July 1975, 151-77.
Close relations exist between the Republic of South Africa and Israel.

1449 FIRST, RUTH; STEELE, JONATHAN; and GURNEY, CHRISTABEL. The South African Connection: Western Investment in Apartheid. Harmondsworth: Penguin, 1973. London: Temple Smith, 1972. 352 p., index, tabls.
The role of foreign investment in South Africa.

1450 FISCHER, GEORGES. "La Non-Reconnaissance du Transkei."
      Annuaire Français de Droit International, 22, 1976, 63-76.
            The General Assembly resolution not to recognize Transkei as a sovereign state leaves unclear several legal questions. The ICJ has established some principles, but the Special Committee against Apartheid should observe nations' behavior on this.

1451 FISHEL, MURRAY I. "The International Aspects of South West Africa: The Historical Prospective." Journal of Human Relations, 13, Second Quarterly, 1965, 196-207.

1452 FRANCK, THOMAS M. et al. Policy Papers on the Legitimacy of Mandatory Sanctions by the United Nations against Rhodesia. New York: New York University, Center for International Studies, 1968. 36 p., bibl.
            Contains essays by T. Franck and L. B. Sohn.

1453 FRANKO, LAWRENCE G. "The European Connection: How Durable?" In: Bissell, R. E. and Crocker, C. A., (eds.). South Africa into the 1980s. Boulder, Col., 1979, 187-208.

1454 FRYER, A. K. "South Africa, a Target for Foreign Policy: A Footnote to J. D. B. Miller's 'Notes'." Australian Outlook, 26, 2, August 1972, 216-21.
            By appropriate foreign policy inputs, states could influence South African events in line with the principles such states purport to sustain.

1455 GADIEL, D. L. "Australian Commercial Policy and South Africa's Economic Isolation." Economic Record, September 1974, 381-404.

1456 GALTUNG, JOHAN. "On the Effects of International Economic Sanctions with Examples from the Case of Rhodesia." World Politics, 19, 3, April 1967, 378-416, figs.
            A theory on the effect of economic sanctions which suggests that internal consolidation followed at some time by disintegration should be expected in Rhodesia.

1457 GANN, L. H. "Portugal, Africa, and the Future." Journal of Modern African Studies, 13, 1, March 1975, 1-18.

1458 _____ and DUIGNAN, PETER. South Africa: War, Revolution, or Peace? Stanford: Hoover Institution, 1978. 85 p.

1459 GENTILI, A. "L'Africa Australe fra Radicalizzazione e Stabilizzazione." Politica Internazionale, 7, July 1977, 27-34.
            "Southern Africa between Radicalization and Stabilization." The US and the USSR can assist Africa through the provision of economic pressure on the white-dominated states.

1460 GERVASI, SEAN. "Breakdown of the U. S. Embargo." In: Western Mass. Association of Concerned African Scholars, (ed.). U. S. Military Involvement in Southern Africa. Boston, 1978, 133-56.
"Extensive evidence exposes the continued funneling of military supplies and equipment to South Africa and even to the illegal Smith regime."

1461 _____. "The Strategy of the Neo-colonialist Powers in Southern Africa." Black Liberator, 1, 4, 1972, 225-30.

1462 _____ and TURNER, JAMES. "The American Economic Future in Southern Africa: An Analysis of an Agency for International Development Study on Zimbabwe and Namibia." Journal of Southern African Affairs, 3, 1, January 1978, 85-98.

1463 GESHEKTER, CHARLES L. "Independent Mozambique and Its Neighbors: Now What?" Africa Today, 22, 3, July-September 1975, 31-36.
Comments on the future relations between Mozambique and its neighbors and sees the possible formation of a web of South African neocolonialism.

1464 GEYSER, O. Detente in Southern Africa. Bloemfontein: Institute for Contemporary History, 1976. 59 p., bibl. [Focus on Politics, 2].
A strongly positive view of South Africa's Africa policy. The author stresses the Soviet threat to Africa and the successes of South Africa's Africa policies.

1465 GIFFORD, LORD. "The Liberation Struggle in Mozambique and the Outside World." In: Stokke, O. and Widstrand, C., (eds.). The UN-OAU Conference on Southern Africa, Vol. II. Uppsala, 1973, 89-96.
Gives a short historical background to the struggle, the Portuguese reaction, and the response of the international community.

1466 GINIEWSKI, PAUL. "Le Dialogue s'amorce entre Pretoria et l'Afrique Noire." Revue Française d'Etudes Politiques Africaines, 70, October 1971, 36-43.

1467 _____. "Un Nouveau Problème International: Le Sort du Sud-Ouest Africain." Politique Etrangère, 30, 3, 1965, 280-6.
Author recommends a regrouping of the territories of Southern Africa to prevent chaos.

1468 _____. "South Africa and the Defence of the Cape Route." NATO's Fifteen Nations, 17, 2, April/May 1972, 24-6.
Increasing Soviet and Chinese penetration of Southern Africa "make African strategic geography a matter of new and increased topical significance . . . ."

1469 GITELSON, S. A. "The Transformation of the Southern African Subordinate State System." Journal of African Studies, 4, 4, Winter 1977, 367-91, tabl.

1470 GLANTZ, MICHAEL H. and EL-KHAWAS, MOHAMED A. "On the Liberation of African Liberation Movements." In: Weinstein, W., (ed.). Chinese and Soviet Aid to Africa. New York, 1975, 202-21, tabl.
Liberation movement "ties with external powers have often embroiled these movements in the larger [world] cleavages, even to the extent that their revolutionary effort becomes 'stalled'."

1471 GOGUEL, A.-M. "L'Impact de l'Afrique du Sud sur les Pays de l'Océan Indien." Revue Française d'Etudes Politiques Africaines, 97, January 1974, 41-64.
South Africa in Indian Ocean politics.

1472 GOLDBLATT, I. The Mandated Territory of South West Africa in Relation to the United Nations. Cape Town: Struik, 1961. 67 p.
Description, analysis, and criticism of South Africa's policies and actions with respect to South West Africa.

1473 GOOD, ROBERT C. "Rhodesia: Towards a New Policy Context." Africa Today, 18, 4, October 1971, 3-16.

1474 _____. U. D. I.: The International Politics of the Rhodesian Rebellion. London: Faber and Faber. Princeton: Princeton University, 1973. 368 p., bibl., ill., index, map.
Author was US Ambassador in Lusaka from 1965 until 1968. The book is partly memoir, partly documentary research, and partly interview-based.

1475 GRUNDY, KENNETH W. "Anti-Neocolonialism in South Africa's Foreign Policy Rhetoric." In: Shaw, T. M. and Heard, K. A., (eds.). Cooperation and Conflict in Southern Africa. Washington, D. C., 1977, 351-64.
"This paper is designed to discuss and to stimulate discussion about the present and future international role of the Republic of South Africa and especially of South Africa's role in the global capitalist order."

1476 _____. Confrontation and Accommodation in Southern Africa: The Limits of Independence. Berkeley: University of California, 1973. 360 p., bibl., ill., index, tabls.
A study of the international relations of the Southern Africa regional sub-system. Guerrilla warfare and South Africa's Africa polices are examined.

1477 _____. "Host Countries and the Southern African Liberation Struggle." Africa Quarterly, 10, 1, April-June 1970, 15-24.

1478 GRUNDY, K. W. "Intermediary Power and Global Dependency: The Case of South Africa." International Studies Quarterly, 20, 4, December 1976, 553-80.

Utilizing data on South Africa, the author seeks to demonstrate the ambiquities and policy alternatives confronting intermediary states in the context of global politics and economics. A dependency model of international relations between industrial states and the less-developed countries is sketched.

1479 _____. "Regional Relations in Southern Africa and the Global Political Economy." In: DeLancey, M. W., (ed.). Aspects of International Relations in Nigeria. Bloomington, Ind., 1979, 90-125.

"Conditions of dominance and submission, super and subordination, imposition, penetration, accommodation, and resistance characterize regional and global affairs regarding Southern Africa."

1480 _____. "The 'Southern Border' of Africa." In: Widstrand, C. G., (ed.). African Boundary Problems. New York, 1969, 119-60, figs., maps, tabls.

A subsystem analysis of the international relations of Southern Africa.

1481 GUELKE, ANIMESH. "Africa as a Market for South African Goods." Journal of Modern African Studies, 12, 1, 1974, 69-88, tabls.

Political factors outweigh economic factors in Africa-South Africa relations.

1482 GUPTA, A. "Arms, African States and the Commonwealth." Economic and Political Weekly, 6, 14, April 1971, 747-52.

The only way in which African states can influence events in South Africa is to make South Africa a cold war issue.

1483 _____. "Collapse of the Portuguese Empire and the Dialectics of Liberation of Southern Africa." International Studies, 14, 1, January-March 1975, 1-20.

How will Portuguese decolonization affect white power in South Africa? Will South Africa respond with military intervention or by extending its economic domination of the region?

1484 _____. "Southern Africa: The Internal Variables." IDSA Journal, 9, 4, April-June 1977, 339-48.

The South Africa situation is thoroughly tied to international politics and economics. What futures are possible for the South African problem?

1485 GUTTERIDGE, W. "The Coming Confrontation in Southern Africa."
Conflict Studies, 15, August 1971, 1-19.
Several factors could upset the apparent stability of
the South African regime.

1486 GYGLI, P. "Power-political Position and Strategic Problems
of the Republic of South Africa." South Africa International,
6, 4, 179-87.
The Western nations must act more strongly to assist
South Africa against Soviet, Chinese, and black African external attacks and terrorist and communist internal attacks.

1487 HALDERMAN, J. W. "Some Legal Aspects of Sanctions in the
Rhodesian Case." International Comparative Law Quarterly,
17, 3, July 1968, 672-705.
Do the Rhodesian sanctions aid or injure the development of a system to maintain peace and security?

1488 HALL, RICHARD, (ed.). South-West Africa (Namibia): Proposals for Action. London: African Bureau, 1970. 45 p.
Various writers on separate aspects, historical and
current, of the Namibia problem. The mandate period, the
actions of the ICJ, and the UN are among the topics. Only
legal apects are considered.

1489 HALLETTI, ROBIN. "The South African Intervention in Angola,
1975-1976." African Affairs, 77, 308, July 1978, 347-86.
Attempts to piece together the actual course of events
from recently available material, and points out continued
gaps in our knowledge of these events owing to much additional information remaining secret.

1490 HARKAVY, ROBERT E. "The Pariah State Syndrome." Orbis, 21,
3, Fall 1977, 623-49.
Israel, South Africa, Taiwan, and South Korea are
treated as pariah states. The four may come together in
some sort of mutual aid or defense agreement.

1491 HARVEY, CHARLES. "The Effect of External Pressure on British
Firms Investing in South Africa." In: Shaw, T. M. and
Heard, K. A., (eds.). Cooperation and Conflict in Southern
Africa. Washington, D. C., 1977, 394-407, tabl.
"This paper is mainly an analysis of recent efforts in
the UK to get British firms to raise the wages of their
black employees in South Africa."

1492 HERSOV, BASIL E. "Opportunity and Responsibility." South
Africa International, 8, 4, April 1978, 177-96.
Essay examines foreign policies of and public views on
South Africa in US, UK, West Germany, and France. South
Africa's image is deteriorating but "an emerging clarity of
assessment" is also taking place.

1493 HILL, C. R. "Unilateral Declaration of Independence and South African Foreign Policy." Journal of Commonwealth Political Studies, 7, 2, July 1969, 96-103.
UDI has caused no major changes in South Africa's foreign policy, although it has caused South African leaders to clarify the policy.

1494 HIRSCHMANN, DAVID. "Pressures on Apartheid." Foreign Affairs, 52, 1, October 1973, 168-79.
Three pressures are discussed - from Bantustan leadership; from urban blacks; and from external groups, organizations, and businesses.

1495 _____. "Southern Africa: Détente?" Journal of Modern African Studies, 14, 1, March 1976, 107-26.
Detente has led to only a very limited success in improving communication between South Africa and Zambia.

1496 HODDER-WILLIAMS, RICHARD and HUGO, PIERRE. "Verligtheid and Change in South Africa: Stirrings in the Afrikaans Intellectual Community." Round Table, 263, July 1976, 261-70.
Changes in political thought are occurring among Afrikaaners and this may affect domestic and foreign policies of the RSA.

1497 HORWITZ, RALPH. "South Africa: The Background to Sanctions." Political Quarterly, 42, 2, April/June 1971, 163-76.
Using the Parsonian theory of the social system, Horwitz argues that economic sanctions may delay the ending of apartheid.

1498 HOUGH, M. "Selected Excerpts Regarding South Africa's Position on Nuclear Energy." Politikon, 1, 2, December 1974, 67-73.
What possible economic, political, or military advantages would South Africa gain from nuclear energy? Particular emphasis on military uses.

1499 HOWE, R. W. "War in Southern Africa." Foreign Affairs, 48, 1, October 1969, 150-65.
Guerrilla wars in Mozambique, Angola, Rhodesia, South Africa, and Namibia and some of the external connections are discussed.

1500 HÜBNER-DICK, GISELA and SEIDELMANN, REIMUND. "Simulating Economic Sanctions and Incentives: Hypothetical Alternatives of United States Policy on South Africa." Journal of Peace Research, 15, 2, 1978, 153-74.

1501 HULTMAN, TAMI and KRAMER, REED. "United States Investments in Southern Africa." In: El-Khawas, M. A. and Kornegay, F. A., Jr., (eds.). American-Southern African Relations. Bibliographic Essays. Westport, Conn., 1975, 115-37, bibl.
    Assesses recent works on the role of US capital in Southern Africa.

1502 HYAM, RONALD. The Failure of South African Expansion. London: Macmillan. New York: Africana, 1972. 219 p., bibl., ill., index, maps.
    South Africa's recent policies of developing closer relations with neighboring states have antecedents that reach far back into South African history.

1503 HYNNING, CLIFFORD J. "The Future of South West Africa: A Plebiscite?" American Journal of International Law, 65, 4, September 1971, 144-7.

1504 INGHAM, DEREK. "If Rhodesia Crumbles: Towards a Commonwealth Presence?" Round Table, 256, October 1974, 451-5.
    Suggests a Commonwealth role in the peaceful solution of the Rhodesian crisis.

1505 INTERNATIONAL DEFENCE AND AID FUND. The Rhodesia-Zambia Border Closure Jan.-Feb. 1973. London: IDAF, 1973. 26 p.
    The 1973 closure of the border by Rhodesia attracted international attention. The failure of the closure was a major, but perhaps temporary, defeat for the Smith regime.

1506 J. N. "Constantes et Contradictions de l'Afrique du Sud." Revue de Défense Nationale, 26, July 1970, 1190-202.

1507 JACOBS, WALTER D. South Africa Looks Outward. New York: American-African Affairs Association, 1968. 13 p.

1508 JANKE, P. "Southern Africa: End of Empire." Conflict Studies, 52, December 1974, 1-23.
    Instability in most African states tends to draw the attention of their governments away from South Africa. The detente policy of Zambia and South Africa may be successful, therefore.

1509 JEEVES, ALAN H. "The Problem of South Africa." International Journal, 26, 2, Spring 1971, 418-32.
    The South Africa question remains a strong threat to continued Commonwealth existence.

1510 ———. "South Africa and the Politics of Accommodation." International Journal, 30, 3, Summer 1975, 504-17.

1511 JEEVES, A. H. "Südafrika und Seine Nachbarn. Bemühungen der Regierung Vorster um eine Neuausrichtung der Afrika-Politik." Europa-Archiv, 30, 13, July 1975, 429-35.

1512 JOHNS, SHERIDAN W. "Opposition in Southern Africa: Segments, Linkages and Cohesion." In: Shaw T. M. and Heard, K. A., (eds.). Cooperation and Conflict in Southern Africa. Washington, D. C., 1977, 184-227.
   Includes information on Botswana, Lesotho, and Swaziland and their interactions with other states andnationalist movements in Southern Africa.

1513 _____. "Südafrikas Diplomatische Öffnung nach Norden." Europa-Archiv, 26, 22, November 1971, 783-94.
   The significance and relative success/failure of South Africa's attempts to improve relations with the black-ruled states of Africa.

1514 JONES, DAVID. Aid and Development in Southern Africa: British Aid to Botswana, Lesotho, and Swaziland. London: Croom Helm, 1977. 313 p., index, tabls.
   "British aid to these countries in this period emerges as virtuous though sometimes frustratingly neutral. Despite enormous potential for 'leverage' . . . the general effect of British aid was to permit the recipients to do more of what they would have done anyway."

1515 KAPUNGU, LEONARD T. "Economic Sanctions in the Rhodesian Context." In: El-Ayouty, Y. and Brooks, H. C., (eds.). Africa and International Organization. The Hague, 1974, 103-26, annex.
   The legal case for the UN sanctions against Rhodesia and a brief note on the effects of sanctions.

1516 _____. The United Nations and Economic Sanctions against Rhodesia. Lexington, Mass.: Lexington Books, 1973. 155 p., index, map, tabls.
   Research was conducted in Denmark, West Germany, France, Italy, Malawi, the Netherlands, Sweden, Switzerland, Zambia, the UK, and the USA. These states were "chosen for their special relationship in the enforcement of economic sanctions against Rhodesia."

1517 KEMP, GEOFFREY. "South Africa's Defence Programme." Survival, 14, 4, July/August 1972, 158-61.
   "Despite persistent talk of the Soviet naval threat to the Cape Route, South Africa still invests most of her defense effort towards the threat from within Southern Africa."

1518 KITAZAWA, YOKO. From Tokyo to Johannesburg: A Study of Japan's Growing Economic Links with the Republic of South Africa. New York: Interfaith Center for Corporate Responsibility, 1975. 47 p.
Japan's trade with South Africa is significant in the consideration of any economic sanctions against the Republic.

1519 KLARE, MICHAEL and PROKOSCH, ERIK. "Evading the Arms Embargo: How the U. S. Arms South Africa and Rhodesia." In: Western Mass. Association of Concerned African Scholars, (ed.). U. S. Military Involvement in Southern Africa. Boston, 1978, 157-71.
US business continues to trade with Rhodesia and South Africa in spite of the sanctions. The author presents a case study of aircraft sales as an example.

1520 KORNEGAY, FRANCIS A. "The Impact of Post-coup Portuguese Africa on the Southern Africa Sub-system." Current Bibliography on African Affairs, 8, 1, 1975, 43-53.
A review essay.

1521 KUNERT, DIRK. "South Africa in a Hostile World." South Africa International, 9, 1, July 1978, 1-10, 37-43.
The Third World is aligned with the Soviet Union against South Africa. South African raw materials are the Soviet goal and they follow a realistic policy to get those resources, unlike the West and its escapist policy.

1522 KUNIN, V. "South Africa: Imperialism and Neocolonialist Bridgehead." International Affairs (Moscow), 1, January 1971, 40-4.
Economic expansionism of South Africa and its ties to the imperialist powers.

1523 LADOUCEUR, PAUL. "Canadian Humanitarian Aid for Southern Africa." In: Anglin, D.; Shaw, T.; and Widstrand, C., (eds.). Canada, Scandinavia and Southern Africa. Uppsala, 1978, 85-104, tabls.
The author is Chief Planning Officer of the Commonwealth Division of the Canadian International Development Agency (CIDA).

1524 LANDELL-MILLS, P. M. "The 1969 Southern African Customs Union Agreement." Journal of Modern African Studies, 9, 2, August 1971, 263-81, tabls.
In 1969 Botswana, Lesotho, Swaziland, and South Africa signed a new customs union agreement to replace one signed in 1910.

1525 LANDIS, ELIZABETH S. "Namibia: Impending Independence?" In: Carter, G. M. and O'Meara, P., (eds.). Southern Africa in Crisis. Bloomington, 1977, 163-99.
A review of the international aspects of the Namibian situation.

1526 _____. "Namibia: Legal Aspects." In: Stokke, O. and Widstrand, C., (eds.). The UN-OAU Conference on Southern Africa, Vol. II. Uppsala, 1973, 107-15.
Asserts that the basic legal question now confronting the international community is how to implement the June 1971 Opinion of the International Court of Justice.

1527 _____. "South African Policy." Bulletin of the Atomic Scientists, 26, 3, March 1970, p. 47.

1528 LANGDON, STEVEN. "The Canadian Economy and Southern Africa." In: Anglin, D.; Shaw, T.; and Widstrand, C., (eds.). Canada, Scandinavia and Southern Africa. Uppsala, 1978, 15-27, tabls.
Trade connections and investment links between Canada and South Africa cause a strong status quo orientation in Canada's Southern Africa policy.

1529 LAPCHICK, RICHARD E. The Politics of Race and International Sport: The Case of South Africa. Westport, Conn,: Greenwood, 1975. 268 p., bibl., index, tabls.
A major aspect of the international campaign against apartheid has been the attempt to isolate South Africa from the world of sport. The author analyzes this campaign up to 1970 and presents a brief update to 1974.

1530 LAWRIE, G. G. "South Africa's World Position." Journal of Modern African Studies, 2, 1, March 1964, 41-54.
South Africa's foreign policy is basically a status quo policy, but there are aspects of imperialism as well.

1531 LEGASSICK, MARTIN. "The Southern African Bloc: Integration for Defense or Expansion." Africa Today, 15, October/November 1968, 9-12.
South African, Rhodesian, and Portuguese cooperation in Southern Africa.

1532 LEGUM, COLIN. "'National Liberation' in Southern Africa." Problems of Communism, 24, 1, January/February 1975, 1-20.
External aid from Communist states to African liberation movements does not mean that these movements are puppets of the Communist states. Once the territories become independent, the influence bought with foreign aid may be more significant.

1533   LEGUM, C. "The Problems of the Land-locked Countries of Southern Africa in the Confrontation between Independent Africa, South Africa, Rhodesia and Portugal." In: Cervenka, Z., (ed.). Land-locked Countries of Africa. Uppsala, 1973, 165-81.

1534   _____. "Southern Africa: The Politics of Détente." Yearbook of World Affairs, 30, 1976, 14-29.
What are the effects in the politics of the Southern African sub-system of the change of regime in Portugal?

1535   _____. Southern Africa: The Secret Diplomacy of Detente; South Africa at the Crossroads. London: Collings, 1975. 91 p., tabls.
Analysis of the effects of the Portuguese revolution and its consequent changes in colonial rule in Africa on the politics of white-ruled Southern Africa.

1536   _____. Southern Africa: The Year of the Whirlwind. New York: Africana, 1977. 72 p.
A review of domestic and international politics in Southern Africa, 1975-1976. Several relevant documents are reproduced here.

1537   _____. Vorster's Gamble for Africa: How the Search for Peace Failed. London: Rex Collings, 1976. 127 p.
Implications of the Soweto riots and Transkei's independence, the failure of détente, and related documents are the major topics. Most of this is a reprint from Africa Contemporary Record.

1538   _____ and HODGES, TONY. After Angola: The War over Southern Africa. London: Rex Collings. New York: Africana, 1976. 85 p.
Most of this is reprinted from Africa Contemporary Record. Intervention by the major powers in Angola and the implications for Rhodesia are the main subjects.

1539   _____ and LEGUM, MARGARET. South Africa: Crisis for the West. New York: Praeger, 1964. 333 p., fig., index, map.
Most of this volume is devoted to domestic aspects, but see Part 2, "International Power and South Africa" which is an examination of intervention and sanctions.

1540   _____ and _____. "South Africa in the Contemporary World." Issue, 3, 3, Fall 1973, 17-27, tabls.
Looks at South Africa in the African and international context after 25 years of apartheid, and finds that internal contradictions have been sharpened, and external relations have declined. Calls for an "effective black challenge" as the only means of producing meaningful change.

1541 LEIGHTON, NEIL O. "A Perspective on Fundamental Change in Southern Africa: Lusaka - Before and Beyond." Africa Today, 23, 3, July-September 1976, 17-24, tabl.
Kissinger's Lusaka speech, although advocating majority rule in Rhodesia, carefully avoids mentioning internal change in South Africa.

1542 LEISTNER, GERALD M. E. Co-operation for Development in Southern Africa. Pretoria: Africa Institute, 1972. 30 p., bibl., tabls. [Occasional Paper, 33].
"The functioning of South Africa's modern economy promotes and helps to sustain development in neighbouring countries."

1543 _____. "Economic Co-operation in Southern Africa." South Africa International, 6, 1, July 1975, 35-44, bibl.
Economic cooperation with South Africa could provide important benefits for the states of Southern Africa.

1544 _____. "South Africa: Economic Interests in Africa." South African Journal of African Affairs, 2, 1972, 32-46.

1545 _____. South Africa's Development Aid to African States. Pretoria: Africa Institute, 1970. 34 p., bibl.
Purposes and forms of South Africa's aid to her neighbors with brief descriptions of programs in Lesotho, Malawi, Angola, Botswana, Mozambique, Rhodesia, and Swaziland.

1546 "Lesotho: Aid or Alms." Bulletin of the Africa Institute of South Africa, 11, 10, 1973, 374-8.
"A South African assessment of Lesotho's external development assistance, the role of South Africa, and relations between Lesotho and the Republic."

1547 LEYS, ROGER. "Scandinavian Development Assistance to Botswana, Lesotho and Swaziland." In: Anglin, D.; Shaw, T.; and Widstrand, C., (eds.). Canada, Scandinavia and Southern Africa. Uppsala, 1978, 47-64, tabls.
"The main purpose of this paper is to provoke a debate about the strategy and content of Scandinavian aid to the 'rand' states."

1548 LITVAK, LAWRENCE; DeGRASSE, ROBERT; and McTIGUE, KATHLEEN. South Africa: Foreign Investment and Apartheid. Washington, D. C.: Institute for Policy Studies, 1979. 104 p.
Argues that investment in South Africa supports apartheid. A list of US banks and corporations with investments and operations in the Republic is included.

1549 LOCKE, R. and MAZRUI, A. A. "The Chinese Model and the Soviet Model in Eastern and Southern Africa. Struggles for Liberation and Development." Asian Forum, 9, 1, Winter/Spring 1976/1977, 1-15.

1550 MACKLER, IAN. Pattern for Profit in Southern Africa. New York: Atheneum, 1975. 100 p., bibl., index.
  The author contends that the US supported South Africa in its efforts to incorporate South West Africa from 1946 through 1950. A major factor in this US stand was the growing investment of US firms in Southern Africa. The author draws parallels to the present situation in the Republic and its relations with the USA.

1551 MADISON AREA COMMITTEE ON SOUTHERN AFRICA. South Africa and Israel. Madison, Wisc.: The Committee, 1971. 43 p.

1552 MAGNUSSON, AKE. Sverige-Sydafrika: En Studie av en Ekonomisk Relation. Uppsala: Scandinavian Institute of African Studies, 1974.
  Sweden-South Africa: A Study of an Economic Relation.

1553 _____. "Sveriges Förbindelser med Sydafrika." Internasjonal Politikk, 2, April/June 1975, 219-40.
  "Sweden's Relations with South Africa." The author examines economic ties, the wage conditions of Swedish firms operating in South Africa, and actions concerning South Africa that are taking place in Sweden.

1554 _____. Swedish Investments in South Africa. Uppsala: Scandinavian Institute of African Studies, 1974. 57 p., tabls. [Research Report, No. 23].
  This is a condensed version of the author's major study, Svergie-Sydafrika. In addition to Swedish investments, Sweden's foreign policy for South Africa and proposals for its modification are discussed.

1555 MAGUBANE, B. "The South African Problem as a View of Imperialism." Journal of Black Studies, 3, 1, September 1972, 75-94.
  South Africa is part of a global problem of imperial domination.

1556 MAKGETLA, NEVA S. and SEIDMAN, ANN. "U. S. Transnational Corporations' Involvement in South Africa's Military-Industrial Complex." In: Western Mass. Association of Concerned Scholars, (ed.). U. S. Military Involvement in Southern Africa. Boston, 1978, 197-220.
  US MNC involvement in South Africa in recent years has been an important factor in the development of that country's military-industrial complex.

1557 MALINGA, PHINEAS. "Imperialism Struggles to Save Southern Africa." African Communist, 62, 1975, 34-40.

1558 MANNING, C. A. W. "South Africa's Racial Policies. A Threat to Peace?" In: Rhoodie, N., (ed.). South African Dialogue. Philadelphia, 1972, 590-611.
Asks whether these policies are really a threat to peace or whether the critics of South Africa and the enemies of the West are using the alleged threat as a pretext for obtaining anti-South African action at the UN.

1559 MARCO, EDMOND A. "Dialogue in Africa." Optima, 21, 3, September 1971, 106-17, ill., map.
Holds that dialogue between black and white leaders is preferable to the use of force and that peace in Africa can only be achieved and maintained by a deliberate keeping open of doors. But the concept is threatening the existence of the OAU.

1560 MARIÑAS OTERO, L. "La Unión Aduanera y Monetaria de Africa del Sur." Revista de Política Internacional, 143, January/February 1976, 187-96.
Although South Africa is the dominant member of the Customs and Monetary Union of South Africa, the other members (Botswana, Lesotho, and Swaziland) maintain independent foreign policies.

1561 MARQUARD, LEO. A Federation of Southern Africa. London: Oxford University, 1971. 142 p., index, map.
"The purpose of this book is to examine the possibilities of federalism in Southern Africa. It is by no means a blueprint for federation."

1562 _____. Our Foreign Policy. Johannesburg: South African Institute of Race Relations, 1969. 25 p.
The author argues that apartheid must be abandoned if South Africa is to have peace and security with its neighbors.

1563 _____. The Peoples and Policies of South Africa. New York: Oxford University, 1969. 266 p.
Although the largest portion of this volume deals with domestic affairs, there is some consideration of foreign policy, in particular with respect to Botswana, Lesotho, Swaziland, and South West Africa.

1564 MASON, P. "South Africa and the World: Some Maxims and Axioms." Foreign Affairs, 43, 1, October 1964, 150-64.
Apartheid is unacceptable. The US and Great Britain should pressure South Africa to change this policy through peaceful means before violence is the only alternative.

1565 McCRYSTAL, L. P. "Angola and Mozambique: Prospects for Economic Cooperation in Southern Africa." South African Journal of African Affairs, 2, 1972, 85-92.

1566 McHENRY, DEAN. *United States Firms in South Africa: Study Project on External Investment in South Africa and Namibia.* Bloomington, Ind.: African Studies Program, Indiana University, 1975. 40 p., tabls.

1567 McKINNELL, ROBERT. "Sanctions and the Rhodesian Economy." *Journal of Modern African Studies*, 7, 4, December 1969, 599-81.
"This article attempts to analyze the affect of sanctions thus far and, secondly, to examine the long-run growth prospects of an economy which, it is expected, will continue to be subject to sanctions in the future."

1568 MELI, F. "Nationalism and Internationalism in South African Liberation." *African Communist*, 57, 1974, 42-55.

1569 MENAUL, STEWART. "The Security of the Southern Oceans: Southern Africa the Key." *NATO's Fifteen Nations*, 17, 2, April/May 1972, 40-6.
"Southern Africa is the key to the security of NATO's lines of communications." South Africa is the key to maintaining these communications.

1570 MERWE, E. J. VAN DER. "The Customs Union Agreement between Botswana, Lesotho, Swaziland and the Republic of South Africa." *South African Journal of African Affairs*, 2, 1972, 64-75, tabls.
A customs agreement reached in 1910 was replaced in 1970 by a new agreement, herein described, between the Republic of South Africa, Botswana, Lesotho, and Swaziland.

1571 METROWICH, F. R. *Rhodesia: Birth of a Nation.* Pretoria: Africa Institute, 1969. 168 p., ill.
Rhodesia's position in international relations is included in this volume.

1572 MIDDLEMAS, KEITH. *Cabora Bassa: Engineering and Politics in Southern Africa.* London: Weidenfeld and Nicolson, 1975. 367, p., bibl., ill., index.
This volume places the Cabora Bassa Dam project in Mozambique in the wider context of Southern African politics.

1573 MILLER, J. D. B. "South Africa's Departure." *Journal of Commonwealth Political Studies*, 1, 1, November 1961, 56-74.
What are the conclusions about the existence of the Commonwealth that can be drawn from the withdrawal of South Africa from the organization?

1574 MILLER, ROGER L. *The Economic Impact of U. S. Restrictions on Trade with Rhodesia: A Preliminary View.* Coral Cables: Center for Advanced International Studies, University of Miami, 1974. 29 p.

1575 MINGST, KAREN A. "Southern Africa's Commodity Trade with the United States: The Impact of Political Change." Africa Today, 24, 2, April-June 1977, 5-28, fig., tabls.
Development of four scenarios of the future of commodity trade between Southern Africa and the US suggests re-examination of US policy.

1576 MINTER, WILLIAM. "U. S. Policy in Angola and Mozambique." Africa Today, 23, 3, July-September 1976, 55-60.
It is unlikely that the United States will stop trying to neutralize the effect the Angolan and Mozambique struggles are having in Southern Africa.

1577 MINTY, ABDUL S. "Scandinavia, Canada and the Arms Embargo." In: Anglin, D.; Shaw, T.; and Widstrand, C., (eds.). Canada, Scandinavia and Southern Africa. Uppsala, 1978, 37-49.
The author is the Director of the Anti-Apartheid Movement, London.

1578 MLAMBO, ESHMAEL. Rhodesia: The Struggle for a Birthright. London: C. Hurst, 1972. 333 p.
Although the larger part of this volume is concerned with internal events, there are chapters on the international aspects, too. The UN, the OAU, the Commonwealth, and British policy are examined.

1579 MOHAN, JITENDRA. "South Africa and the Suez Crisis." International Journal, 16, 4, Autumn 1961, 327-57.
Reactions in South Africa to the Suez crisis were a reflection of attitudes on apartheid. Views of members of Parliament, the press, and the Jewish community are described.

1580 MOLTENO, ROBERT V. Africa and South Africa: The Implications of South Africa's "Outward-looking" Policy. London: Africa Bureau, 1971. 28 p.
"Is it in the interests . . . of African states to depart from the declarations of the United Nations and the Organization of African Unity, which favour the isolation of South Africa?"

1581 _____. "South Africa's Expansion Policy." Reality, 3, 5, November 1971, 4-6; 3, 6, January 1972, 14-7; 4, 1, March 1972, 16-20.

1582 _____. "South Africa's Forward Policy in Africa." Round Table, July 1971, 329-45.
South African attempts to improve relations with other African states have gone under a variety of names.

1583 MORRISON, RODNEY J. "Apartheid and International Monetary Reform." Review of Politics, 32, 3, July 1970, 338-46.

1584 MTSHALI, B. VULINDLELA. Rhodesia: Background to Conflict.
      New York: Hawthorn Books, 1967. 255 p., bibl., index.
           Although much of this book is devoted to the domestic
      history of Rhodesia, there are chapters on UN action and
      African states' policies on Rhodesia.

1585 MUBAKO, S. V. "The Rhodesian Border Blockade of 1973 and
      the African Liberation Struggle." Journal of Commonwealth
      and Comparative Politics, 12, 3, November 1974, 297-312.
           The increasing conflict between Zambia and Rhodesia and
      the effects of that conflict upon Rhodesia should indicate
      to the settlers the wisdom of a negotiated settlement.

1586 MUDGE, G. A. "Domestic Policies and United Nations Activ-
      ities: The Cases of Rhodesia and the Republic of South
      Africa." International Organization, 21, 1, Winter 1967,
      55-78.
           UN criticism of South Africa and Rhodesia has not had
      the desired effects. Indeed, such criticism has led to the
      confirmation of extremist governments in power.

1587 MUFUKA, KENNETH N. "Reflections on Southern Rhodesia: An
      African Viewpoint." Africa Today, 24, 2, April-June 1977,
      51-5.
           Kissinger's 1976 policy is similar to the 1969 "peaceful
      settlement" position, and is rejected by most Africans.

1588 MUGUMBO, AGRIPPAH T. "The Emergence and Collapse of South
      Africa's Black African Policy." Journal of East African
      Research and Development, 5, 1, 1975, 19-36.

1589 _____. "The Rise and Fall of 'Pax Suid-Afrika': A His-
      torical Analysis of South Africa's Black African Foreign
      Policy." Kenya Historical Review, 3, 1, 1975, 127-45. Also
      in: Ojuka, A. and Ochieng, W., (eds.). Politics and Leader-
      ship in Africa. Kampala, 1975, 272-95.
           "This essay looks into the origins of this outward
      movement, focusing especially on the period from the mid-
      1960s onwards."

1590 _____. "Zimbabwe, Detente and the Strategy of Deceit."
      Africa Today, 25, 2, April-June 1978, 45-55.

1591 MUNGER, EDWIN S. "John Vorster and the United States."
      American Universities Field Staff Reports, Central and South-
      ern Africa, 12, 2, 1968, 1-9, ill.
           Vorster's foreign policy is influenced by his lack of
      knowledge (and travel) of areas outside of South Africa.

1592 _____. Notes on the Formation of South African Foreign
      Policy. Pasadena: Castle, 1965. 102 p.

1593 MUNGER, E. S. "Rhodesia: Republic in Gestation." American Universities Field Staff Reports, Central and Southern Africa Series, 13, 4, 1969, 1-30.
Considers the effects of sanctions on Rhodesia.

1594 _____. "South Africa: Are There Silver Linings?" Foreign Affairs, 47, 2, January 1969, 375-86.
Analysis of South Africa's Africa policy.

1595 NEHWATI, FRANCIS. "Economic Sanctions against Rhodesia." In: Stokke, O. and Widstrand, C., (eds.). The UN-OAU Conference on Southern Africa, Vol. II. Uppsala, 1973, 151-61, tabls.
Assesses the effectiveness of the sanctions, and concludes they have not been as successful as was expected.

1596 NKOSI, Z. "South African Imperialism." African Communist, 30, 1967, 25-39.
Examination of South Africa's relations with other African states.

1597 NOEL, M. "L'Evolution des Relations Economiques entre la France et l'Afrique du Sud." Revue Française d'Études Politiques Africaines, 14, February 1972, 43-57, tabl.

1598 NOLUTSHUNGU, SAM C. South Africa in Africa: A Study in Ideology and Foreign Policy. New York: Africana, 1975. 329 p., bibl., index, maps, tabls.
"A thorough examination of South Africa's policies towards the rest of the continent in the post-war period." Author is a black South African political scientist.

1599 NOONE, BREAM. Australian Economic Ties with South Africa. North Fitzroy, Victoria: International Development Action, 1973. 47 p., bibl., fig., ill., maps, tabls.

1600 NUSCHELER, F. "Der Britisch-Rhodesische Verfassungskonflikt". Verfassung und Recht in Übersee, 6, 4, 1973, 407-27.
International sanctions are a failure in the Rhodesian case and there is no alternative for black nationalists but to defeat by violent means the colonialists.

1601 NYATHI, V. M. "South African Imperialism in Southern Africa". African Review, 5, 4, 1975, 451-71.
Dynamics, mechanisms, and effects of South African imperialism.

1602 NYERERE, J. K. "Rhodesia in the Context of Southern Africa." Foreign Affairs, 44, 3, April 1966, 373-86.
Britain's relations with black Africa depend upon her resolution of the Rhodesian rebellion.

1603 OBOZUWA, A. UKIOMOGBE. The Namibian Question: Legal and Political Aspects. Benin City: Ethiope Publishing, 1973. 256 p., bibl., index.
A legal analysis of the role of the ICJ.

1604 OLIVIER, GERITT C. "Die Publiek, die Parlement, die Burokrasie en Buitelandse Beleid met Speciale Verwysing na Suid-Afrika." Politikon, 3, 1, June 1976, 2-27.
"Public, Parliament, Bureaucracy and Foreign Policy with Special Reference to South Africa." This case study indicates that the peculiar nature of the foreign policy process means that normal democratic processes and procedures cannot be followed.

1605 _____. "South African Foreign Policy." In: Worrall, D., (ed.). South Africa: Government and Politics. Pretoria, 1971, 285-331.

1606 OSTROWSKY, JÜRGEN. "Militärische Kooperation Bundesrepublik-Sudafrika. Zu einem Memorandum des Auswärtigem Amtes." Blätter für Deutsche und Internationale Politik, 22, 5, May 1977, 574-92.
"Military Cooperation between the Federal Republic and South Africa: On a Memorandum of the Ministry of Foreign Affairs." In spite of UN resolutions to the contrary, Germany continues to sell arms to South Africa.

1607 PACHAI, BRIDGAL. The International Aspects of the South African Indian Question, 1860-1971. Cape Town: C. Struik, 1971. 318 p., bibl., index.
The Indian population of the Republic of South Africa has been a factor in South Africa's international relations. Several chapters deal with the UN role.

1608 PARK, STEPHEN. Business as Usual: Transactions Violating Rhodesian Sanctions. Washington: Carnegie Endowment for International Peace, 1973. 54 p.

1609 PENN, J. "South Africa: Background and Strategic Value." South Africa International, 7, 1, July 1976, 41-56.
South Africa, like Israel, is a bastion of the free world in a highly strategic location, but South Africa is poorly treated by the West. The USA should take a more positive policy toward South Africa.

1610 PERERA, L. H. HORACE. "Non-governmental Action in Support of the Victims of Colonialism and Apartheid in Southern Africa." In: Stokke, O. and Widstrand, C., (eds.). The UN-OAU Conference on Southern Africa, Vol. II. Uppsala, 1973, 17-24.
Condemns the record of some governments regarding colonialism and apartheid, and calls for non-violent action by non-governmental organizations.

1611 PETERSON, CHARLES W. "The Military Balance in Southern Africa." In: Potholm, C. P. and Dale, R., (eds.). Southern Africa in Perspective. New York, 1972, 298-317.
"This chapter will analyze . . . six topics: (1) terrain and climate considerations; (2) the logical combatants: the white-controlled areas; (3) the logical combatants: the African states; (4) the probable neutrals; (5) combat events, 1961-1969; and (6) the future course of armed struggle."

1612 PHILLIPS, K. "The Prospects for Guerrilla Warfare in South Africa." International Relations, 4, 1, May 1972, 108-17.
External intervention is one of three variables that must be examined to determine the liklihood of guerrilla warfare in South Africa.

1613 POSER, G. "Soviet Sea Power, Southern Africa and World Equilibrium." South Africa International, 7, 2, October 1976, 73-80.
The West must take the side of South Africa to prevent Soviet expansion.

1614 POTHOLM, CHRISTIAN P. "The Limits of Systemic Growth: Southern Africa Today." In: Shaw, T. M. and Heard, K. A., (eds.). Cooperation and Conflict in Southern Africa. Washington, D. C., 1977, 118-43.
The author examines various subsystem models in an attempt "to indicate the kind of system which seems to be evolving in the area."

1615 _____. "Southern Africa's Scenarios Revisited: A Framework for Speculation." Plural Societies, Autumn 1974, 63-9.
The author presents his views on future developments in Southern Africa.

1616 _____ and DALE, RICHARD, (eds.). Southern Africa in Perspective; Essays in Regional Politics. New York: Free Press, 1972. 418 p., bibl., index, tabls.
Contains 23 essays on individual countries and relationships between them.

1617 QUESTER, GEORGE H. "Paris, Pretoria, Peking . . . Proliferation?" Bulletin of the Atomic Scientists, October 1970, 12-6.
"South Africa . . . is not likely to defy the world's consensus on proliferation. . . . Pretoria would have nothing to gain by making nuclear weapons of its own."

1618 RAMAMURTHI, T. G. "Southern Africa and the Indian Ocean." India Quarterly, 28, 4, October-December 1972, 341-6.
Decolonization has stopped in Southern Africa because the West does not want to endanger its strategic interests in South Africa.

1619 REED, DOUGLAS. The Battle for Rhodesia. New York: Devin-Adair, 1967. 150 p.
A plea for international support for the government of Ian Smith in an effort to prevent World War III.

1620 REITSMA, HENDRIK J. A. "South Africa and the Red Dragon: A Study in Perception." Africa Today, 23, 1, January-March 1976, 47-67.
South Africa perceives international communism as an enemy and has become increasingly worried about China's efforts to establish itself as an important factor in African affairs.

1621 RHOODIE, ESCHEL M. "Southern Africa: Towards a New Commonwealth?" In: Potholm, C. P. and Dale, R., (eds.). Southern Africa in Perspective. New York, 1972, 276-97, tabls.
The author speaks very favorably of the possibilities of economic union in Southern Africa, based on past experiences of cooperation between black and white in the region.

1622 RHOODIE, NIC, (ed.). South African Dialogue: Contrasts in South African Thinking on Basic Race Issues. Philadelphia: Westminister. Johannesburg: McGraw-Hill, 1972. 620 p., index.
The major part of this book is concerned with South Africa's domestic problems, but three essays on the international ramifications of these problems are included. See references to essays by J. Barratt, C. Manning, and D. Worral.

1623 RIPPON, GEOFFREY. "The Importance of South Africa." Survival, 12, 9, September 1970, 292-7.
"Should there ever be another major conventional war . . . then the sea route around South Africa will be of vital importance."

1624 ROGERS, BARBARA. "Namibia: Economic and Other Aspects." In: Stokke, O. and Widstrand, C., (eds.). The UN-OAU Conference on Southern Africa, Vol. II. Uppsala, 1973, 117-36.
Proposes economic measures to be taken by the UN and others for ending the South African occupation of Namibia. Includes an appendix listing multinational companies operating in Namibia which have subsidiaries in Africa.

1625 ———. "South Africa's Fifth Column in the United States." Africa Report, 22, 6, November/December 1977, 14-7, ill.
South Africa wages a multifaceted campaign to maintain support for its policies among US citizens.

1626 ———. South Africa's Stake in Britain. London: Africa Bureau, 1971. 36 p., bibl.
South African and British trade and commerce with each other.

1627 ROGERS, B. "'Sunny South Africa': A Worldwide Propaganda Machine." Africa Report, 22, 5, September/October 1977, 2-8.

1628 ROPP, KLAUS VON DER. "Freiden oder Krieg im Südlichen Afrika." Aussenpolitik, 28, 4, 1977, 437-54.
"Peace or War in South Africa?" US, Soviet, and EEC policies in Southern Africa will prevent a peaceful solution to the problems that exist there.

1629 _____. "Southern Africa after Portugal's Withdrawal." Aussenpolitik, 27, 1, 1976, 84-102. In German edition: 27, 1, 1976, 80-97.
The independence of Angola and Mozambique has caused changes in South African, OAU, UN, and major power policies for Southern Africa.

1630 _____. "Das Veränderte Kräftspiel im Süden Afrikas." Aussenpolitik, 26, 1, 1975, 56-72.
"The Changed Balance of Powers in Southern Africa." Analysis of the effects of the independence of Angola and Mozambique on the international relations of Southern Africa.

1631 "Round Table: The Dilemma of Foreign Investment in South Africa." American Journal of International Law, 65, 4, September 1971, 293-322.

1632 ROUSSEAU, P. E. "Détente in Southern Africa." South Africa International, 6, 3, January 1976, 117-23.
The countries in Southern Africa are learning that economic cooperation is better than political confrontation.

1633 ROUX, DOMINIQUE DE. "Le Mozambique et la Bataille de l'Afrique Australe." Défense Nationale, 29, October 1973, 115-22.

1634 SALAH-BEY, A. "L'Afrique et l'O. I. T. - Evaluation Politique." Revue Algérienne des Sciences Juridiques, Politiques et Economiques, 1, January 1964, 100-27.
Analysis of the problem of South Africa's membership in the ILO.

1635 SCHEIBER, MICHAEL T. "Apartheid under Pressure: South Africa's Military Strength in a Changing Political Context." Africa Today, 23, 1, January-March 1976, 27-45, tabls.
Throughout the evolution of a new system in South Africa, the military will be the key force in holding the country together. South Africa is well equipped to meet any military threats from neighboring African states, and attemps to force the South African government to change its policies through embargo or collective international efforts have been relatively unsuccessful.

1636 SCHRIRE, R. "South Africa and the World: Which Way Now?" International Affairs Bulletin, 2, 1, 1978, 31-40.

1637 SCHROEDER, B. "Les Rélations entre l'Afrique du Sud et la Rhodésie." Revue Française d'Études Politiques Africaines, 10, 116, August 1975, 36-55.
South Africa has successfully moved Rhodesia from a position of British domination to one of South African domination.

1638 SCOTT, MICHAEL. "Report on a Study Project on External Investment in South Africa and Namibia." Objective: Justice, 7, 4, October-December 1975, 31-3.

1639 SEGAL, RONALD, (ed.). Sanctions against South Africa. London: Penguin, 1964. 272 p.
A collection of papers and commentaries from the International Conference on Economic Sanctions against South Africa.

1640 _____ and FIRST, RUTH. South-West Africa - Travesty of Trust. London: Andre Deutsch, 1967. 352 p., index, tabls.
This volume contains the papers presented at a conference at Oxford in 1966. Topics include the international status of South West Africa, South African rule in the territory, the decision of the ICJ, and proposition for international pressure to cause change.

1641 SEIDMAN, ANN and SEIDMAN, NEVA. South Africa and U. S. Multinational Corporations. Westport, Conn.: Lawrence Hill, 1978. 251 p., maps, tabls.
This study of US economic interests in South Africa includes chapters on minerals, oil companies, manufacturing, labor unions, and financiers. South Africa's foreign policy and relations with neighboring states are examined. Chapters are devoted to Namibia, Mozambique and Angola, Swaziland-Lesotho-Botswana, Zimbabwe, and Zambia.

1642 _____ and _____. "Southern African Contradictions: Part I, The Role of U. S. Based Multinational Corporations (MNCs)." Contemporary Crises, 1, 3, July 1977, 261-87.
South Africa as a regional imperialist sub-center and the role of US capital and MNCs in South Africa.

1643 _____ and _____. "Southern African Contradictions: Part II, South Africa's Outward Reach." Contemporary Crises, 1, 4, October 1977, 371-401.
The neocolonial control of Southern Africa by South Africa is described. Namibia, Botswana, Lesotho, Swaziland, Mozambique, Angola, and Zambia are examined in this context.

1644 SEILER, JOHN. "South African Perspectives and Responses to External Pressure." Journal of Modern African Studies, 13, 3, September 1975, 447-68.

Explores the pattern of South African responses to external pressures in terms of the basic motivation and dynamics of the country's decision-making process. Rejects the use of the "laager model" as misleading, and focuses instead on the world perspectives of the decision-makers, as evidenced by their "definition of the situation."

1645 _____, (ed.). Southern Africa Since the Portuguese Coup. Boulder, Col.: Westview, 1980. 272 p., index, maps.

Essays examine Cuban and Soviet roles; the regional policies of South Africa, Mozambique, and Zaire; the role of black African states in the liberation struggle; and white South African reactions to recent events.

1646 SETAI, BETHUEL. "Prospects for a Southern African Common Market." Journal of African Studies, 1, 3, Fall 1974, 310-34, tabls.

The prospects for a common market are not good until South Africa solves its internal problems.

1647 SHAMUYARIRA, N. M. "A Revolutionary Situation in Southern Africa." African Review, 4, 2, 1974, 159-79.

South Africa's strategic importance influences US and NATO attitudes on the liberation movements. The OAU and some other superpower must support the liberationists in case of Western European or US opposition.

1648 SHAW, TIMOTHY M. "International Organizations and the Politics of Southern Africa: Towards Regional Integration or Liberation?" Journal of Southern African Studies, 3, 1, October 1976, 1-19.

A wide-ranging discussion of the international relations of Southern Africa. Author concludes that violence is almost inevitable.

1649 _____. "The International Sub-system of Southern Africa: Introduction." In: Cervenka, Z., (ed.). Land-locked Countries of Africa. Uppsala, 1973, 161-4.

Author recommends the adoption of an "issue area framework" for sub-system analysis.

1650 _____. "Introduction to Southern Africa as a Regional Subsystem." In: Shaw, T. M. and Heard, K. A., (eds.). Cooperation and Conflict in Southern Africa. Washington, D. C., 1977, 1-15.

The introduction to a collection of essays presented at a conference at Dalhousie University in 1973.

1651 SHAW, T. M. "Oil, Israel and the OAU: An Introduction to the Political Economy of Energy in Southern Africa." Africa Today, 23, 1, January-March 1976, 15-26.
   Afro-Arab tensions will increase the probability of Israel and South Africa being able to exploit African oil.

1652 _____. "The Political Economy of Technology in Southern Africa." In: Shaw, T. M. and Heard, K. A., (eds.). Cooperation and Conflict in Southern Africa. Washington, D. C., 1977, 365-79.
   "This essay suggests that the hegemony of South Africa in Southern Africa is a function of its dependence on the international capitalist economy . . . . However, the perpetuation of its regional dominance is related not only to the indulgence of the great powers . . . its regional power is also dependent on its economic growth and world demand for its resources."

1653 _____. "South Africa's Military Capability and the Future of Race Relations." In: Mazrui, A. A. and Patel, H. H., (eds.). Africa in World Affairs: The Next Thirty Years. New York, 1973, 37-61, tabls.
   The author analyzes the developing diplomatic, economic, and military conflict between white South Africa and black Africa. He suggests that the conflict may be seen as rich vs. poor as well as black vs. white.

1654 _____. "Southern Africa: Cooperation and Conflict in an International Sub-System." Journal of Modern African Studies, 12, 4, 1974, 633-55.
   A review article with suggestions for future research.

1655 _____. "Southern Africa: Dependence, Interdependence and Independence in a Regional Subsystem." In: Shaw, T. M. and Heard, K. A., (eds.). Cooperation and Conflict in Southern Africa. Washington, D. C., 1977, 81-97, tabls.
   "Patterns of interaction in the region differ between issue areas in terms of both their mode and actors."

1656 _____ and HEARD, KENNETH A., (eds.). Cooperation and Conflict in Southern Africa: Papers on a Regional Subsystem. Washington: University Press of America, 1977. 279 p., bibl., figs., map, tabls.
   Contains eighteen essays organized into four sections. See references under Bowman, Barratt, Shaw, Boyd, Potholm, Keppel-Jones, Schlemmer, Johns, Anglin, Thahane, Rothchild/ Curry, Grundy, Konczacki, Harvey, Spence, Hill, and Cervenka.

1657 SHEPHERD, G. W., JR. Anti-Apartheid: Transnational Conflict and Western Policy in the Liberation of South Africa. Westport, Conn.: Greenwood, 1977. 246 p., indices.
   Private international organizations against apartheid.

1658 SIERRA, SERGIO. "Racism and Fascism in the Southern Atlantic." African Communist, 74, 1978, 100-9.
"A member of the Central Committee of the Communist Party of Uruguay discusses the growing links between the Vorster regime and the fascist dictatorships in Latin America.

1659 SIMON, BERNARD. "1976: South Africa's Troubled Year." Rivista di Studi Politici Internazionali, 44, 1, January-March 1977, 43-58.
South Africa's strategic position has been weakened by events in Angola and Mozambique. Soviet influence may next be felt in SWA.

1660 SITHOLE, MASIPULA. "Rhodesia: An Assessment of the Viability of the Anglo-American Proposals." World Affairs, 141, 1, Summer 1978, 71-81.

1661 SJOLLEMA, BALDWIN. "The World Council of Churches: Policies and Programmes in Support of the Liberation Struggle in Southern Africa." In: Stokke, O. and Widstrand, C., (eds.). The UN-OAU Conference on Southern Africa, Vol. II. Uppsala, 1973, 25-32.
Summarizes the position of the WCC in regard to the race question in general and Southern Africa in particular. Includes general and specific information.

1662 SKJELSBAEK, K.; SEMB, H.; and STRØMME, S. A. "Det Okonomiske Samkvem Mellom Sor-Afrika og Norge." Internasjonal Politikk, 1, January-March 1977, 77-103.
"Economic Relations between South Africa and Norway." A detailed survey of economic relations between the two states.

1663 SMIT, P. "South Africa and the Indian Ocean: The South African Point of View." In: Cottrell, A. J. and Burrell, R. M., (eds.). The Indian Ocean: Its Political, Economic, and Military Importance. New York, 1972, 269-92, map, tabls.
"An analysis of the economic significance of the Indian Ocean for South Africa as well as of the military and political consequences that might arise for South Africa from the increasing Soviet penetration of the Indian Ocean and, from that Ocean, of East and Southern Africa." Also see the essay by L. W. Bowman in this volume.

1664 SMOCK, D. R. "The Forgotten Rhodesians." Foreign Affairs, 47, 3, April 1969, 532-44.
The interests and policies of Zambia, South Africa, the Rhodesian minority government, and Britain are described, but the author stresses that the interests of the majority of Rhodesia's citizens are neglected by all of these parties.

1665 SOUBEYROL, JACQUES. "L'Action Internationale contre l'Apartheid." Revue Générale de Droit International Public, 69, 2, April-June 1965, 326-69.
Political and legal actions against apartheid are assessed; these actions have had little effect upon apartheid.

1666 _____. "L'Afrique Australe en Accusation." Annee Africaine 1972, 1972, 228-80.
A review of events and policies in Southern African international relations.

1667 _____. "Afrique Noire et Afrique Australe." Annee Africaine 1970, 1970, 101-42.
The UN campaign against Portugal, South Africa, and Rhodesia has had only limited effects.

1668 _____. "La Coercition Internationale contre les Etats d'Afrique Australe." Annee Africaine 1974, 1974, 206-47.
A review of events and policies in Southern African international relatins.

1669 _____. "Les Sanctions Internationales contre les Territoires Gouvernés par les Minorités Blanches en Afrique Australe." Annee Africaine 1973, 1973, 137-41.

1670 _____. "Société Internationale et Afrique Australe en 1975." Annee Africaine 1975, 1975, 136-86.
A general review of great power and international organization activities aimed at changing the apartheid system.

1671 "South Africa." Armed Forces Journal International, June 1973, 21-35, ill., map, tabls.
Series of short articles concerning the policy of apartheid, South Africa's strategic military position, and her specific armaments (army, navy, and air force).

1672 "South African Aid to Africa." Bulletin of the Africa Institute of South Africa, 11, 4, 1973, 127-38.
Description of the African economic aid programs of the Republic of South Africa.

1673 "The South African Gold Agreement: Its Effect upon the Control of Liquidity in the International Monetary System." New York University Journal of International Law and Politics, 3, 2, Winter 1970, 358-75.

1674 SOUTH AFRICAN INSTITUTE OF RACE RELATIONS. South Africa in Africa: An Evaluation of Detente. Johannesburg: The Institute, 1976.

1675 "Southern Africa: Problems and U. S. Alternatives: A Guide to Discussion, Study and Resources." Intercom, 70, September 1970, 2-70.
This special issue is designed as a teaching aid for secondary school use in the USA.

1676 SPENCE, JOHN E. "Détente in Southern Africa: An Interim Judgement." International Affairs, 53, 1, January 1977, 1-16.
Rhodesia, South Africa, and the Transkei in international relations.

1677 _____. "Nuclear Weapons and South Africa: The Incentives and Constraints on Policy." In: Shaw, T. M. and Heard, K. A., (eds.). Cooperation and Conflict in Southern Africa. Washington, D. C., 1977, 408-28.
Although South Africa's nuclear capabilities and policies have not been considered of importance, a number of reasons now exist that make analysis of this subject of interest.

1678 _____. The Political and Military Framework. London: Africa Publications Trust, 1975. 114 p.
South African foreign relations with emphasis on military and economic relations. The defense of the Indian Ocean is also considered.

1679 _____. Republic under Pressure: A Study of South African Foreign Policy. London: Oxford University, 1965. 132 p. tabls.
"A remarkable feature of post-war international relations has been the steady decline in stature of South Africa." This study includes chapters on South Africa in Africa, South Africa in international organizations, and linkages between domestic and external politics.

1680 _____. "South Africa and the Defense of the West." Survival, 13, 3, March 1971, 78-84. Also see: Round Table, 241, January 1971, 15-24.
"The security element in this emotive debate has more often been the subject of assertion rather than argument and it has not usually been given the analysis it deserves. The article . . . analyzes the defence aspects."

1681 _____. "South African Foreign Policy: Changing Perspectives." World Today, 34, 11, November 1978, 417-25.

1682 _____. South African Foreign Policy in Today's World. Johannesburg: South African Institute of International Affairs, 1975. 16 p.

1683 SPENCE, J. E. "South African Foreign Policy: The 'Outward Movement'." In: Potholm, C. P. and Dale, R., (eds.). Southern Africa in Perspective. New York, 1972, 46-58.
Description and analysis of South Africa's attempts after the mid-1960s to build better relations with African states, especially in the southern part of the continent.

1684 _____. "Southern Africa's Uncertain Future: Adjusting to a New Balance of Power." Round Table, 258, April 1975, 159-65.
Changes must occur in South Africa's foreign policy because the regional balance of power has been upset by the revolution in Portugal.

1685 _____. "The Strategic Significance of South Africa." Yearbook of World Affairs, 27, 1973, 134-52.
South Africa's control of the South Atlantic and the Indian Ocean make it of strategic importance to the USA and France, especially with the growth since 1970 of the Soviet presence in the Indian Ocean.

1686 _____. The Strategic Significance of Southern Africa. London: Royal United Service Institution, 1970. 48 p.
A review of British-South African relations.

1687 _____. "Tradition and Change in South African Foreign Policy." Journal of Commonwealth Political Studies, 1, 2, May 1962, 136-52.
After demarcating the foreign policy objectives of the Republic, the author examines the effects of the Nationalist Party take-over in 1948 and its apartheid policy on the attainment of those objectives.

1688 SPRACK, JOHN. Rhodesia: South Africa's Sixth Province: An Analysis of the Links between South Africa and Rhodesia. London: International Defence and Aid Fund, 1974. 87 p., bibl., maps.

1689 SPRING, MARTIN C. Confrontation: The Approaching Crisis between the United States and South Africa. Johannesburg: Valiant, 1977. 181 p.
A firm statement of South African national interests and how these clash with US interests.

1690 STAUB, H. O. "Südafrika: Problem und Herausforderung." Europäische Rundschau, 5, 2, 1977, 33-42.
"South Africa: Problem and Challenge." International economic relations, boycotts, and apartheid.

1691 STEINHART, EDWARD I. "Shylock and Prospero: Anti-Semitism, Zionism and South African Ideology." Ufahamu, 4, 3, Winter 1974, 35-56.

1692   STEPHENSON, G. V. "The Impact of International Economic Sanctions on the Internal Viability of Rhodesia." Geographical Review, 65, 3, July 1975, 377-89, maps, tabls.
An analysis of the differential impact of sanctions on the populations of Rhodesia.

1693   STEVENS, RICHARD P. "South Africa and Independent Black Africa." Africa Today, 17, 3, May/June 1970, 25-32.
South Africa's dialogue policy and visit of members of Malawi's government to the Republic.

1694   STEWARD, ALEXANDER. The World, the West and Pretoria. New York: David McKay, 1977. 308 p., index.
This mixture of South African history and the Republic's place in international relations is basically a defense of South African policies. US and British policy, the UN, and the Angolan crisis are considered.

1695   STOCKHOLM INTERNATIONAL PEACE RESEARCH INSTITUTE. Southern Africa: The Escalation of a Conflict. New York: Praeger, 1976. 235 p., bibl.

1696   STOKKE, OLAV. "Sør-Afrikas Utadrettede Strategi." Internasjonal Politikk, 4, October-December 1972, 619-57.
"South Africa's Forward Strategy." A description of South Africa's Africa policies and the possibilities of change in the domestic and international policies of that state.

1697   _____. " Utviklingsbistand og Frigøring Sbevegelser." Internasjonal Politikk, 4, 1970, 398-422.
"Development Assistance and Liberation Movements." An examination of moral and legal factors involved in the apparent conflict between the desire to aid liberation movements and the principle of non-interference in the domestic affairs of a sovereign state.

1698   _____ and WIDSTRAND, CARL, (eds.). The UN-OAU Conference on Southern Africa, Oslo, 9-14 April 1973. Uppsala: Scandinavian Institute of African Studies, 1973. Vol. I, 275 p. Vol. II, 346 p.
Government, international organization, and liberation movement representations and individual experts took part in this conference. Volume I reports recommendations and summaries of committee meetings. Volume II contains several papers presented at the conference.

1699   STRACK, HARRY R. Sanctions: The Case of Rhodesia. Syracuse, N. Y.: Syracuse University, 1978. 296 p., bibl., index, map, tabls.
Analysis of Rhodesia's efforts to evade sanctions and consideration of interdependency and international organization.

1700   STULTZ, NEWELL M. "The Politics of Security: South Africa under Verwoerd, 1961-1966." Journal of Modern African Studies, 7, 1, April 1969, 3-20.
        South African foreign policy becomes "qualitatively different" after the establishment in 1961 of the Republic.

1701   SUCKLING, J. "Foreign Investment and Domestic Savings in the Republic of South Africa, 1957-1972." South African Journal of Economics, 43, 3, September 1975, 315-21, figs., tabls.

1702   SUTCLIFFE, R. B. "The Political Economy of Rhodesian Sanctions." Journal of Commonwealth Political Studies, 7, 2, July 1969, 113-25.
        Sanctions have failed because the burden of their effect has fallen upon politically weak groups in Rhodesian society - white farmers and rural Africans.

1703   SYLVESTER, ROBERT. "Observation of the U. S. Arms Embargo." In: Western Mass. Association of Concerned African Scholars, (ed.). U. S. Military Involvement in Southern Africa. Boston, 1978, 221-43.
        "Government rhetoric seems designed to conceal a complex, sometimes contradictory set of procedures which have not stopped the continued flow of arms to South Africa."

1704   TEPLINSKIJ, L. B. "Izrail i JUAR: Soucastniki Zagovora Protiv Prav Narodov." Sovetskoe Gosudarstvo i Pravo, 7, 1971, 55-62.
        "Israel and the Republic of South Africa: Accomplices in the Plot against the Peoples' Rights." Collusion between Israel and South Africa is a factor in the oppression of all of the peoples of Africa.

1705   THAHANE, TIMOTHY T. "Financial Communications and Labour Transactions in Southern Africa." In: Shaw, T. M. and Heard, K. A., (eds.). Cooperation and Conflict in Southern Africa. Washington, D. C., 1977, 290-311, tabls.
        "Southern Africa should offer the best prospects for intra-regional co-operation in Africa. However, this is not the case . . . . Why is this so?"

1706   TIEWUL, S. A. "Apartheid: Steps toward International Control." Zambia Law Journal, 6, 1974, 101-27, tabls.

1707   TU'MAH, JŪRJ. "Janūb Ifrīqiyā wa Isrā'īl." Shu'ūn Filastīn-īyah, 28, December 1973, 85-97.
        Relations between South Africa and Israel.

1708 TURNER, BIFF. "A Fresh Start for the Southern African Customs Union." African Affairs, 70, July 1971, 269-76.

Botswana, Lesotho, and Swaziland have been in a customs union with South Africa since 1910. The agreement was renewed in 1969.

1709 UNITED STATES. HOUSE COMMITTEE ON FOREIGN AFFAIRS. SUBCOMMITTEE ON AFRICA. Critical Developments in Namibia. Washington: Government Printing Office, 1974. 350 p., maps.

In addition to testimony presented in front of the US House Sub-Committee on African Affairs, this document contains numerous reprints of articles, documents, and other items on the external and internal factors involved in the Namibia situation.

1710 UTETE, C. M. BOTISO. "Détente in Southern Africa." Black World, 24, 7, May 1975, 30-6, map.

"Rhodesia could not have survived the combination of African guerrilla pressure and UN-imposed economic sanctions had not South Africa come to the aid of the government.

1711 VANDENBOSCH, AMRY. South Africa and the World: The Foreign Policy of Apartheid. Lexington: University of Kentucky, 1970. 303 p., index.

Earlier periods of South African history are examined in the opening chapters, but more than half of the volume is devoted to analysis of South Africa's foreign relations since 1948.

1712 VÄYRYNEN, RAIMO. "South Africa: A Coming Nuclear-weapon Power?" Instant Research on Peace and Violence, 7, 1, 1977, 34-47.

German, US, and French companies and governments have been actively transferring various types of nuclear technology to S. Africa. South Africa's motives are examined and the transfers are analyzed.

1713 VENTER, DENIS, (ed.). International Relations in Southern Africa. Johannesburg: South African Institute of International Affairs, 1974. 44 p.

1714 VILLIERS, B. DE. "African Détente and Development Aid." Internationales Afrika Forum, 11, 11/12, November/December 1975, 601-4.

South African foreign policy and its aid to African states.

1715 _____. "Southern Africa and the Future." South Africa International, 5, 4, April 1975, 189-97.

Recent UN actions and the independence of Mozambique cause the South African government to wish to reach settlement with black Africa.

1716 WALL, PATRICK, (ed.). The Southern Oceans and the Security of the Free World. London: Stacey International, 1977. 226 p., index, maps.
Various essays, mainly by military officers, analyze Soviet policy and threats to security in the Indian Ocean-Southern Atlantic. South Africa's role and position receives special attention.

1717 WEBB, COLIN DE B. "The Foreign Policy of the Union of South Africa." In: Black, J. E. and Thompson, K. W., (eds.). Foreign Policies in a World of Change. New York, 1963, 425-50, bibl, map.
'The Commonwealth, the UN, Africa, communism, and economic relations are the major topics.

1718 WEISBORD, ROBERT G. "The Dilemma of South African Jewry." Journal of Modern African Stuides, 5, 2, September 1967, 233-41.
The South African Jewish community is in a difficult position as a white group in a racist society.

1719 "Who is Arming the Republic of South Africa and Why?" International Afffairs, 1, January 1969, 82-7.

1720 WILLIAMS, MICHAEL and PARSONAGE, MICHAEL. "Britain and Rhodesia: The Economic Background to Sanctions." World Today, 29, 9, September 1973, 379-88, tabls.

1721 WILMER, S. "Ten Years of Vain Negotiations." Instant Research on Peace and Violence, 2, 1972, 67-72. Also in: East African Journal, 9, 2, February 1972, 26-9.
Ten years of negotiation have provided no benefits for the black population of Rhodesia.

1722 WINDRICH, ELAINE. "Rhodesia: The Road from Luanda to Geneva." World Today, 33, 3, March 1977, 101-11.
US policy for Rhodesia is considered.

1723 WINSTON, JOHN. "Réflexions sur l'Intervention Sud-Africaine en Angola." Revue Française d'Etudes Politiques Africaines, 142, October 1977, 60-9.
South African intervention (and its failure) in Angola has caused severe losses for its foreign policy. The intervention was undertaken in support of US interests.

1724 WOLDRING, K. "The Prospect of Federalisation in Southern Africa." African Review, 3, 3, 1973, 453-78.
True federation with all races having access to the central government may be a solution to South Africa's problem.

1725 WOLDRING, K. "South Africa's Africa Policy Reconsidered." African Review, 5, 1, 1975, 77-93.
   South Africa's Africa policy is doomed to failure because its major purpose is the defense of the status quo in South Africa.

1726 WOODS, DONALD. "South Africa's Face to the World." Foreign Affairs, 56, 3, April 1978, 521-8.
   South Africa wages a huge propaganda campaign to maintain the support of the peoples and governments of the industrial democracies. Without that support, a black majority government would soon replace the Nationalist Party government. Author is a former South African journalist.

1727 WORRAL, DENIS. "South Africa's Reaction to External Criticism." In: Rhoodie, N., (ed.). South African Dialogue. Philadelphia, 1972, 562-89.
   Examines the influence of outside criticism on internal attitudes (official and unofficial) and polices, and asserts that the thrust of change in South Africa has been in the opposite direction from that demanded by the country's critics.

1728 ZACKLIN, RALPH. "Challenge of Rhodesia. Toward an International Policy." International Conciliation, 575, November 1969, 5-72.
   Sanctions, their effects and their enforcement.

1729 _____. The United Nations and Rhodesia: A Study in International Law. New York: Praeger, 1974. 202 p., index.
   A study of the theory and reality of international sanctions. The author concludes that sanctions are a necessary procedure if a just solution to the Rhodesian problem is to be found.

1730 ZARTMAN, I. WILLIAM. "The African States as a Source of Change." In: Bissell, R. E. and Crocker, C. A., (eds.). South Africa into the 1980s. Boulder, Col., 1979, 107-32.

The USSR, the PRC, the UK, and France: Relations with Africa

1731 ABIR, MORDECHAI. "The Contentious Horn of Africa." Conflict Studies, 24, June 1972, 5-19.
Analysis of the interests of Russia, China, Israel, and Egypt in the Eritrean conflict.

1732 _____. "Red Sea Politics." In: Conflicts in Africa. London: International Institute for Strategic Studies, 1972. 25-41, map.
The views of the USA, USSR, and PRC are considered.

1733 ADANALIAN, ALICE A. "The Horn of Africa." World Affairs, 131, 1, April-June 1968, 38-42.

1734 ADIE, IAN. "China's Return to Africa." South Africa International, 3, 4, April 1973, 191-212.
China's activities in Africa are a part of its competition with the USA and the USSR. Economic relations with African states and aid to Southern African liberation movements are examined.

1735 ADIE, W. A. C. "Africa on China's Stage." Politikon, 2, 1, June 1975, 62-73.
An analysis of China's policy for the Third World, in which Africa plays a role of special importance.

1736 _____. "China and Africa Today." Race, 5, 4, April 1964, 3-25.

1737 _____. "China's Year in Africa, 1974." Bulletin of the Africa Institute of South Africa, 13, 3, 1975, 95-101.

1738 _____. "Chinese Policy towards Africa." In: Hamrell, Sven and Widstrand, C. G., (eds.). The Soviet Bloc, China and Africa. Uppsala, 1967, 43-63.
China's views of Africa, Asia, and Latin America are similar to the early view Stalin held of China.

1739 AHMAD, S. S. "Nigeria-China Relations. An Approach to Positive Neutrality." Pakistan Horizon, 26, 1, 1973, 48-54.
    The opening of relations between Nigeria and the PRC in 1971 indicates that Nigeria has a truly independent policy.

1740 AJIBOLA, W. A. "The British Parliament and Foreign Policy Making: A Case Study of Britain's Policy towards the Nigerian Civil War." Nigerian Journal of Economic and Social Studies, 16, 1, March 1974, 129-47, tabls. Also in: "British Parliament and the Nigerian Crisis: Some Background Notes." Quarterly Journal of Administration, 8, 3, April 1974, 305-23.
    In-depth interviews with 20 M. P.s, interviews with all relevant ministers, questionaires from 120 M. P.s, and content analysis of relevant debates were used in this study. Author concludes that debates had almost no effect on government's policy.

1741 _____. "The British Pressure Groups and the Nigerian Civil War: Some Background Notes." Quarterly Journal of Administration, 12, 1, October 1977, 67-84.
    A continuation of the material presented in the author's 1974 article in this journal.

1742 ALBRIGHT, DAVID E. "Soviet Policy." Problems of Communism, 27, 1, January/February 1978, 20-39, ill.
    "Priorities, general approach, objectives, and operating style that have shaped Soviet policy in the recent period."

1743 ALEXANDRE, PIERRE. "Francophonie: The French and Africa." Journal of Contemporary History, 4, 1, January 1969, 117-26.

1744 ALLEN, PHILIP M. "Francophonie Considered." Africa Report, 13, 6, June 1968, 6-11. See also: "The Mystique of Francophonie." In: Tandon, Y., (ed.). Readings in African International Relations, Vol. I. Nairobi, 1972, 278-89.
    Argues that it is the mystique of the French language and culture which forms the essential basis of francophonie, and suggests that such a basis is rather superficial and perhaps even transient.

1745 AMMI-OZ, MOSHE. "Les Impératifs de la Politique Militaire Française en Afrique Noire à l'Epoque de la Décolonisation." Revue Française d'Etudes Politiques Africaines, 134, February 1977, 65-89.
    The experiences of World War II have shown the importance of Africa to French security, but it was also necessary to recognize African nationalism and pressure for decolonization. Military assistance, defense agreements, and other arrangements have been concluded to reconcile these needs.

1746 AMOA, G.-K. "Relations between Africa and Europe in Historical Perspective." University of Ghana Law Journal, 13, 1/2, 1976, 7-58.

1747 "Angola after Independence: The Struggle for Supremacy." Conflict Studies, 64, November 1975, 1-15.
A major cause of the Angolan civil war is the Soviet drive for strategic influence in Africa.

1748 ANSPRENGER, FRANZ. "Die Bedeutung von de Gaulles Politik für das Überleben Balkanisierter Staaten des Ehemals Französischen Afrikas." Politische Vierteljahresschriff, 2, 1970, 473-515.
French policy toward the francophone states must be considered a success. Discusses military aid, economic aid, private investment, and cultural aid of France in Africa.

1749 _____ and CZEMPIEL, ERNEST-OTTO. Südafrika in der Politik Grossbritanniens und der USA. Kaiser, Mainz: Matthias-Grünewald, 1977. 164 p., bibl.
Summaries in French and English are included in this foreign policy study.

1750 ANTI-TAYLOR, WILLIAM. "China through African Eyes." Race, 5, 4, April 1964, 48-51.

1751 APALIN, G. "Peking and the 'Third World'." International Affairs (Moscow), December 1972, 28-34.
A Soviet analysis of the Third World policy of the PRC, viewed principally as a policy to isolate the USSR.

1752 ATTWOOD, WILLIAM M. The Reds and the Blacks. New York: Harper and Row, 1967. 341 p., index, map.
A former US Ambassador to Kenya and Guinea describes what he saw of "Soviet and Chinese efforts to penetrate and subvert Africa."

1753 AUSTIN, DENNIS. Britain and South Africa. London: Oxford University, 1966. 191 p., bibl., maps, tabls.
"The primary aim of this study . . . has been to try and 'measure the extent of British interests in South Africa, and the degree to which they are likely to influence United Kingdom policy towards the Republic'." Topics include the UN, the OAU, Rhodesia, and the High Commission territories.

1754 _____. "Britain and Southern/Central Africa." In: University of Southampton. The Indian Ocean in International Politics. Southampton, 1973, 147-52.
An assessment of British attitudes and interests.

1755 BAHR, E. "Die Leitbilder der Europäischen Politik in Afrika." Aussenpolitik, 11, 3, March 1960, 173-81.
"Main Types of European Policy in Africa." Each colonial power has a different Africa policy.

1756 BAILEY, MARTIN. "Tanzania and China." African Affairs, 74, 294, January 1975, 39-50, tabls.
The evolution of Sino-Tanzanian relations. Major factors are Chinese aid and the view of China as a model for development. Chinese aid has increased the independence of Tanzania's foreign policy.

1757 BARBER, JAMES. "The Impact of the Rhodesian Crisis on the Commonwealth." Journal of Commonwealth Political Studies, 7, 2, July 1969, 83-95.
Three issues are considered: Britain's central role in the Commonwealth, race relations, and splits with the Commonwealth.

1758 BARRATT, JOHN. The Angolan Conflict: Internal and International Aspects. Braamfontein: South African Institute of International Affairs, 1976. 21 p., map.

1759 BELL, J. BOWYER. "Endemic Insurgency and International Order: The Eritrean Experience." Orbis, 18, 2, Summer 1974, 427-50.
International politics may play a role in insurgency, but there must be a domestic basis for disorder.

1760 _____. The Horn of Africa: Strategic Magnet in the Seventies. New York: Crane, Russak, 1973. 52 p., bibl., map.
An "analysis of the present and potential strategic importance of the Horn." Internal trends, regional stress, and major power interests are considered.

1761 _____. "Strategic Implications of the Soviet Presence in Somalia." Orbis, 19, 2, Summer 1975, 402-11.
Similar to the author's 1973 book.

1762 BENNETT, VALERIE PLAVE. "Soviet Bloc-Ghanaian Relations Since the Downfall of Nkrumah." In: Weinstein, Warren, (ed.). Chinese and Soviet Aid to Africa. New York, 1975, 120-41, tabls.
An examination of the effects of a regime change in one member of a two-party relationship upon that relationship and the reactions of the Soviets to the disappearance of radical leaders in Africa.

1763   BERG, E. J. "The Economic Basis of Political Choice in
        French West Africa." American Political Science Review, 54,
        2, June 1960, 391-405.
             A classic analysis of the economic ties between France
        and the French colonies.

1764   BERNER, W. "Peking und der Kongolesische Partisanenkrieg."
        Ost Probleme, 18, 20, 1966, 610-20.
             Analysis of rebel movements in Zaire and the role of
        PRC aid to such groups.

1765   BISSELL, RICHARD E. "Southern Africa: Testing Détente."
        In: Kirk, G. and Wessell, N., (eds.). The Soviet Threat.
        New York, 1978, 88-98.
             "The passage of three years of prolonged confrontation
        between the Soviet Union and the United States in southern
        Africa . . . is casting doubt on the wisdom of linking the
        future of détente to political and military developments in
        distant regions such as southern Africa."

1766   BOURGI, A. "Aspects Actuels de la Coopération Franco-
        Africaine." Annuaire du Tiers Monde, 1, 1975, 188-204.
             The French are forced to rethink their Africa policy
        and to renegotiate their agreements with African states
        because of decisions made by Madagascar and Mauritania.

1767   BRAUN, D. "The Indian Ocean in Afro-Asian Perspective."
        World Today, 28, 6, June 1972, 249-56.
             US Indian Ocean policy must consider the desires of the
        littoral states.

1768   BRETTON, HENRY. Patron-Client Relations: Middle Africa and
        the Powers. New York: General Learning, 1971. 24 p., bibl.
             "In all political, economic, and military respects, the
        region remains consigned to passivity, if not impotence."

1769   BROMHARD, DAVID. "Conflicts in Trade Unions." In: Wein-
        stein, Warren, (ed.). Chinese and Soviet Aid to Africa.
        New York, 1975, 283-7.
             The author, an employee of the AFL-CIO, writes on Soviet
        and Chinese efforts to infiltrate African labor movements.

1770   BRZEZINSKI, ZBIGNIEW K., (ed.). Africa and the Communist
        World. Stanford, Cal.: Stanford University for the Hoover
        Institution, 1963. 272 p., index, map, tabls.
             There are six essays on China, the Soviet Union, Eastern
        Europe, and Yugoslavia.

1771   _____. "Conclusion: The African Challenge." In:
        Brzezinski, Z., (ed.). Africa and the Communist World.
        California, 1963, 204-29.

1772 BURRACK, D. "Der Kreml Revidient seine Afrika-Politik: Die Haltung der UdSSR zur Militärregimen in Schwarzafrika." Ost-Europa, 22,3, March 1972, 191-8.
"The Attitude of the USSR towards Military Regimes in Black Africa." The Soviets have learned to coexist with states with which there is no ideological agreement.

1773 _____. "Moskau und Peking im Wettlauf um Afrika." Aussenpolitik, 23, 5, May 1972, 279-85.
"Moscow and Peking in Competition for Africa." Economic aid and support for liberation movements are the main elements in Sino-Soviet conflict in Africa.

1774 _____. "Peking's Einfluss in Schwarzafrika." Aussenpolitik, 23, 1, January 1972, 21-9.
"Peking's Influence in Black Africa." Largely a consideration of Sino-Soviet competition.

1775 BUTLITSKY, A. "Defeat of Imperialism in Angola." Social Sciences, 8, 2, 1977, 94-101.

1776 CAMARA, SYLVAIN S. Le Conflit Franco-Guinéen 1958-1971. Paris: Fondation Nationale des Sciences Politiques, 1975. 338 p., bibl.

1777 _____. La Guinée sans la France. Paris: Fondation Nationale des Sciences Politiques, 1976. 299 p., bibl., tabl.

1778 _____. "Les Origines du Conflit Franco-Guinéen." Revue Française d'Etudes Politiques Africaines, 114, June 1975, 31-47.
The conflict between Guinea and France was not the result of the personalities of De Gaulle and Touré; the conflict was caused as a matter of the policies of Touré and De Gaulle. The conflict served broader aims of both men and their governments.

1779 CAMPBELL, JOHN C. "Soviet Policy in Africa and the Middle East." Current History, 73, 430, October 1977, 100-4.

1780 CARHART, THADDEUS. "China's African Policies." Stanford Journal of International Studies, 10, Spring 1975, 238-53.

1781 CARTIGNY, S. "Bibliographie sur la Francophone." Etudes Internationales, 5, 2, June 1974, 399-427.
Contained in a special issue on international cooperation between French-speaking states.

1782 CASTAGNO, A. A. "The Horn of Africa and the Competition for Power." In: Cottrell, A. J. and Burrell, R. M., (eds.). The Indian Ocean. New York, 1972, 155-80.
Brief review of competition for control of the area.

1783   CATTELL, DAVID T. "The Soviet Union Seeks a Policy for Afro-Asia." In: London, K., (ed.). New Nations in a Divided World. New York, 1963, 163-79.
        Soviet policy for Africa lacks coherence and planning, much like that of the USA. Soviet chances for hegemony in Africa are not very strong.

1784   CERVENKA, ZDENEK. The Nigerian War, 1967-1970. Frankfurt am Main: Bernard and Graefe, 1971. 459 p., bibl., index.
        This history of the war contains chapters on the role of international observers, press coverage of the war, and "The Great Powers and the Nigerian War." There is an extensive bibliography.

1785   CHANG, TUNG TSAI. "Chinese Communist Diplomatic Setbacks in Africa." Issues and Studies, 2, 1, October 1965, 11-8.
        A brief analysis in a journal published in Taiwan.

1786   CHELLI, MILENA. "L'Evolution Historique de l'Organisation de Recherches sur l'Afrique en URSS." Revue d'Etudes Comparatives Est-Ouest, 8, 1, March 1977, 165-80.
        A brief review of Soviet research facilities and directions of research.

1787   CHRISTIE, MICHAEL J. The Simonstown Agreement: Britain's Defence and the Sale of Arms to South Africa. London: Africa Bureau, 1970. 23 p.
        British-South African relations, political and military.

1788   CIGNOUX, C. J. "El Communismo y Africa." Politika Internacional, 50/51, July-October 1960, 79-96.
        Description of attempts of the communist states to gain influence in Africa.

1789   CLARK, C. "Soviet and Afro-Asian Voting in the United Nations General Assembly, 1946-1965." Australian Outlook, 24, 3, December 1970, 296-308.
        An analysis of the Afro-Asian bloc in the UN and the success of the Soviet Union in gaining the support of this bloc.

1790   CLEMENS, W. C., JR. "Soviet Policy in the Third World in the 1970s: Five Alternative Futures." Orbis, 13, 2, Summer 1969, 476-501.

1791   COHN, HELEN DESFOSSES. "Soviet-American Relations and the African Arena." Survey, 19, 1, Winter 1973, 147-64.
        A comparison of US and Soviet policy. Author sees a large degree of parallelism.

1792 COHN, H. D. Soviet Policy toward Black Africa: The Focus on National Integration. New York: Praeger, 1972. 316 p.
"This study analyzes Soviet theories regarding the nation-building process in black Africa." Some discussion of the effects of such theories on Soviet policy decisions.

1793 _____. "The Study of Communist Powers in Africa: The Problem of Perspective." Studies in Comparative Communism, 6, 3, Autumn 1973, 301-9.
A review of B. Larkin's 1971 book on China and Africa.

1794 "The Cold War in the Tropics: Communist Pressure on the African Mind." Round Table, 201, December 1960, 15-21.
Soviets are organizing front organizations in Africa and they are willing to work with neutralist and nationalist movements.

1795 CONFINO, MICHAEL and SHAMIR, SHIMON, (eds.). The USSR and the Middle East. Jerusalem: Israel Universities. London: John Wiley, 1973. 441 p.
Although no essays are directly concerned with Soviet policy in sub-Saharan Africa, several essays do contain information on the Sudan and the Persian Gulf that are of interest.

1796 COOLEY, JOHN K. East Wind over Africa: Red China's African Offensive. New York: Walker, 1965. 246 p., bibl., index, tabls.
The PRC has decided to make Africa a major arena of its international relations. This volume describes China's objectives, tactics, successes, and failures.

1797 CORBETT, EDWARD M. The French Presence in Black Africa. Washington: Black Orpheus, 1972. 209 p., bibl., index, tabls.
Sociocultural, political, economic, aid, and military factors are analyzed in this study of France's relations with Africa.

1798 CORNEVIN, R. "La France et l'Afrique Noire." Etudes Internationales, 1, 4, December 1970, 88-101.
In spite of relations with South Africa and Portugal, France has been successful in developing relations with black African states.

1799 COTTRELL, ALVIN J. and BURRELL, R. M., (eds.). The Indian Ocean: Its Political, Economic, and Military Importance. New York: Praeger, 1972. 370 p., index, maps, tabls.
A collection of papers presented at a 1971 conference in Washington. See references under Bowman, Castagno, Smit, and Wolde-Mariam.

1800 CROCKER, CHESTER A. "France's Changing Military Interests." Africa Report, 13, 6, June 1966, 16-41. See also: "Evolution of France's Military Role in Africa." In: Tandon, Y., (ed.). Readings in African International Relations, Vol. I. Nairobi, 1972, 300-20.
 Does France have a post-colonial military role in Africa? Is it a legitimate role?

1801 CRONJE, SUZANNE. The World and Nigeria: The Diplomatic History of the Biafran War. London: Sidgwick and Jackson, 1972. 409 p., ill., index, maps.
 The author was deeply involved as a supporter of the Biafran effort. Here, she describes British, French, US, Soviet, and African involvement - with emphasis on Britain's role.

1802 CROZIER, BRIAN. The Soviet Presence in Somalia. London: Institute for the Study of Conflict, 1975. 20 p., bibl., map. [Conflict Studies, 54].
 The emphasis is on Soviet relations with Somalia.

1803 DALLIN, ALEXANDER. "The Soviet Union: Political Activity." In: Brzezinski, Z., (ed.). Africa and the Communist World. Stanford, Cal., 1963, 7-48.
 Ideological perspectives, strategy, techniques, and prospects are considered. Also see the companion essay by Erlich and Sonne.

1804 DAVIDSON, BASIL. "L'Afrique Recolonisée?" Temps Modernes, 27, 297, April 1971, 1801-24.
 Africa's relations with Europe and the USA are discussed.

1805 DAY, JOHN. "A Failure of Foreign Policy: The Case of Rhodesia." In: Leifer, Michael, (ed.). Constraints and Adjustments in British Foreign Policy. London, 1972, 150-71.
 Author examines "constraints" on British policy and concludes that these are not sufficient to explain Britain's failure.

1806 DE LA TORRE, S. "Aspectos de la Politica China en las Luchas de Liberación de Africa." Revista de Política Internacional, 149, February 1977, 147-73.
 Chinese policy for Africa is considered to be a factor promoting world revolution.

1807 DECRAENE, PHILIPPE. "De Gaulle's Africa Policy." Africa Quarterly, 10, 3, October-December 1970, 204-16.
 Analysis of the factors underlying De Gaulle's Africa policies.

1808 DECRAENE, P. "Le Voyage de M. Pompidou en Afrique Noire: l'Ouverture devrait de Sommais prendre le Pas sur la Continuité." Revue Française d'Etudes Politiques Africaines, 63, March 1971, 84-99.

1809 DELANCEY, MARK W. "The Africa Policies of the United States, the Soviet Union and the People's Republic of China." Africa (Rome), 33, 1, March 1978, 97-116.
A bibliographic essay.

1810 DELAVIGNETTE, R. "L'Evolution de la Politique des Puissances Coloniales après 1945." Revue d'Economique Sociale, September 1960, 11-22.
A description of the international influences leading toward decolonization and a forecasting of the types of relationships that may exist between ex-colonizer/ex-colonized state after independence. Concentrates on France and its colonies.

1811 DELMAS, CLAUDE. "L'Afrique dans les Enjeux de la Politique Mondiale." Revue Générale, 12, December 1974, 69-89.
The place of Africa in communist global strategy.

1812 _____. "De Belgrade à Addis-Abéba." Revue Générale, 4, April 1978, 71-87.
A general comment on détente and human rights with some analysis of Soviet policy in Africa.

1813 _____. "Les Enjeux des Conflits dans l'Afrique Divisée." Revue Générale, 8/9, August/September 1978, 67-82.
Africa has become a theater of the on-going cold war-détente era. Moscow sees détente as a time to make advances in Africa. African conflicts thus are cold war conflicts.

1814 DESFOSSES, HELEN. "Naval Strategy and Aid Policy: A Study of Soviet-Somali Relations." In: Weinstein, Warren, (ed.). Chinese and Soviet Aid to Africa. New York, 1975, 183-201, tabls.
Examination of Soviet policy "reveals a much less dynamic and successful policy" in the Indian Ocean. Written prior to the 1977/78 changes in alliance patterns in the Horn of Africa.

1815 DESHPANDE, G. P. and SHINDE, B. E. "China and the Liberation Movements in Africa." IDSA Journal, 9, 4, April-June 1977, 421-6.
China supports liberation movements for ideological reasons derived from the Stalinist perspective on world affairs.

1816 DESMARESCAUX, JOSEPH. "La Visite du Président de la République en Afrique Noire." Revue de Défense Nationale, 27, April 1971, 659-62.

1817 DESSENS, P. "Le Litige du Sahara Occidental." Maghreb-Machrek, 71, 1976, 26-46.
French, German, and American interests have played and continue to play roles in Western Saharan affairs.

1818 DIRNECKER, B. "The Patrice Lumumba Friendship University in Moscow." Modern World, 2, 1963/1964, 128-43.
Organization of the University and its role in Soviet Third World strategy.

1819 DORESSE, JEAN. "Les Nouveaux Accords Culturels Franco-Ethiopiens." Revue Juridique et Politique, 23, 3, July-September 1969, 383-402.

1820 DOVE-EDWIN, GEORGE. "Développons Nos Relations Bilatérales." Nigeria Demain, 51, March 1978, 2-3.
The author is Nigeria's ambassador to France.

1821 DU BOIS, VICTOR D. "To Die in Burundi: Part II, Foreign Reactions." American Universities Field Staff Reports, Central and Southern Africa, 16, 4, 1972, 1-12, ill.
"Perhaps the most ominous result, however, was the relative passivity of the outside world in the face of that tragic knowledge [of genocide in Burundi]."

1822 _____. "Former French Black Africa and France." American Universities Field Staff Reports, West Africa Series, 16, 2, May 1975, 1-11; 16, 3, June 1975, 1-12, ill., tabls.
Part I is "The Continuing Ties" and Part II is "Toward Disengagement."

1823 DUBOUAYS, JEAN-MARIE. "L'Union Soviétique et la Corne de l'Afrique." Défense Nationale, May 1978, 43-50.
The Soviet-Cuban intervention in Ethiopia is part of a grand plan to gain control over both ends of the Red Sea.

1824 DUNCAN, W. RAYMOND, (ed.). Soviet Policy in Developing Countries. Waltham, Mass.: Ginn-Blaisdell, 1970. 350 p.
Of the thirteen essays, two are directly relevant to Africa. Robert Legvold discusses Soviet views of Africa and W. Scott Thompson examines Soviet-Ghanaian relations.

1825 EBINGER, CHARLES K. "External Intervention in Internal War: The Politics of the Angolan Civil War." Orbis, 20, 3, Fall 1976, 669-700.

1826 EISEMANN, PIERRE M. "Rhodésie, une Libération Nationale auto-déterminée?" Annuaire du Tiers Monde, 2, 1976, 333-41.
Exterior pressures are important determinants of the behavior of the nationalists and of the Smith regime.

1827 ELAIGWU, J. ISAWA. "The Nigerian Civil War and the Angolan Civil War. Linkages between Domestic Tensions and International Alignments." Journal of Asian and African Studies, 12, 1-4, January-October 1977, 215-35.
   Great power involvement in domestic conflicts has become a major factor increasing the intensity of such conflicts.

1828 ELIAS, T. O. "The Commonwealth in Africa." Modern Law Review, 31, May 1968, 284-304.
   A Nigerian author here considers the influence of the British model on African domestic politics and the effects of continuing Commonwealth membership.

1829 EL-KHAWAS, MOHAMED A. "Africa, China and the United Nations." African Review, 2, 2, 1972, 277-87.
   Analysis of UN votes shows that African states did not work together as a bloc on the China admission issue.

1830 _____. "China's Changing Policies in Africa." Issue, 3, 1, Spring 1973, 24-8.
   A historical survey of China's policies.

1831 _____. "The Development of China's Foreign Policy toward Africa, 1955-1972." Current Bibliography on African Affairs, 6, Spring 1973, 129-39, tabl.
   Motivations behind and impact of China's policies.

1832 FARER, TOM J. "Ethiopia: Soviet Strategy and Western Fears." Africa Report, 23, 6, November/December 1978, 4-9.

1833 _____. War Clouds on the Horn of Africa: A Crisis for Detente. New York: Carnegie Endowment for International Peace, 1976. 157 p., bibl., maps.
   The author includes discussion of the domestic and local international conflicts, but the emphasis is on Soviet policy and potential conflict with Western policy.

1834 FEDORENKO, NIKOLAI. "The Soviet Union and African Countries." Annals of the American Academy of Political and Social Science, 354, July 1964, 1-8.
   A statement on Soviet African policy.

1835 FERREIRA REIS, A. C. "Africa: Complemento da Europa." Revista Brasileira de Política Internacional, 6, 21, March 1963, 70-87.
   Africa's position as an inferior partner in the Eurafrique concept.

1836 FESSARD DE FOUCAULT, B. "Indépendances Nationales et Coopération Franco-Africaine." Etudes, April 1973, 535-59.
   The history of relations between France and the independent states of Africa.

1837 FLORY, MAURICE. "Aspects Culturels de la Coopération Francophone." Etudes Internationales, 5, 2, June 1974, 244-51.
In a special issue, "La Coopération Internationale entre Pays Francophones."

1838 FRANCESCHINI, PAUL-JEAN. "La Politica Africana della Francia." Affari Esteri, 35, July 1977, 434-47.
"France's African Policy." The Eurafrique concept and France's relations with Africa.

1839 FREEBERNE, M. "Racial Issues and the Sino-Soviet Dispute." Asian Survey, 5, 8, August 1965, 408-16.
China is using race issues in its conflicts with the Soviet Union.

1840 FUKUDA, HARUKO. Britain in Europe: Impact on the Third World. London: MacMillan, 1973. 194 p., index, tabls.
Britain has joined the EEC and this will entail "a radical reform of Britain's relations with the Third World, with Africa, and with the other members of the Commonwealth.

1841 FYODOROV, V. "USSR-Angola: Friendship and Solidarity." International Affairs (USSR), 12, December 1976, 75-8.

1842 GANDOLFI, A. "Les Accords de Coopération en Matière de Politique Etrangère entre la France et les Nouveaux Etats Africains et Malgache." Revue Juridique et Politique d'Outre-Mer, 17, 2, April-June 1963, 202-19.
The initial period of relations between France and her ex-African colonies was modeled after the French-Tunisian agreements.

1843 GENTILI, ANNA MARIA. "L'Africa e la Politica delle Grandi Potenze." Mulino, 252, July/August 1977, 588-605.
"Africa and the Policies of the Big Powers." The USA, the ex-colonial powers, the USSR and the PRC are considered.

1844 GINSBURGS, GEORGE. "The Soviet View of Chinese Influence in Africa and Latin America." In: Rubinstein, Alvin Z., (ed.). Soviet and Chinese Influence in the Third World. New York, 1975, 197-220.
The Soviet view of Chinese aims is that China wishes to discredit the Soviet Union, other socialist states, the USA, and the states of Western Europe. The author describes the various tactics that the Soviets believe the PRC uses to accomplish these ends.

1845 GLAGOW, RAINER. "Das Rote Mer: Eine Neue Konfliktregion? Politik und Sicherheit im Afro-Arabischen Grenzbereich." Orient, 18, 2, June 1977, 16-50; 18, 3, September 1977, 25-68.
"The Red Sea: A New Region of Conflict? Politics and Security in the Afro-Arab Border Area."

1846 GOOD, KENNETH. Western Domination in Africa. Syracuse, N. Y.: Program of Eastern African Studies, Syracuse University, 1972. 71 p.
Western domination in Ethiopia, Zaire, Kenya, Gabon and the roles of Portugal and South Africa are analyzed.

1847 GREIG, IAN. The Communist Challenge to Africa: An Analysis of Contemporary Soviet, Chinese and Cuban Policies. London: Foreign Affairs, 1977. 306 p., bibl., maps.

1848 GREÑO VELASCO, JOSE ENRIQUE. "Estrategia y Política en el Atlántico Sur." Revista de Política Internacional, 148, November/December 1976, 19-43.
Soviet gains in the South Atlantic have caused Brazil and Argentina to alter their strategic policies. A NATO for this area is a possibility, but the US position is not clear.

1849 GRIFFITH, WILLIAM E. "Soviet Policy in Africa and Latin America: The Cuban Connection." In: Griffith, W. E., (ed.). The Soviet Empire: Expansion and Détente. Lexington, Mass., 1976, 337-44.
A brief description of events.

1850 GROMYKO, ANATOLY. "The October Revolution and Africa's Destiny." International Affairs, 9, September 1977, 95-103.

1851 GUILLEREZ, B. "Djibouti: Centre Névralgique de la Corne de l'Afrique." Défense Nationale, February 1976, 79-87.
The strategic importance of Djibouti, and the roles of Somalia, Ethiopia, China, the USA, the USSR, and France.

1852 GUITON, R. J. "Die Neuesten Bemühungen Pekings um Afrika." Europa-Archiv, 19, 6, March 1964, 193-202.
"Peking's Recent Efforts in Africa." Although the Chinese have many purposes, the major factor in their African activities seems to be the competition with the Soviets.

1853 GUPTA, ANIRUDHA. "Ugandan Asians, Britain, India and the Commonwealth." African Affairs, 73, July 1974, 312-24.
The Asians who remained in Uganda after independence, most of whom were expelled by the Amin government, have been a problem for Commonwealth relations.

1854 GUPTA, VIJAY. "The Ethiopia-Somalia Conflict and the Role of the External Powers." Foreign Affairs Reports, 27, 3, March 1978, 39-57.
A chronological review of events.

1855 HAHN, WALTER F. and COTTRELL, ALVIN J. Soviet Shadow over Africa. Miami: Center for Advanced International Studies, University of Miami, 1976. 105 p.

1856 HALL, RICHARD and PEYMAN, HUGH. The Great Uhuru Railway: China's Showpiece in Africa. London: Gollancz, 1976. 208 p., ill.
This study concentrates more on the international relations aspects of the railway than technical matters or the story of its construction.

1857 HAMRELL, S. and WIDSTRAND, C. G., (eds.). The Soviet Bloc, China and Africa. Uppsala: Scandinavian Institute of African Affairs, 1967. 173 p.
Contains seven essays (plus discussion) presented in 1963 at the Scandinavian Institute. See references by Legum, Morison, Adie, Ansprenger, Müller, and Lowenthal.

1858 HATCH, J. "South African Crisis in the Commonwealth." Journal of International Affairs, 15, 1, January 1961, 68-76.
The effects of South Africa's expulsion from the Commonwealth.

1859 HEINZLMEIR, HELMUT. "Der Konflikt am Horn von Afrika." Aus Politik und Zeitgeschichte, 16, April 1978, 3-16.
The Soviet Union is using the long-standing conflicts in the Horn as a means of increasing its influence in the region.

1860 HENRIKSON, THOMAS H. "Angola and Mozambique: Intervention and Revolution." Current History, 71, 421, November 1976, 153-7.

1861 _____. [Book Review]. Journal of Modern African Studies, 14, 2, June 1976, 350-3.
Review of China's African Revolution by Alan Hutchison, Chinese and Soviet Aid to Africa edited by Warren Weinstein, and Soviet and Chinese Influence in the Third World edited by Alvin Z. Rubinstein.

1862 HEVI, E. J. An African Student in China. London: Pall Mall. New York: Praeger, 1963. 220 p.
An African student's reactions to his life in China. "It is an argument: my argument against communism in Africa."

1863 HODGES, TONY. "The Struggle for Angola: How the World Powers Entered a War in Africa." Round Table, 262, April 1976, 173-84.
The origins and development of US, Soviet, and South African involvement in the Angolan Civil War.

1864 HOLMES, JOHN. "The Impact on the Commonwealth of the Emergence of Africa." International Organization, 16, 2, Spring 1962, 291-302. Also in: Padelford, N. J. and Emerson, R., (eds.). Africa and World Order. New York, 1963, 23-35. Also in: Tandon, Y., (ed.). Readings in African International Relations, Vol I. Nairobi, 1972, 258-68.

1865 HUGHES, ANTHONY J. "News Analysis: Britain's Africa Policy." Africa Report, 19, 2, March/April 1974, 2-5.

1866 HULL, RICHARD W. "China in Africa." Issue, 3, Fall 1972, 49-50.

1867 HURBON, LAËNNEC. "Dialogue entre l'Afrique et l'Europe: De la Reconnaissance des Différences comme Chemin de la Solidarité." Presénce Africaine, 82, 1972, 94-110.
Report on a conference held in Brazzaville in February 1972 to discuss various aspects of relations between Europe and Africa.

1868 HUTCHISON, ALAN. "China and Africa." Afrika Spectrum, 10, 1, 1975, 5-12.

1869 _____. "China in Africa: A Record of Pragmatism and Conservatism." Round Table, 259, July 1975, 263-71.
China's Africa policy is highly successful and she is clearly seen as a leader of the Third World.

1870 _____. China's African Revolution. London: Hutchinson, 1975. 313 p., ill., index, maps.
China's goal in Africa is not power or the export of revolution; it is friendship.

1871 IDENBURG, P. J. "Sovjet-Ruslands Belangstelling voor Afrika." International Spectator, 14, 1, January 1960, 3-18.
"Soviet Russia's Interest in Africa." Soviet views of African events are becoming more realistic and accurate.

1872 INGRAM, DEREK and WALKER, ANDREW. "Commonwealth Conference 1977: Racial Conflict, Economic Challenges and the Problem of Uganda." Round Table, 267, July 1977, 215-28.
African problems dominated the 1977 Commonwealth Conference.

1873 ISAACS, H. R. "Color in World Affairs." Foreign Affairs, 47, 2, January 1969, 235-50.
The author analyzes the importance of race in US, Soviet, and Chinese foreign policy thinking and behavior.

1874 ISMAEL, TAREQ Y. "The People's Republic of China and Africa." Journal of Modern African Studies, 9, 4, December 1971, 507-29.
China's relations with Africa begin in 1955, but since 1959 the Sino-Soviet competition has been the main feature of China's Africa policy.

1875 JACOBS, WALTER D. Report from Southeast Africa: The Indian Ocean Cockpit; Communist Activity in the Indian Ocean and Penetration of the African Continent. New York: American-African Affairs Association, 1971.

1876 JAMES, MARTIN and VANNEMAN, PETER. "Soviet Intervention in the Horn of Africa: Intensions and Implications." Policy Review, 5, Summer 1978, 15-36.

1877 JENSEN, PETER F. Soviet Research on Africa: With Special Reference to International Relations. Uppsala: Scandinavian Institute of African Studies, 1973. 68 p., bibl. [Research Report, No. 19].
   The author includes analysis and extensive discussion of the views of Soviet writers. Three major topics are examined: relations among African states, relations between Africa and imperialist states, and relations between African and the socialist states. Authors analyzed are Gromyko, Fokeyev, Etinger, and Kollontay.

1878 _____. "Sovjetisk Afrikaforskning, Med Saerlight Henblik pa Internationale Relationer." Økonomi og Politik, 49, 1, 1975, 73-86.
   "Soviet Research on Africa, Particularly in International Relations." Soviet study of African international relations is divided into three groupings - interafrican relations, Africa's relations with imperialist states, and Africa's relations with socialist states.

1879 JESMAN, CZESLAW. "The Roots of Chinese Policy in Africa." Race, 5, 4, April 1964, 26-34.
   A review of T. Filesi, Le Relazione della Cina con l'Africa nel Medio-Evo.

1880 JOHNS, S. "The Comintern, South Africa and the Black Diaspora." Review of Politics, 37, 2, April 1975, 200-34.

1881 JUKES, GEOFFREY. The Indian Ocean in Soviet Naval Policy. London: International Institute for Strategic Studies, 1972. 30 p. [Adelphi Paper, 87].
   Fear of Soviet expansion has been exaggerated.

1882 _____. "Soviet Policy in the Indian Ocean." In: Mcc-Gwire, M.; Booth, K.; and McDonnell, J., (eds.). Soviet Naval Policy : Objectives and Constraints. New York, 1975, 307-18.
   Soviet policy in this area appears to be basically an antisubmarine and anticarrier policy.

1883 JUNDANIAN, B. F. "Great Power Interaction with Africa: An Africanist's Perspective." Studies in Comparative Communism, 6, 3, Autumn, 1973, 319-25.
   A review of Cohn, Soviet Policy toward Black Africa and Larkin, China and Africa, 1949-1970.

1884 KAISER, K. "The Interaction of Regional Subsystems: Some Preliminary Notes on Recurrent Patterns and the Role of the Superpowers." World Politics, 21, 1, October 1968, 84-107.
An investigation and theoretical discussion of the relationships between regional subsystems.

1885 KANET, ROGER E. "Soviet Attitudes toward the Search for National Identity and Material Advancement in Africa." Vieteljabresberichte, 36, June 1969, 143-56.
A review of changing Soviet views of Africa.

1886 _____. "The Soviet Union and the Developing Countries: Policy or Policies?" World Today, 31, 8, August 1975, 338-46.
Soviet aid depends upon the strategic importance of a potential recipient, its value as a market or as a source of raw materials, and the strength of US or Chinese influence in the state.

1887 _____, (ed.). The Soviet Union and the Developing Nations. Baltimore: Johns Hopkins University, 1974. 302 p., index, tabls.
Contains several essays of a general nature and others on Asia, Latin America, and the Middle East. Only one, by A. J. Klinghoffer, deals directly with Africa.

1888 _____. "The Soviet Union and the Third World: A Case of Ideological Revision." Rocky Mountain Social Science Journal, 6, 1, April 1968, 109-16.
A review of changing Soviet views of the Third World.

1889 _____ and BAHRY, DONNA, (eds.). Soviet Economic and Political Relations with the Developing World. New York: Praeger, 1975. 242 p., figs., index, tabls.
Contains twelve essays presented at the First International Slavic Conference, 1974. General essays plus case studies on Asia and the Middle East are included.

1890 KANIEWICZ, G. "Nouvelle Politique Soviétique en Afrique Tropicale." Revue de l'Action Populaire, 159, June 1962, 690-6.
Soviet economic aid is aimed at increasing the strength of the public sector and to increase the size of the proletariat through aid to industrialization.

1891 KAUSHIK, DEVENDRA. "Super Powers in the Indian Ocean." Political Science Review, 10, 3/4, July-December 1971, 63-82.

1892 KEITA, RAPHAËL T. L'Evolution des Relations Franco-Maliennes depuis 1960. Toulouse: Institut d'Etudes Politiques, 1969. 39 p.
Analysis of the various types of cooperation between France and Mali.

1893 KIRKWOOD, KENNETH. Britain and Africa. Baltimore: Johns Hopkins, 1965. 235 p.
"Attention has been given to . . . more fundamental general issues and to a consideration of British-African relationships over time."

1894 KJØLBERG, ANDERS. "Striden om Afrikas Horn." Internasjonal Politikk, 1, 1978, 77-98.
"The Horn of Africa: Conflicts and Actors." The author provides a description of the crisis in the Horn to exemplify more general lessons on East-West conflict in the Third World.

1895 KLINGHOFFER, ARTHUR J. Soviet Perspectives on African Socialism. Rutherford, N. J.: Fairleigh Dickinson University, 1969. 276 p., bibl., index.
"This study is concerned with the Soviet view of socialism in sub-Saharan Africa during the years 1955-1964."

1896 _____. "The Soviet Union and Africa." In: Kanet, R., (ed.). The Soviet Union and the Developing Nations. Baltimore, 1974, 51-77.
"The Soviet Union has come to behave more as a great world power seeking influence and strategic, economic and military position and less as the center of a revolutionary movement aimed at overthrowing African governments and installing Communist regimes."

1897 KUHLEIN, CONRAD. "Die Auswirkungen der Revolution in Äthiopien auf die Lage am Horn von Afrika." Europa-Archiv, 33, 5, March 1978, 135-44.
"The Effects of the Revolution in Ethiopia on the Situation in the Horn of Africa." The political stability of Ethiopia is the key to the future of the Horn.

1898 LARKIN, BRUCE D. "China and Africa." Africa Today, 22, 3, July-September 1975, 61-6.
Review of China's Policy in Africa 1958-71 by Alaba Ogunsanwo.

1899 _____. "China and Africa: A Prospective on the 1970s." Africa Today, 18, 3, July 1971, 1-11.

1900 _____. China and Africa, 1949-1970: The Foreign Policy of the People's Republic of China. Berkeley: University of California, 1971. 268 p., bibl., index, tabls.
Larkin "argues that Peking's long-term aim of world revolution is maintained despite accommodation of African governments which are not challenged by serious radical opposition."

1901 _____. "Interest and Ideology: A Soviet Perspective." Studies in Comparative Communism, 6, 3, Autumn 1973, 310-8.
A review of H. D. Cohn's book on the Soviet Union and Africa.

1902   LARRABEE, S. "Moscow, Angola and the Dialectics of Détente." World Today, 32, 5, May 1976, 173-82.
       Soviet intervention in Angola raises doubts about détente.

1903   LAVRENTYEV, ALEXANDER. "Indian Ocean: The Soviet Perspective." Africa Report, 20, 1, January/February 1975, 46-9, ill.
       Cites a stepped-up Western military presence in the Ocean, calling it "aircraft-carrier diplomacy," and sees it as the traditional reaction by the West whenever it must defend its political and economic interests. Contrasts this approach with peaceful Soviet policies.

1904   LAWRIE, G. G. "Britain's Obligations under the Simonstown Agreements. A Critique of the Opinion of the Law Officers of the Crown." International Affairs, 47, 4, October 1971, 708-28.
       The manner in which these agreements between Britain and South Africa are interpreted will have an important effect upon the sale of British arms to South Africa.

1905   LEBEDEV, IGOR A. "Soviet Policy Considerations and the Indian Ocean." Australian Outlook, 31, 1, April 1977, 133-41.

1906   LEE, M. "La Politique Britannique à l'Egard de l'Afrique Noire." Etudes Internationales, 1, 4, December 1970, 102-9.
       British policy has no solution for its problems in Southern Africa.

1907   LEGUM, COLIN. "The African Environment." Problems of Communism, 27, 1, January/February 1978, 1-19, ill., map.
       African attitudes toward the USSR and changes in those attitudes in recent years are examined.

1908   _____. "Le Incognite del Corno d'Africa." Affari Esteri, 36, October 1977, 721-9.
       Analysis of the effects of Soviet intervention in Ethiopia on the politics of the Horn, especially with regard to moderate Arab states, Ethiopia, and Somalia.

1909   _____. "Realities of the Ethiopian Revolution." World Today, 33, 8, August 1977, 305-12.
       The Russians are active in this area and the power balance is altering.

1910   _____. "The Soviet Union, China and the West in Southern Africa." Foreign Affairs, 54, 4, July 1976, 745-62, map.
       The Sino-Soviet rivalry is more important to those states than their rivalry with the USA.

1911   LEGVOLD, ROBERT. "Moscow's Changing View of Africa's Revolutionary States." Africa Report, 14, 3/4, March/April 1969, 54-8.

1912   LEGVOLD, R. "Soviet and Chinese Influence in Africa." In: Rubinstein, A. Z., (ed.). Soviet and Chinese Influence in the Third World. New York, 1975, 154-75.
          Soviet and Chinese efforts in Africa have diminished as their confidence in their ability to influence there have diminished.

1913   _____. Soviet Policy in West Africa. Cambridge: Harvard University, 1970. 372 p., bibl., index, map.
          Guinea, Ghana, Ivory Coast, Mali, Nigeria, and Senegal were selected for analysis in this comparative study because "these six countries represent in their differing foreign and domestic attitudes the full spectrum of countries the Soviet Union has dealt with in Black Africa." The author makes comparisons between states, between time periods, and between theory and practice.

1914   _____. "The Soviet Union's Strategic Stake in Africa." In: Whitaker, J. S., (ed.). Africa and the United States: Vital Interests. New York, 1978, 153-86.
          "The first half on my essay deals with basic aspects of the Soviet-American competition." The second is an analysis of Soviet military activity and the Soviet view of its role in Africa.

1915   LELLOUCHE, PIERRE and MOISI, DOMINIQUE. "French Policy in Africa: A Lonely Battle against Destabilization." International Security, 3, 4, Spring 1979, 108-33, tabl.
          "Change, in French African policy, has been manifested in and affected by ferment within Africa, the new French military doctrine (elaborated since 1974-75), and the personality of the president. The continuities arise from earlier Gaullist policies. It is the intention of this article to examine, primarily through a study of political/security issues in sub-Saharan Africa, the implications of these policy changes within the context of the Gaullist continuities."

1916   LESSING, PIETER. Africa's Red Harvest. New York: John Day, 1962. 207 p., bibl., index.
          Analysis of Soviet and Chinese communist inroads in Africa.

1917   LEVY, B.-H. and HERTZOG, G. "Réflexions sur la Guerre d'Angola." Revue Française d'Etudes Politiques Africaines, 125, May 1976, 79-115.
          Superpower interest in Angola is examined.

1918   LEWIS, ROY. "Commonwealth Africa and the Enlarged Community." Round Table, October 1971, 515-24.
          The political, cultural, and economic impact of the anticipated British entry into the Common Market.

1919 LHUISARD, A. "La Chine et l'Afrique." Défense Nationale, June 1974, 89-105.
China wishes to align with the Third World against the superpowers.

1920 LINIGER-GOUMAZ, MAX. Eurafrique. Bibliographie Générale. Geneva: Les Editions du Temps, 1970. 160 p., indexes.
Contains 1300 references arranged under 5 major headings. Author, subject/geographic, and chronological indexes aid in the use of this work.

1921 _____. L'Eurafrique, Utopie ou Réalité? Yaoundé: Editions CLE, 1972. 111 p.
The Swiss author analyzes several variants of the Eurafrique concept, stressing the role of Eurafrique in a world dominated by the superpowers. Is it possible or desirable to create a third force?

1922 _____. "L'U. R. S. S., La Chine Populaire et L'Afrique - Essai Bibliographie." Geneve-Afrique, 8, 2, 1969, 69-80.

1923 LIPTON, MERLE. "British Arms for South Africa." World Today, 26, 10, October 1970, 427-34.
The British decision to resume arms sales to South Africa has caused strong negative reactions at home and abroad.

1924 LOWENTHAL, RICHARD. "China." In: Brzezinski, Z., (ed.). Africa and the Communist World. Stanford, 1963, 142-203.
Communist China's special claim to kinship with the ex-colonial world "made her not only a uniquely qualified ally but also a potentially dangerous rival for Soviet penetration of the African continent."

1925 _____. "The Sino-Soviet Split and Its Repercussions in Africa." In: Hamrell, S. and Widstrand, C. G., (eds.). The Soviet Bloc, China and Africa. Uppsala, 1964, 131-45.
The author attempts "to draw the outlines of the general framework of Sino-Soviet Relations" and the African implications of those relations.

1926 MALECOT, G. R. "Raisons de la Présence Française à Djibouti." Revue Française d'Etudes Politiques Africaines, 85, January 1973, 38-53.
France retained this territory because of its strategic importance.

1927 MARIÑAS OTERO, LUIS. "Las Conferencias Presidentiales Francoafricanas." Revista de Política Internacional, 148, November/December 1976, 59-73.
An example of continuing French domination of its ex-African colonies.

1928 MARIÑAS OTERO, L. "La Cumbre Franco-Africana de Dakar."
Revista de Política Internacional, 152, July/August 1977,
225-31.
Security, military cooperation, and economic relations
were the main topics of the French-African summit meeting in
Dakar, 1977.

1929 MATTHIES, VOLKER. China and Africa. Hamburg: Institut für
Asienkunde, 1969. 78 p., bibl.
Africa and the PRC, foreign and economic relations.

1930 _____. "Chinesische Afrikapolitik." Gegenwartskunde,
19, 1, 1970, 57-62.

1931 _____. "Somalia-ein Sowjetischer 'Satellitenstaat' im
Horn von Africa?" Varfassung und Recht in Übersee, 9, 4,
1976, 437-56.
"Somalia, a Soviet 'Satellite' in the Horn of Africa?"
"No," is the author's answer.

1932 MAXEY, K. "Labour and the Rhodesian Situation." African
Affairs, 299, April 1976, 152-62.
The domestic environment in Britain forced the Labour
Government to take unfortunate policy decisions in respect
of Rhodesia.

1933 MAYA, HENRY. "The Imperialist Threat to Africa - I: Uganda."
African Communist, 45, 1971, 37-44.

1934 MAZRUI, ALI A. The Anglo-African Commonwealth: Political
Friction and Cultural Fusion. Oxford: Pergamon, 1967.
163 p., index.
A variety of topics emerge in this consideration of
"Britain's third Commonwealth." British relations with
Africa, Africa in the Commonwealth, the EEC-Britain-Africa,
Rhodesia, and Pan-Africanism are the most obvious of these.
See "The African Conquest of the British Commonwealth," a
selection from this book in Y. Tandon, ed., Readings in
African International Relations.

1935 _____. "The Bolsheviks and the Bantu: From the October
Revolution to the Angolan Civil War." Survey, 22, 3/4,
Summer/Autumn 1976, 288-306.
The Soviet Union has been far ahead of the USA in under-
standing Africa, as evidenced by recent events in Southern
Africa.

1936 MCLANE, CHARLES B. Soviet-African Relations. London:
Central Asian Research Centre, 1974. 190 p., bibl., tabls.
[Soviet-Third World Relations, Vol. 3].
A brief introductory chapter is followed by 35 chapters
on Soviet relations with individual states.

1937 MCLANE, C. B. "Soviet Doctrine and the Military Coups in Africa." International Journal, 21, 3, Summer 1966, 298-310.
 The overthrow of Nkrumah is causing a rethinking by Soviet policy-makers. In the future, ideology will play a lesser role and traditional concepts of power politics a greater role in their policy decisions.

1938 McLENNAN, BARBARA N. "Britain and the New Africa." Il Politico, 34, 4, December 1969, 728-38.

1939 MERLE, M. "La Communauté Franco-Africaine." Revue de l'Action Populaire, 139, June 1960, 667-78.
 The dynamics of decolonization led from a French federation concept to a more confederal arrangement between France and the ex-colonies. See this author's 1961 essay.

1940 _____. "Mort ou Renouveau de la Communauté Franco-Africaine." Revue de l'Action Populaire, 151, September-October 1961, 919-28.
 The confederal arrangements described in this author's 1960 article failed to work, and here he describes the new relationships that have developed between France and the ex-colonies and between the ex-colonies.

1941 METROWICH, F. R. Africa and Communism: A Study of Successes, Setbacks and Stooge States. Johannesburg: Voortrekkerpers, 1967. 261 p., map.

1942 METZLER, JOHN J. "Peiping in Africa: Prospects and Portents for Communist China's Role in the Subsaharan Zone." Issues and Studies, 14, 8, August 1978, 72-84.

1943 MIKSCHE, F. O. "Pseudonationalismus aus Schwarzafrika." Aussenpolitik, 11, 11, November 1960, 733-42.
 The African states are weak and inadequate to repel communism unless the Western powers provide aid and utilize an energetic policy to safeguard their interests.

1944 MORISON, DAVID L. "Africa's Would-be Mentors." Race, 5, 4, April 1964, 52-60.
 A review of Z. Brzezinski, (ed.), Africa and the Communist World.

1945 _____. "Soviet Policy towards Africa." In: Hamrell, S. and Widstrand, C. G., (eds.). The Soviet Bloc, China and Africa. Uppsala, 1964, 30-42.
 Soviet Africa policy operates at two levels - that of the African governments and that of the African public.

1946 _____. "The U. S. S. R. and Africa: Commitment and Isolation." Race, 13, 3, January 1972, 337-46.
 Essays the failures of Soviet policy in Africa, 1958-68.

1947 MORISON, D. "The USSR and Africa in 1972." Africa Quarterly, 12, 1, April-June 1972, 2-8.
Russian Africa policy appears to be based on pragmatic rather than ideological considerations. However, this appearance may be misleading.

1948 _____. The Union of Soviet Socialist Republics and Africa. London: Oxford University, 1964. 124 p.
Soviet aims and attitudes, African studies in the USSR, and Soviet views of individual African countries are the major topics of this volume.

1949 MOSS, ROBERT. "On Standing Up to the Russians in Africa." Policy Review, 5, Summer 1978, 97-104.

1950 MULIRA, J. "The Role of the Soviet Union in the Decolonization Process of Africa: From Lenin to Brezhnev." Mawazo, 4, 4, 1976, 26-35.

1951 NATUFE, O. I. "Nigeria and Soviet Attitudes to African Military Regimes, 1965-70." Survey, 22, 1, Winter 1976, 93-111.
Soviet foreign policy for Africa is predicated more upon concerns of national interest than on considerations of ideology.

1952 N'DONGO, SALLY. La 'Coopération' Franco-Africaine. Paris: François Maspero, 1972. 139 p.
Largely a study of the Senegalese workers in France, but overall an indictment of French neocolonial rule in Africa.

1953 NEILSON, WALDEMAR. The Great Powers and Africa. New York: Praeger, 1969. 431 p., index, tabls.
Major sections on US, European, and communist states' relations with Africa.

1954 NELSON, DANIEL. "Sino-African Relationship: Renewing an Ancient Contact." East Asian Review, 4, 1, Spring 1977, 68-89.
China's interest in the non-important continent of Africa is an opportunity to study foreign policy making in a situation without the normal constraints.

1955 NEUHAUSER, CHARLES. Third World Politics: China and the Afro-Asian People's Solidarity Organization. Cambridge: Harvard University, 1968. 99 p.. index.
"This book will attempt to explore the history of the AAPSO and of its relationship to Chinese foreign policy."

1956 NIMER, B. "The Congo in Soviet Policy." Survey, 19, 1, Winter, 1973, 184-210.

Although the Soviets supported the leftist parties and personalities in Congo/Zaire, they have attempted to maintain relations with the government as part of their overall strategy of competition with the USA and the PRC.

1958 NWEKE, G. A. External Intervention in African Conflicts: France and French-speaking West Africa in the Nigerian Civil War, 1967- 1970. Boston: African Studies Center, Boston University, 1976. 77 p., map, tabls.

1959 O'BRIEN, RITA CRUISE. White Society in Black Africa: The French of Senegal. Evanston: Northwestern University. London: Faber and Faber, 1972. 320 p., bibl., ill., index, maps, tabls.

This study provides information on French and Senegalese international relations as well as the role of French people in Senegal from 1900 until the late 1960s. The are chapters on French technical assistance and investments in Senegal.

1960 OBUKHOV, L. "Imperialism's Ideological Subversion in Africa." International Affairs, 6, June 1971, 71-6.

The western states are waging a mass media propaganda war against the African states.

1961 OGUNBADEJO, OYE. "Ideology and Pragmatism: The Soviet Role in Nigeria, 1960-1977." Orbis, 21, 4, Winter 1978, 803-30, tabls.

Viewing Soviet policy as a pendulum, swinging back and forth between ideology and pragmatism: the author assesses the different phases of the policy, their shifting doctrinal bases, and the repercussions for Moscow and Lagos.

1962 OGUNSANWO, ALABA. China's Policy in Africa, 1958-1971. New York: Cambridge University, 1974. 336 p., bibl., index, tabls.

"From the peripheral status it occupied in the middle fifties, China's policy in Africa rapidly achieved a self-propelling and compulsive momentum."

1963 OPPERMANN, T. "'Eurafrika' - Idee und Wirklichkeit." Europa-Archiv, 15, 23, December 1960, 695-707.

Background, origins, and present conditions of the Eurafrique concept and reality.

1964 ORITZ, E. "Las Grandes Potencias y la Crisis de Nigeria." Estudios Internacionales, 3, 1, April-June 1969, 63-9.

1965 OTTAWAY, MARINA. "Soviet Marxism and African Socialism." Journal of Modern African Studies, 16, 3, September 1978, 477-85.
Examines the adaptability of Marxism as an ideology to the African continent.

1966 OVSGANY, I. D. et al. A Study of Soviet Foreign Policy Translated by David Skvirsky. Moscow: Progress, 1975. 379 p.
See Chapter III, "The Soviet Union and the Developing States," 87-162.

1967 PAPP, DANIEL S. "Angola, National Liberation, and the Soviet Union." Parameters: Journal of U. S. Army War College, 8, 1, March 1978, 26-39.

1968 _____. "National Liberation during Détente: The Soviet Outlook." International Journal, 32, 1, Winter 1976/1977, 82-9.
The Soviet belief system impels them to continue support for national liberation movements during this period of détente.

1969 PEDINI, M. "Europe e Africa." Affari Esteri, 23, July 1974, 48-55.
Relations between Europe and Africa, with emphasis on Europe's relations with Africa. The author compares Europe's policies for North Africa and the rest of Africa.

1970 PENDERGAST, W. R. "French Cultural Relations." Chronique de Politique Etrangère, 27, 3, May 1974, 339-56.
Maintains that cultural expansion is a stable and important element within the French repertoire of diplomatic strategies and proceeds to investigate and compare it to the policies of other major powers.

1971 PEPY, M. DANIEL. "France's Relations with Africa." African Affairs, 69, 275, April 155-62.
Brief summary of French relations from 1959 to 1969, with emphasis on recent developments.

1972 PERSON, YVES. "La France et l'Afrique Noire: Histoire d'Une Aliénation." Revue Française d'Etudes Politiques Africaines, 63, March 1971, 55-83.

1973 PLANTEY, ALAIN. "Indépendance et Coopération." Revue Juridique et Politique, Indépendance et Coopération, 31, 4, December 1977, 1079-107.
A review of French relations with its former African colonies.

1974   PLESSIS, JAN DU. <u>The Brezhnev Doctrine and South Africa</u>.
        Pretoria: Foreign Affairs Association, 1977.
        A South African view of Soviet foreign policy.

1975   _____. "Communist Objectives in Africa in the Seventies."
        <u>Bulletin of the Africa Institute of South Africa</u>, <u>12</u>, 10,
        1974, 426-31.

1976   _____. "The Soviet Union's Foreign Policy towards Africa."
        <u>Bulletin of the Africa Institute of South Africa</u>, <u>12</u>, 3,
        1974, 105-15.
        Soviet policy is "closely linked with the success or failure of the terrorist movements."

1977   PRINSLOO, DAAN S. "China's Quest for Africa." <u>Journal of Social and Political Studies</u>, <u>3</u>, 1, Spring 1978, 27-40.

1978   R. B. "China's Impact on Africa - A Summing Up." <u>Race</u>, <u>5</u>, 4, April 1964, 75-82.
        This special issue of <u>Race</u> is devoted to studies of African-Chinese relations.

1979   RABIER, C. <u>and</u> ANGRAND, J. "La Stratégie des Grandes Puissances autour du Territoire Français de Affars et des Issas et de l'Océan Indien." <u>Revue Française de Science Politique</u>, <u>26</u>, 3, June 1976, 521-34, map.
        Although many of the states that border the Indian Ocean are opposed to the growth of Soviet and US forces in the Ocean, those states are too weak and too divided to prevent it.

1980   RATHBONE, RICHARD. "France and Africa." <u>Current History</u>, <u>64</u>, 379, March 1973, 111-3.
        France's domination of its ex-colonies is no longer very strong.

1981   REAU, GUY. "Les Grandes Puissances et l'Océan Indien." <u>Revue de Défense Nationale</u>, 27, October 1971, 1464-80.

1982   REINTON, P. O. "Imperialism and the Southern Sudan." <u>Journal of Peace Research</u>, <u>8</u>, 3/4, 1971, 239-47.
        British imperialism is seen as a major force in the continuing civil war in Sudan.

1983   RETTMAN, R. J. "The Tanzam Rail Link: China's 'Loss-Leader' in Africa." <u>World Affairs</u>, <u>136</u>, 3, Winter 1973/1974, 232-58.
        The West wished to keep Zambia dependent upon South Africa and so refused to support the Tanzam project. The PRC supplied the aid in order to extend its influence, build trade relations, and outdo the West and the Soviets.

1984   RIVKIN, ARNOLD. "Arms for Africa?" Foreign Affairs, 38, 1, October 1959, 84-94.
         The author proposes that no military aid should be given to African states and that the major powers should guarantee the continent, except Egypt, against aggression and subversion.

1985   ROBBS, PETER. "Africa and the Indian Ocean." Africa Report, 21, 3, May/June 1976, 41-5, ill., map.
         Strategic survey of political and military developments in East and South African countries bordering on the Indian Ocean, in the context of Soviet-Western confrontation.

1986   ROBERTS, NIGEL S. "Africa and the Six Crises of the Commonwealth." World Review, 13, 3, October 1974, 23-31.

1987   ROCHOU, JEAN-CLAUDE. Et Si Nous Faisons la Révolution Eurafricaine. Paris: La Pensée Universelle, 1972. 203 p.
         A somewhat romantic view of Euro-African relations and world politics.

1988   ROPP, K. VON DER. "Die Franko-Afrikanischen Beziehungen." Aussenpolitik, 25, 4, 1974, 461-76.
         Economic and military relations between France and Africa have decreased, but cultural relations have increased. There is some information on the large French commercial firms such as SCOA and CFAO.

1989   ROUCEK, J. S. "La Colisión de los Mundos Comunistas en Africa." Revista de Política Internacional, 78, March/April 1965, 15-70.
         Soviet and Chinese interpretations of the application of Marxism and Leninism to Africa are different, but the real basis for their competition in Africa is the protection and advancement of their national interests.

1990   _____. "The Indian Ocean in Global Geopolitics." International Review of History and Political Science, 8, 4, November 1971, 57-77.
         A brief historical review of the post-World War II period.

1991   _____. "The Sino-Soviet Conflict in Its Racial Aspects." Revue du Sud-Est Asiatique Extrême-Orient, 1, 1970, 59-83.
         Racism is major factor in the Sino-Soviet competition for influence in Africa.

1992   ROYAL AFRICAN SOCIETY. Europe and Africa: Trends and Relationships. London: Royal African Society, 1978. 49 p.
         Contains four essays on relations between Europe and Africa.

1993   RUBINSTEIN, ALVIN Z. "Soviet Policy in the Third World Perspective." Military Review, 58, 7, July 1978, 2-9.
        Soviet attempts to gain positions of influence in Africa and Asia have not been successful; the client states seem to follow their interests rather that those of the USSR even after large amounts of Soviet assistance have been granted.

1994   _____. "The Soviet Union's Imperial Game in Africa." Optima, 26, 3, 1977, 114-25, ill., map.
        Soviet foreign policy for Africa.

1995   RUBIO GARCIA, LEANDRO. "Unión, Comunidad y Cooperación: Förmulas en un Proceso de Descolonización." Revista de Política Internacional, 152, July/August 1977, 215-24; 153, September/October 1977, 205-18.
        "Union, Community and Cooperation: Formulas in a Process of Decolonization." The three stages of France's relations with its ex-colonies in Africa.

1996   SABOURIN, LOUIS. "La Coopération entre Pays Francophones dans une Perspective Globale." Etudes Internationales, 5, 2, June 1974, 195-207.
        In a special issue, "La Coopération Internationale entre Pays Francophones."

1997   SAKARAI, LAWRENCE J. "The Imperialists in Africa." Africa Quarterly, 16, 1, July 1976, 29-59.

1998   SALVINI, GIANPAOLO. "Il Conflitto Etiopico." Aggiornamenti Sociali, 28, 12, December 1977, 707-18.
        "The Ethiopian Conflict." A discussion of the various conflicts that make up the Ethiopian conflict and the motives and actions of the various states (African and non-African) that are involved.

1999   SAMUELS, MICHAEL A., (ed.). The Horn of Africa. New Brunswick, N. J.: Transaction, 1978. 96 p.
        A collection of short essays analyzing several of the international aspects of the crisis in the Horn. Cuban, Soviet, US, Israeli, Iranian, and Saudi policies and interests are discussed.

2000   SCALAPINO, R. A. "Sino-Soviet Competition in Africa." Foreign Affairs, 42, 4, July 1964, 640-54.
        A description of Soviet and Chinese tactics and purposes in their competition for influence in Africa.

2001   SCHATTEN, FRITZ. Communism in Africa. New York: Praeger, 1966. 352 p., index.
        The progress of the USSR and the PRC in the take-over of Africa. Case studies on Guinea, Ghana, and Mali are included. Africa is portrayed as a center for cold war competition.

2002   SCHMIDT, R. "Angola: Ein Internationaler Konflikt."
        Aussenpolitik, 27, 4, 1976, 460-71.
             Description of the international aspects of the Angolan
        Civil War, with emphasis on the implications for interna-
        tional politics of the Soviet involvement in that conflict.

2003   SCHOELL, F. L. "Les Influences Extra-Européennes en Afrique
        au Sud du Sahara." Revue Economique et Sociale, September
        1960, 40-53.
             The author discusses mainly Soviet and US influences
        and techniques for expanding their influence in Africa.

2004   SCHÜTZE, W. "Bilanz und Perspektiven der Französisch-Afri-
        kanischen Gemeinschaft." Europa-Archiv, 15, 5, March 1960,
        155-66.
             "Balance Sheet and Perspectives of the Franco-African
        Community." The French-speaking African states may move
        from a position of autonomy in a French community to one of
        independence.

2005   SCHWAB, PETER. "Cold War on the Horn of Africa." African
        Affairs, 77, 306, January 1978, 6-20.
             Argues that the recent radicalization of political
        systems in Ethiopia and Somalia generated a response amongst
        a multiplicity of states in and out of Africa that led to
        the crisis on the Horn of Africa and the Red Sea.

2006   SECK, A. "Logique d'une Philosophie: Les Nouveaux Accords
        Franco-Sénégalais." Ethiopiques, 2, April 1975, 8-17.

2007   SERVOISE, E. "Die Weiterentwicklung der Französisch-Afri-
        kanischen Gemeinschaft." Europa-Archiv, 16, 7, April 1961,
        149-62.
             "The Evolution of the French African Community." An
        analysis of the dynamic situation of relations between
        France and its former colonies in Africa.

2008   SHAW, TIMOTHY M. [Book Review] Journal of Modern African
        Studies, 15, 2, June 1977, 324-7.
             Review of Freedom Railway: China and the Tanzania-
        Zambia Link, by Martin Bailey, and The Great Uhuru Railway:
        China's Showpiece in Africa, by Richard Hall and Hugh Peyman.

2009   SIBEKO, ALEXANDER. "The Battle for Angola." African Com-
        munist, 62, 1975, 41-51.

2010   SIM, J. P. "Soviet Naval Presence in the Indian Ocean."
        Australian Outlook, 31, 1, April 1977, 185-92.

2011   SIRCAR, PIRBATI K. "The Great Uhuru (Freedom) Railway:
        China's Link to Africa." China Report, 14, 2, March/April
        1978, 15-23, tabls.

2012 SKURNIK, W. A. E. "Africa and the Superpowers." Current History, 71, 421, November 1976, 145-8.

2013 SLAWECKI, L. M. S. "The Two China's in Africa." Foreign Affairs, 41, 2, January 1963, 398-409.
    Nationalist China is winning the competition with the PRC for African support at the UN. In time, Africa will be an important market for Taiwan's industries.

2014 SMOLANSKY, O. M. "Soviet Policy in the Middle East and Africa." Current History, 75, 440, October 1978, 113-6.

2015 SPENCE, J. E. "British Policy towards the High Commission Territories." Journal of Modern African Studies, 2, 2, July 1964, 221-45.
    The article deals mainly with British colonial policy, but there is also information on the conflicts between Britain and South Africa over the three territories.

2016 "A Statement of the Soviet Government: Text of a Soviet Government Statement on Recent Aggressive NATO Activity in Africa." International Affairs, 8, August 1978, 141-3.

2017 STENT, ANGELA. "The Soviet Union and the Nigerian Civil War: A Triumph of Realism." Issue, 3, 2, Summer 1973, 43-8.
    Analysis of changing Soviet attitudes and policies for Africa.

2018 STEVENS, CHRISTOPHER. "Africa and the Soviet Union." International Relations, 3, 12, November 1971, 1014-25.
    A general review with some emphasis on Soviet relations with Ghana and Nigeria.

2019 _____. "The Soviet Union and Angola." African Affairs, 75, 299, April 1976, 137-51.
    Soviet involvement in Angola marks an important maturation in Soviet policy since the Congo experience fifteen years earlier.

2020 _____. The Soviet Union and Black Africa. New York: Holmes and Meier, 1976. 276 p., tabls.
    Economic, political, and military relations between 1953 and 1972. Case studies of Ghana, Guinea, Mali, Nigeria, and Kenya are included and there are chapters on trade, aid, and strategic considerations.

2021 SURET-CANALE, JEAN. "Difficultés du Neo-colonialisme Français en Afrique Tropicale." Revue Canadienne des Etudes Africaines, 8, 2, 1974, 211-34.

2022 TANDON, YASHPAL. "The Internationalization of the Civil War: Lessons from the Congo, Nigeria and Vietnam." In: Mazrui, A. A. and Patel, H., (eds.). Africa in World Affairs: The Next Thirty Years. New York, 1973, 63-78.
    The author uses a "teleological" approach to identify the purposes of internationalization and to draw lessons from the three cases.

2023 "Tanzam-Railway: Breakthrough for Red China." Bulletin of the Africa Institute of South Africa, July 1972, 237-41.
    The Tanzam Railway Project is China's Trojan Horse in Africa.

2024 TARABRIN, E. A. The New Scramble for Africa. Moscow: Progress, 1974. 319 p., tabls.
    A Soviet writer's view of the struggle between the Western imperialist powers for control of Africa.

2025 TAU, B. "The Imperialist Threat to Africa: 2. Guinea." African Communist, 45, 1971, 45-50.

2026 TEILLAC, JEAN. "Europe et Afrique: Vers un Nouveau Contrat?" Défense Nationale, 30, February 1974, 59-66.

2027 THOMAS, T. "Angola, Südafrika und die NATO." Blätter für Deutsche und Internationale Politik, 21, 2, February 1976, 139-51.
    The USA, France, and Germany in the NATO context have economic and strategic needs for preservation of the South African regime.

2028 THOMPSON, W. SCOTT. "The Communist Powers and Africa." Orbis, 16, 4, Winter 1973, 1066-9.
    A review of Larkin, China and Africa and Legvold, Soviet Policy in West Africa.

2029 _____ and SILVERS, BRETT. "South Africa in Soviet Strategy." In: Bissell, R. E. and Crocker, C. A., (eds.). South Africa into the 1980s. Boulder, Col., 1979, 133-58.

2030 TING, D. "La Chine et les Etats d'Afrique Noire." Annuaire du Tiers Monde, 1, 1975, 205-22.
    The PRC's Africa policy has met with outstanding success.

2031 TORRE, SEVANDO DE LA. "Aspectos de la Política China en las Luchas de Liberacion de Africa." Revista de Política Internacional, 149, January/February 1977, 147-74.

2032 TOUSCOUZ, J. "La 'Normalisation' de la Coopération Bilatérale de la France avec les Pays Africains 'Francophones' (Aspects Juridiques)." Etudes Internationales, 5, 2, June 1974, 208-25.

2033 TRIVIERE, LEON. "La Chine et l'Expansion Soviétique en Afrique (1977)." Mondes Asiatiques, 12, Winter 1977/1978, 289-300.
   China's emphasis on developing its relations with Africa is based mainly on a desire to prevent the expansion of Soviet influence on the continent.

2034 TUNTENG, P-KIVEN. "France-Africa: Plus ça Change." Africa Report, 20, 4, July/August 1974, 2-6.

2035 UNIVERSITY OF SOUTHAMPTON. The Indian Ocean in International Politics. Southampton: Department of Extra-Mural Studies, University of Southampton, 1973. 236 p.
   See references to essays by D. Austin, D. Braun, and C. Mitchell.

2036 URFER, SYLVAIN. "Coopération Sino-Tanzanienne." Afrique Contemporaine, 15, 83, January/February 1976, 12-6, tabl.
   China's successes in its relations with Tanzania have greatly enhanced the reputation of the PRC among Third World states.

2037 VALENTA, JIRI. "The Soviet-Cuban Intervention in Angola, 1975." Studies in Comparative Communism, 11, 1/2, Spring/Summer 1978, 3-33.
   The Soviet-Cuban intervention in Angola in 1975 was an important contributing factor to the temporary freeze in Soviet-American relations in 1976. Nevertheless, in the future, we can expect the Soviets to continue to exploit the opportunities available in unstable African politics.

2038 VAN DE MEERSSCHE, P. "Algeriie en Afrika in de Politiek van De Gaulle." Internationale Spectator, 15, 9, May 1961, 243-77.

2039 VANNEMAN, PETER. "Soviet National Security Policy in Southern Africa and the Indian Ocean: The Case of Mozambique." Politikon, 3, 1, June 1976, 42-50.
   Soviet aims and methods in the Indian Ocean are examined. Mozambique is the focus of Soviet activity and Soviet gains there may greatly alter the balance of power in this area.

2040 _____ and JAMES, M. "Soviet Thrust into the Horn of Africa: The Next Targets." Strategic Review, 6, 2, Spring 1978, 33-40.

2041 VAUDIAUX, J. "L'Evolution Politique et Juridique de la Coopération Franco-Africaine et Malgache." Revue Générale de Droit International Public, 74, 4, October-December 1970, 922-68.
   French-African relations are based on inequality, but there is some very slight movement toward equality.

2042   VENGROFF, R. "Neo-colonialism and Policy Outputs in Africa." *Comparative Political Studies*, 8, 2, July 1975, 234-50.
        An empirical test of the reality of neocolonialism suggests that military and foreign policy are especially subject to external control. Two different patterns of neocolonialism exist.

2043   VERLET, MARTIN. "L'Afrique, l'Impérialisme et Giscard d'Estaing." *Cahiers du Communisme*, 53, 5, May 1977, 94-109, map.

2044   VILLIERS, CAS DE. "China's Decade in Africa." In: Africa Institute of South Africa, (ed.). *Africa in the Seventies*. Pretoria, 1974.

2045   VOLGHIN, A. "Africa in Peking's Foreign Policy." *International Affairs*, September 1969, 26-32.
        A Soviet view of the Africa policy of the PRC.

2046   WALL, PATRICK, (ed.). *The Indian Ocean and the Threat to the West: Four Studies in Global Strategy*. London: Stacey International, 1975. 198 p., ill., index, maps.
        Four essays draw similar conclusions. As Britain has withdrawn from the Indian Ocean, the Soviets and the Chinese have moved in. The area is important because of oil shipments in the Ocean and South Africa is a key to the security of the area.

2047   "Waning French Influence in Africa." *Bulletin of the Africa Institute of South Africa*, 12, 3, 1974, 89-93.
        Suggests that francophone states are becoming increasingly independent of France.

2048   WARIAVWALLA, BHARAT. "Superpowers and the Angola Conflict." *IDSA Journal*, 9, 4, April-June 1977, 404-20.
        Angola was a severe strain for détente but it also showed how little influence the superpowers have over their clients.

2049   WAUTHIER, CLAUDE. "France and Africa: Long Live Neo-colonialism." *Issue*, 2, 1, Spring 1972, 23-6.
        Changing African views of French foreign policy and French reactions to charges of being a neocolonial power.

2050   WEINSTEIN, B. "Francophonie: A Language-based Movement in World Politics." *International Organization*, 30, 3, Summer 1976, 485-507.
        The attempts to improve (and expand) the use of French language through a variety of organizations and techniques in 26 states, many of which are in Africa, may lead to the emergence of a new cultural force that could affect the way states interact.

2051 WEINSTEIN, WARREN. "Chinese Policy in Central Africa: 1960-63." In: Weinstein, Warren, (ed.). Chinese and Soviet Aid to Africa. New York, 1975, 56-82, map, tabls.
Zaire, Rwanda and Burundi are discussed.

2052 WHITE, DOROTHY S. "De Gaulle and Black Africa." Orbis, 13, 4, Winter 1970, 1159-84.
The good relations between France and francophone Africa are in large measure due to the policies of Charles De Gaulle.

2053 WHITEMAN, KAYE. "Pompidou and Africa: Gaullism After De Gaulle." World Today, 26, 6, June 1970, 241-7.

2054 WOLFERS, MICHAEL. "The Chinese Presence in Commonwealth Africa and Its Significance." Commonwealth, 15, 5, October 1971, 109-11.

2055 YAKOBSON, SERGIUS. "The Soviet Union and Ethiopia: A Case of Traditional Behavior." Review of Politics, 25, 3, July 1963, 329-42.
Soviet-Ethiopian relations today (1963) are a continuation of relations developed during the Tsarist period.

2056 _____. "The U. S. S. R. and Ethiopia: A Case of Traditional Behavior." In: London, K., (ed.). New Nations in a Divided World. New York, 1963, 180-92.
The Soviet policy for Ethiopia is a refurbished version of pre-Revolution Russian policy.

2057 YEH, PO-T'ANG. "Peiping's Policy towards Africa as Viewed from the Independence of Guinea-Bissau and Mozambique." Issues and Studies, 10, 15, December 1974, 2-12.
A major turning point in the PRC's Africa policy was reached in 1970. Aid was greatly increased and a larger, broader spectrum of recipients was established. The PRC sees its support of African racist war as part of a global strategy.

2058 YOUNGER, K. "Reflections on Africa and the Commonwealth." World Today, 18, 3, March 1962, 121-9. Also in: Tandon, Y., (ed.). Readings in African International Relations. Nairobi, 1972, 268-71.
Pan-Africanism may prove to be a competitor against the Commonwealth. If the Commonwealth is to survive, it must improve its performance in assisting in the practical needs of its members.

2059 _____. "Wandlungen der Britischen Haltung gegen über Afrika." Europa-Archiv, 16, 10, May 1961, 241-8.
"Changes in the British Attitude toward Africa." Britain and its allies now wish Africa to remain neutral.

2060 YU, GEORGE T. China and Tanzania: A Study in Cooperative Interaction. Berkeley: Center for Chinese Studies, University of California, 1970. 100 p. [China Research Monographs].

Trade, aid, military assistance, China as a model of development, and Chinese objectives are analyzed. Several documents are reprinted.

2061 _____. "China and the Third World." Asian Survey, 17, 11, November 1977, 1036-48.

Chinese policies for Asia, Africa, and Latin America are examined individually and as parts of a Third World policy.

2062 _____. "China in Africa." Yearbook of World Affairs, 24, 1970, 125-37, tabl

Studies China's utilization of select foreign policy instruments in its interaction with other states, with a focus on those of Africa.

2063 _____. China's Africa Policy: A Study of Tanzania. New York: Praeger, 1975. 200 p., bibl., ill., index, tabls.

An updating of his earlier monograph. Yu includes a chapter on Swedish aid to Tanzania for purposes of comparison. Chinese aid is viewed as true cooperation, rather than a political tool.

2064 _____. "China's Failure in Africa." Asian Survey, 6, 8, August 1966, 461-8.

Anti-imperialism, anti-Soviet Unionism, and Third World solidarity are the major objectives of the PRC.

2065 _____. "China's Impact." Problems of Communism, 27, 1, January/February 1978, 40-50, ill., tabls.

"Examines the nature of Sino-Soviet rivalry on the continent and its implications for Soviet-African relations."

2066 _____. "China's Role in Africa." Annals of the American Academy of Political and Social Science, 432, July 1977, 96-109.

After a brief review of earlier stages of China's Africa policy, the author analyzes present Chinese policy. Africa occupies an important position in China's overall foreign policy.

2067 _____. "Chinese Rivalry in Africa." Race, 5, 4, April 1964, 35-46, tabls.

ROC vs. PRC in efforts to woo friends in Africa.

2068 _____. "Dragon in the Bush: Peking's Presence in Africa." Asian Survey, 8, 12, December 1968, 1018-26.

"Really meaningful interaction with China involves fewer than six states, including Guinea, Mali, and Tanzania."

2069 YU, G. T. "Peking's African Diplomacy." Problems of Communism, 21, 2, March/April 1972, 16-24, ill.
"Tanzania and the Congo (B) have constituted primary focuses of Chinese policy in Africa . . . . This article will analyze and compare these two important foreign-policy ventures of the Chinese in Africa."

2070 _____. "Sino-African Relations: A Survey." Asian Survey, 5, 7, July 1965, 321-32.
The PRC stresses self-determination, self-reliance, and Afro-Asian unity in its Africa policy. A large variety of foreign policy tools are used in the implementation of this policy, but success is limited by the growth of African nationalism and the PRC's lack of exportable resources.

2071 _____. "Working on the Railroad: China and the Tanzania-Zambia Railway." Asian Survey, 11, 11, November 1971, 1101-17.
The railway project funded by the PRC is examined as a foreign policy and behavior case study.

2072 ZEA, L. "La Revolución de los Pueblos Africanos." Cuadernos Americanos, 20, 5, September/October 1961, 9-21.
The African struggle for independence is made more difficult by the Soviet-American conflict and the conflict of interests between the colonial powers.

The USA:
Relations with Africa

2073 ABOUL-ENEIN, MOHAMMED I. M. "The United States Reaches a Deadlock in Namibia." Revue Egyptienne de Droit International, 31, 1975, 123-56.

2074 ADELMAN, KENNETH L. "The Black Man's Burden." Foreign Policy, 28, Fall 1977, 86-109.
Carter's policy for Southern Africa is inadequate for black rule there is not a human rights solution. An equitable partition of South Africa might be a solution, but the whites are as determined as the Israelis to maintain a homeland.

2075 AFRICAN-AMERICAN INSTITUTE. CONFERENCE OF AFRICAN AND AMERICAN REPRESENTATIVES. America's Africa Policy. New York: AAI, 1972. 40 p., ill.
Rhodesia, Namibia, South Africa, the Portuguese territories, trade, investment, and aid are each considered in brief reports from a conference conducted in Lusaka.

2076 AFRICAN BIBLIOGRAPHIC CENTER. AF-LOG: African Interests of American Organizations. Washington: ABC, 1975. 886 p., index. [Current Reading List Series, 11, 2].
A listing of businesses, missions, and other US organizations with interests in Africa.

2077 AMERY, JULIAN. "The Crisis in Southern Africa: Policy Options for London and Washington." Policy Review, 2, Fall 1977, 89-111. Also in: South Africa International, 8, 4, April 1978, 197-208.
The USA and the USSR are the only states that will matter in the solution of the Southern Africa problem.

2078 ANYAKOHA, M. W. "American Foreign Relations with Africa: The Kissinger Years." Current Bibliography on African Affairs, 10, 2, 1977/1978, 147-62.
Brief comments and 283 references.

2079 ARKHURST, FREDERICK S. "Introduction." In: Arkhurst, F. S., (ed.). U. S. Policy towards Africa. New York, 1975, 1-10, map.
This introduction includes commentary on US Africa policy and, in particular, US policy for Southern Africa.

2080 _____, (ed.). U. S. Policy toward Africa. New York: Praeger, 1975. 295 p., map, tabls.
A collection of six papers (and statistical appendix) that were presented as a seminar conducted by the Phelps Stokes Froundation. Authors include I. Wallerstein, H. J. Spiro, G. M. Houser, E. and R. L. West, and W. B. Ofuatey-Kodjoe.

2081 ARSENAULT, RAYMOND. "White on Chrome: Southern Congressmen and Rhodesia 1962-1971." Issue, 2, Winter 1972, 46-57, tabl.
Critical look at the stance of some Southern Congressmen and Senators regarding trade sanctions against Rhodesia.

2082 BAKER, DONALD. "Kissinger-Carter: Two Views of Southern Africa." Africa Institute Bulletin, 15, 8, 1977, 196-202, ill.
"US foreign policy toward Southern Africa is in the process of major transformations, and these changes will have a decided impact in the area."

2083 BAKER, ROSS K. "American Policy toward Africa: Cause for Indictment?" Worldview, 15, 12, December 1972, 18-24.
Those who have argued for a more liberal US Africa policy must review their assumptions.

2084 _____. "The 'Back-Burner' Revisited: America's African Policy." Orbis, 15, 1, Spring 1971, 428-47.
US policy has been symbolic, not substantive. It has been imitative, reactive, and vicarious.

2085 _____. "Towards a New Constituency for a More Active American Foreign Policy for Africa." Issue, 3, 1, Spring 1973, 12-9, tabls.
Identifies five constituencies for an Africa pressure group in the USA.

2086 BALL, GEORGE. "Asking for Trouble in South Africa." Atlantic, 240, 4, October 1977, 43-50. Also in: Rivista di Studi Internazionali, 45, 2, 1978, 261-74. Also in: South Africa International, 8, 3, January 1978, 148-65.
US policy must take into account the domestic realities of South Africa. We must not isolate the South African whites.

2087   BELFIGLIO, V. J. "American Viewpoints on Multinational Development in South Africa." Issues and Studies, 12, 12, December 1976, 55-76.
   The demographic histories of the USA and the RSA are different and US policy should not be based on attempts to draw parallels between those histories. The two states should cooperate.

2088   BENDER, GERALD J. "Angola, the Cubans, and American Anxieties." Foreign Policy, 31, Summer 1978, 3-30.

2089   _____. "La Diplomatie de Kissinger et l'Angola." Revue Française d'Etudes Politiques Africaines, 126, June 1976, 73-95.
   American policy for the past six years has been based on premises totally foreign to the realities of the situation in Angola.

2090   BERNARD, J. P. "Afrique Noire — Etats-Unis. Quelques Aspects de la Diplomatie Américaine Face à la Décolonisation et Après." Res Publica, 12, 2, 1970, 217-37.
   American policy has four major characteristics: stress key states, leave the other states to the ex-colonial power, support for regional groupings, and bring African students to American universities.

2091   BEZBORUAH, MONORANJAN. U. S. Strategy in the Indian Ocean. New York: Praeger, 1977. 288 p.
   The US is attempting to fill a vacuum in the Indian Ocean, but has exaggerated the Soviet build-up there.

2092   BIENEN, HENRY. "U. S. Foreign Policy in a Changing Africa." Political Science Quarterly, 93, 3, Fall 1978, 443-64.

2093   BISSELL, RICHARD E. "United States Policy in Africa." Current History, 73, 432, December 1977, 193-5.

2094   BIXLER, RAYMOND W. The Foreign Policy of the United States in Liberia. New York: Pagent, 1957. 143 p., bibl.
   A history of US relations with Liberia from the founding of Liberia until the early 1950s.

2095   BÖGE, W. "Dekolonisation und Amerikanische Aussenpolitik: Eine Analyse der Politischen Beziehungen Angola-USA, 1945-1975." Afrika Spectrum, 10, 3, 1975, 219-31.
   The main feature of US policy in Angola has been the lack of any long term plan. Portuguese attitudes were never understood by American policy makers. In the end, both the Portuguese and the Africans were hostile to the US and so the US ended with no role in decolonization at all.

2096 BOWEN, MICHAEL; FREEMAN, GARY; and MILLER, KAY. Passing by: The United States and Genocide in Burundi. Washington: Carnegie Endowment for International Peace, 1973. 49 p.
US policy for Africa during the massacres in Burundi is critically examined.

2097 BOWMAN, LARRY W. South Africa's Outward Strategy: Foreign Policy Dilemma for the United States. Athens, O.: Center for International Studies, Ohio University, 1971. 25 p. map, tabl. [Papers in International Studies, Africa Series, No. 13].
Similar to the author's essay in International Affairs, 1971.

2098 ———. "South Africa's Southern Strategy and Its Implications for the United States." International Affairs, 47, 1, January 1971, 19-30.
A Southern Africa subsystem has emerged and this poses problems for US policy makers.

2099 ———. "Southern Africa Policy for the Seventies." Issue, 1, 1, Fall 1971, 25-6.
A examination of Nixon's Africa policy.

2100 BOWMAN, L. G. "Vesten og det Sørlige Africa: Strategisk Betinget Tilnaerming." Internasjonal Politikk, 2, April-June 1974, 301-13.
The evolution of US policy toward Southern Africa with emphasis upon Nixon's policy.

2101 BUTCHER, GOLER T. "America's New Opportunity." Africa Report, 20, 5, September/October 1974, 17-36.

2102 ———. "The Constituency and the Challenge." Africa Report, 22, 1, January/February 1977, 2-5.

2103 ———. "Reflections on U. S. Policy towards Namibia." Issue, 4, 3, Fall 1974, 59-62.
Recommends that US policy should support self-determination for Namibia. Author is an official of the US Department of State.

2104 CHALLENOR, HERSCHELLE S. Black Africa and U. S. African Policy. Santa Barbara, Cal.: Center for the Study of Democratic Institutions, 1974.
A recording - on tape or record - of a discussion with fellows at the Center on the role of Afro-Americans in influencing US Africa policy. Concept of "Trans-Africanism" is defined.

2105 CHALLENOR, H. S. "The Influence of Black Americans on U. S. Foreign Policy toward Africa." In: Said, A. A., (ed.). Ethnicity and U. S. Foreign Policy. New York, 1978, 139-74.
"Despite an ongoing and recently increasing interest in political and economic developments on the African continent, black Americans historically have exerted little influence on United States policy toward Africa, but . . . their influence is increasing."

2106 CHESTER, EDWARD W. Clash of Titans: Africa and United States Foreign Policy. Maryknoll, N. Y.: Orbis, 1974. 316 p., bibl., index, tabl.
This history of relations between the US and Africa begins at the period of the slave trade and concludes with a summary of post-World War II events. However, the majority of the author's attention is devoted to the earlier periods.

2107 CHETTLE, J. H. "The Evolution of the United States Policy towards South Africa." Modern Age, 16, 3, Summer 1972, 259-70.
The United States has been very antagonistic toward South Africa, but under Nixon a more positive approach began.

2108 CHINWEIZU. The West and the Rest of Us. New York: Random House, 1975. 520 p.
Western imperialism and neo-colonialism, with emphasis on cultural dependence and the role of the USA.

2109 CLARK, DICK. "An Alternative U. S. Policy." Africa Report, 21, 1, January/February 1976, 16-7.

2110 COHEN, B. The Black and White Minstrel Show: Carter, Young and Africa. Nottingham: Bertrand Russell Peace Foundation, 1977. 21 p.

2111 COLA ALBERICH, J. "Periplo Africano del Vice-Presidente Humphrey." Revista de Política Internacional, 95, January/February 1968, 111-22.
In 1967/68 Humphrey traveled to Liberia, Ivory Coast, Ghana, Zaire, Zambia, Ethiopia, Somalia, and Kenya as a symbol of US interest in African affairs.

2112 "Controversy over United States Policy toward Rhodesia: Pro and Con." Congressional Digest, 52, 2, February 1973, 33-64.
U S policy is discussed and arguments on both sides of the Rhodesian chrome issue are presented.

2113 COTTER, W. R. and KARIS, T. "'We Have Nothing to Hide': Contacts between South Africa and the U. S." Social Dynamics, 3, 2, December 1977, 3-14.

2114 COX, COURTLAND. "Western Strategy in Southern Africa." In: Western Mass. Association of Concerned African Scholars, (ed.). U. S. Military Involvement in Southern Africa. Boston, 1978, 39-57.

"United States policies for Southern Africa appear to reflect the conflicting views of successive presidential administrations . . . . These differences are, however, more apparent than real. Underneath, the basic aim is to avoid fundamental reconstruction of the regional political economies."

2115 CROCKER, CHESTER A. "The African Dimension of Indian Ocean Policy." Orbis, 20, 3, Fall 1976, 637-67.

US policy has not been as effective as that of the Soviets in the Indian Ocean area. However, the Angolan Civil War is causing profound rethinking among US policy makers.

2116 _____ and LEWIS, WILLIAM H. "Missing Opportunities in Africa." Foreign Policy, 35, Summer 1979, 142-61.

Although the Carter administration considers Africa as a priority area, its policy is inadequate to its goals. Recommendations for policy changes are made.

2117 CZEMPIEL, E. O. "Die Politik der USA im Südlichen Afrika." Politik und Zeitgeschichte, 30, July 1976, 3-24.

American policy in Southern Africa has supported the wrong side for the period studied.

2118 DALE, RICHARD. "The Implications of Botswana-South African Relations for American Foreign Policy." Africa Today, 16, 1, February/March 1969, 8-12.

2119 DARLINGTON, CHARLES F. and DARLINGTON, ALICE B. African Betrayal. New York: David McKay, 1968. 359 p., ill., index, map.

An ex-US Ambassador to Gabon and his wife recount their impressions of Gabon, US and French Africa policies, and other topics. Chapters are included on AID and the Peace Corps.

2120 DAVIS, JOHN A. "Black Americans and United States Policy toward Africa." Journal of International Affairs, 23, 2, 1969, 236-49.

"Black Americans should influence the formulation and execution of US policy toward African nations."

2121 DAVIS, NATHANIEL. "The Angola Decision of 1975: A Personal Memoir." Foreign Affairs, 57, 1, Fall 1978, 109-24.

The author was US Assistant Secretary for African Affairs in 1975.

2122   DEUTSCH, RICHARD. "Carter's African Record." Africa Report, 23, 2, March/April 1978, 47-9.

2123   _____. "New Wave in Washington." Africa Report, 23, 3, May/June 1978, 39-42.

2124   DIGGS, CHARLES C., JR. "Action Manifesto." Issue, 2, 1, Spring 1972, 52-60.
        Recommendations for US Africa policy by an important black US Congressman.

2125   DINGEMAN, JAMES. "Covert Operations in Central and Southern Africa." In: Western Mass. Association of Concerned African Scholars, (ed.). U. S. Military Involvement in Southern Africa. Boston, 1978, 82-108.
        US destabilizing operations have taken place in Zaire, Ghana, and Angola.

2126   DUGGAN, WILLIAM R. A Socioeconomic Profile of South Africa. New York: Praeger, 1973. 181 p.
        The views of a retired US diplomat based on experience in S. Africa, newspaper accounts, and documents. He recommends US support for the present government in RSA as a means of inducing change.

2127   DUIGNAN, P. and GANN, LEWIS. "A Different View of United States Policy in Africa." Western Political Quarterly, 13, 4, December 1960, 918-23.
        US policy should aim at the prevention of chaos in Africa, and concentrate less on favoring the independence of unviable states and the abandonment of white settlers.

2128   EASUM, DONALD B. "United States Policy toward South Africa." Issue, 5, 3, Fall 1975, 66-72.
        A statement by an official of the US State Department.

2129   EL-KHAWAS, MOHAMED A. "American Involvement in Angola and Mozambique." In: El-Khawas, M. A. and Kornegay, F. A., Jr., (eds.). American-Southern African Relations: Bibliographic Essays. Westport, Conn., 1975, 1-31, bibl.
        Reviews the literature specifically dealing with US policy toward Portuguese Africa.

2130   _____. "American Involvement in Portuguese Africa: The Legacy of the Nixon Years." Ufahamu, 6, 1, 1975, 117-30, tabl.

2131   _____. "United States Foreign Policy toward Africa, 1960-1972." Current Bibliography on African Affairs, July 1972, 407-20.
        US policy is now becoming free of its previous dependence upon the opinions of its NATO allies.

2132　EL-KHAWAS, M. A. "U. S. Foreign Policy towards Angola and Mozambique, 1960-1974." Current Bibliography on African Affairs, 8, 3, 1975, 186-203, tabls.
　　　Asserts that US policy has shown a tendency to ignore the professed American commitment to the principle of national self-determination, in favor of promoting its economic and defense interests in both Europe and Southern Africa.

2133　_____ and COHEN, BARRY, (eds.). The Kissinger Study of Southern Africa: National Security Study Memorandum 39. Westport, Conn.: Lawrence Hill, 1976. 189 p.
　　　Contains a complete copy of NSSM 39 and analysis of the memo in the context of Nixon's and Ford's Southern Africa policies.

2134　_____ and KORNEGAY, F. A., JR., (eds.). American-Southern African Relations: Bibliographic Essays. Westport, Conn.: Greenwood, 1975. 188 p.
　　　Contains bibliographic essays by El-Khawas, Nyang, Rogers, Kornegay, and Hultman/Kramer.

2135　EMERSON, RUPERT. Africa and United States Policy. Englewood Cliffs, N. J.: Prentice-Hall, 1967. 117 p., index, map. [America's Role in World Affairs Series].
　　　An examination of the developing relations between black Africa and the US with chapters on Afro-American attitudes and various other links between the continent and the US.

2136　_____. "American Policy in Africa." Foreign Affairs, 40, 2, January 1960, 303-15.
　　　An early survey of American interests in newly-independent Africa.

2137　_____. "The Character of American Interests in Africa." In: Goldschmidt, W., (ed.). The United States and Africa. New York, 1963, 3-35.
　　　A general introduction to American thoughts and views on Africa with some advice for US foreign policy makers.

2138　_____. "Race in Africa: United States Foreign Policy." In: Shepherd, G. W., Jr., (ed.). Racial Influences on American Foreign Policy. New York, 1970, 165-85.

2139　_____ and KILSON, M. "The American Dilemma in a Changing World: The Rise of Africa and the Negro American." Daedalus, Fall 1965, 1055-84.
　　　Racial domination by whites in the USA and in South Africa is a major challenge to US policy.

2140 ENLOE, CYNTHIA. "Mercenarization." In: Western Mass. Association of Concerned African Scholars, (ed.). U. S. Military Involvement in Southern Africa. Boston, 1978, 109-30.
   Although the US public would object to sending troops to Rhodesia, many Americans are there as "khaki" and "white collar" mercenaries.

2141 FERGUSON, CLYDE and COTTER, WILLIAM R. "South Africa: What Is to Be Done." Foreign Affairs, 56, 2, January 1978, 253-74.
   Suggestions for US policy makers to follow to induce change in South African domestic politics.

2142 FERKISS, VICTOR C. "United States Policy: An American View." In: Lewis, W. H., (ed.). French-speaking Africa: The Search for Identity. New York, 1965, 194-204.
   "In perhaps no other area of the globe are American aims less direct and specific and more simply a function and exemplification of the general ends of United States foreign policy than in French-speaking Africa."

2143 _____. "U. S. Policy in Southern Africa." In: Davis, J. A. and Baker, J. K., (eds.). Southern Africa in Transition. New York, 1966, 285-304.
   "American policy toward Southern Africa is an extension of American policy toward Africa generally - an essentially negative policy . . . ."

2144 FOLTZ, WILLIAM, J. "United States Policy toward Southern Africa: Economic and Strategic Constraints." Political Science Quarterly, 92, 1, Spring 1977, 47-64.
   Why do American governments support white-dominated regimes in Africa? This article attacks the assumption that the USA has any vital economic or political interests in Southern Africa.

2145 FOSTER, BADI G. "United States Foreign Policy toward Africa: An Afro-American Perspective." Issue, 2, 2, Summer 1972, 45-51.
   Afro-American views should play a role in the formulation of US Africa policy.

2146 FROELICH, J.-C. "Les Etats-Unis et l'Afrique Noire." Revue de Défense Nationale, November 1965, 1712-28.
   The closing of the US Embassy in Brazzaville marks an important turning point in US policy and suggests certain problems with US policy.

2147 GANN, L. H. "South Africa and the U. S. Arms Embargo." South Africa International, 8, 3, January 1978, 127-36, tabl.
   The US should reverse its arms embargo and see South Africa as a valuable part of Western strategy and economics.

2148 GAPPERT, GARY. "The Emerging Political Economy of the Indian Ocean and United States Policy toward Africa." <u>Current Bibliography on African Affairs</u>, <u>4</u>, 6, November 1971, 397-418, tabl.
   South Africa's position in the Indian Ocean and recommendations for US policy.

2149 _____ and GARRY, T., (eds.). <u>The Congo, Africa and America</u>. Syracuse, N. Y.: Program of Eastern African Studies, Syracuse University, 1965. 63 p. [Occasional Paper, 15].
   See references to essays by Lefever, Schaufele, and Mazrui.

2150 GOLDSCHMIDT, WALTER, (ed.). <u>The United States and Africa</u>, 2nd. edition. New York: Praeger, 1963. 298 p., map.

2151 GOOD, ROBERT C. "Rhodesia: More of the Same with a Difference." <u>Africa Today</u>, <u>23</u>, 3, July-September 1976, 37-45.
   Discusses options for US policy in a presidential campaign year.

2152 GRAN, GUY. "Zaire 1978: The Ethical and Intellectual Bankruptcy of the World System." <u>Africa Today</u>, <u>25</u>, 4, October-December 1978, 5-24, tabls.
   Investigates why IMF, IBRD, and AID continue to bolster Mobutu's regime although aid is not getting to peasants and debts are not being repaid. Suggests it is due to West's need for cobalt.

2153 GRIFFITH, W. E. "Die Sowetjetisch-Amerikanische Konfrontation im Südlichen Afrika." <u>Europa-Archiv</u>, <u>32</u>, 2, January 1977, 31-40.
   The USA and USSR are struggling to replace the now departed colonial powers as the major influence in Africa. Rhodesia, South West Africa, and RSA are the major points of conflict.

2154 GROMYKO, ANATOLY. "Neo-Colonialism's Manoeuvers in Southern Africa." <u>International Affairs</u>, 12, December 1977, 96-102.

2155 GROSS, E. A. "The Coalescing Problem of Southern Africa." <u>Foreign Affairs</u>, <u>46</u>, 4, July 1968, 743-57.
   The US and the UK should work together to solve the problems of Rhodesia, South Africa, and Portuguese Africa.

2156 GRUNDY, K. W. "The Stanleyville Rescue: American Policy in the Congo." <u>Yale Review</u>, <u>56</u>, 2, Winter 1967, 242-55.
   African reactions to US involvement in the Stanleyville rescue was negative.

2157 GRUNDY, K. W. and FALCHI, J. P. "The United States and Socialism in Africa." Journal of Asian and African Studies, 4, 4, October 1969, 300-14.
To what extent do considerations of the nature of a country's economic system affect US aid policies to that country?

2158 GUPTA, ANIRUDHA. "The Angolan Crisis and Foreign Intervention." Foreign Affairs Reports, 25, 2, February 1976, 20-30.
The USA and the USSR intervened in Angola in their efforts to gain furthur control over world affairs.

2159 GURTOV, M. "Kennedy and Africa." In: Gurtov, M. The United States against the Third World: Antinationalism and Intervention. New York, 1974, 41-81.
"The strong impression Kennedy created of devotion to the principles of self-determination and anticolonialism is contradicted by his, and his advisors', overriding concern about Africa's strategic and economic value."

2160 HALLIDAY, F. "U. S. Policy in the Horn of Africa." Review of African Political Economy, 10, September-December 1977, 8-32.

2161 HANCE, WILLIAM A. "The Case for and against United States Disengagement from South Africa." In: Hance, W. A., (ed.). Southern Africa and the United States. New York, 1968, 105-60, tabls.
Sanctions and disengagement are analyzed. US involvement in Southern Africa is described. Author opposes use of war, sanctions, and economic disengagement in the conditions that prevailed at the time of writing.

2162 _____, (ed.). Southern Africa and the United States. New York: Columbia University, 1968. 171 p., bibl., tabls.
Contains essays by Hance and McKay that are relevant to international relations.

2163 HARSCH, ERNEST and THOMAS, TONY. Angola: The Hidden History of Washington's War. New York: Pathfinder, 1976. 157 p., bibl., ill.

2164 HEATHCOTE, N. "American Policy towards the United Nations Operation in the Congo." Australian Outlook, 18, 1, April 1964, 77-97.
American policy on the Congo question was an attempt to win friends among African governments.

2165 HELMREICH, W. B., (comp.). Afro-Americans and Africa: Black Nationalism at the Crossroads. Westport, Conn.: Greenwood, 1977. 74 p. bibl. [African Bibliographic Center, Special Bibliography Series, New Series, 3].

2166 HERSKOVITS, JEAN, (ed.). "Subsaharan Africa." In: Schlesinger, A. M., Jr., (ed.). The Dynamics of World Power: A Documentary History of United States Foreign Policy 1945-1973, Vol 5. New York, 1973, 539-1231.
     This contains an extensive collection of speeches, newspaper editorials, and other items on US relations with the continent and specific states, including sections on economic policy, Southern Africa, Nigeria, and Zaire.

2167 HILL, ADELAIDE C. and KILSON, MARTIN. Apropos of Africa: Sentiments of Negro American Leaders on Africa from the 1800s to the 1950s. London: Frank Cass, 1969. 390 p.
     This collection of letters, articles, and other statements by black Americans, including Charles C. Diggs on American-African relations, is an important reference for the study of the role of black Americans in the formulation of US Africa policy.

2168 HOTTELET, RICHARD C., (ed.). "The United Nations." In: Schlesinger, A. M., Jr., (ed.). The Dynamics of World Power: A Documentary History of United States Foreign Policy 1945-1973, Vol. 5. New York, 1973, 3-538.
     This extensive collection of speeches and other materials contains sections on the Congo mission, South Africa, sanctions against Rhodesia, and South West Africa.

2169 HOUSER, GEORGE M. "U. S. Policy and Southern Africa." In: Arkhurst, F. S., (ed.). U. S. Policy towards Africa. New York, 1975, 88-130.
     The author is Director of the American Committee on Africa and a founder of the Congress of Racial Equality. Here he argues that US policy in Southern Africa is aimed at maintaining the status quo.

2170 HOWE, RUSSELL W. Along the Africa Shore: An Historical Review of Two Centuries of U.S.- African Relations. New York: Barnes and Noble, 1975. 197 p., ill., index.
     This volume is not "a diplomatic history in the formal sense. [The author has] exercised the newsman's prerogative to select and emphasize . . . . This was made easier by the fact that, for much of the period under review, there was no coherent policy as such to trace."

2171 HUMPHREY, HUBERT H. "Africa's Challenge to U. S. Policy." Africa Report, 21, 4, July/August 1976, 45-7.

2172 IDENBURG, P. J. "Amerika's Beleid voor Afrika." International Spectator, 14, 9, May 1960, 237-49.
     "American Policy for Africa." Until 1956 US policy was dependent upon the views of its NATO allies, but it is now taking a more independent course.

2173 IJERE, M. O. "Economics and African Nationalism." Civilisations, 22, 4, 1972, 547-62.
Among other topics, the author considers the contributions of black Americans to African economic independence.

2174 ISAACMAN, ALLEN and DAVIS, JENNIFER. "United States Policy toward Mozambique Since 1945: 'The Defense of Colonialism and Regional Stability'." Africa Today, 25, 1, January-March 1978, 29-55.
Since WW II the US had actively supported repressive Portuguese regimes which placed US in direct opposition to the aspirations of the Mozambican people. Examines subtle changes in US policy which are still part of a larger American strategy to contain radical change in Southern Africa.

2175 JOHNSON, WALTON R. "Afro-America and Southern Africa: A Reassessment of the Past." Africa Today, 19, 3, Summer 1972, 5-12.
Black American involvement in Southern Africa "has heretofore had little practical consequence in terms of change."

2176 JOHNSON, WILLARD R. et al. "United States Foreign Policy towards Africa." Africa Today, 20, 1, Winter 1973, 15-44, tabls.
Documents US complicity in the continuation of white racism in Southern Africa and argues for changes in US policy.

2177 KANZA, THOMAS R. "African-American Relations: An African Perspective." Africa Report, 21, 4, July/August 1976, 37-9.

2178 KAPLAN, L. S. "The United States, Belgium, and the Congo Crisis of 1960." Review of Politics, 29, 2, April 1967, 239-56.
US Congo policy is indicative of the conflict between US anticolonialism and the US struggle against the USSR.

2179 KEMP, GEOFFREY. "U. S. Strategic Interests and Military Options in Sub-Saharan Africa." In: Whitaker, J. S., (ed.). Africa and the United States: Vital Interests. New York, 1978, 120-52, map.
Argues that a new strategic map of the world has emerged and the importance of the Southern seas is now greater.

2180 KENNAN, GEORGE F. "Hazardous Courses in Southern Africa." Foreign Affairs, 49, 2, January 1971, 218-36. Also in: Roland, Joan G., (ed.). Africa: The Heritage and the Challenge. Greenwich, Conn., 1974, 437-55.
A recommendation from one of the most famous of students of US foreign policy that the United States should favor the status quo in South Africa.

2181 KENNAN, G. F. "What Policies toward Southern Africa?" Current, 127, March 1971, 50-61.

2182 KHAMA, SERETSE. "African-American Relations in the 1970s: Prospects and Problems." Africa Today, 18, 3, July 1971, 25-34.
　　　The author is President of Botswana.

2183 KISSINGER, HENRY A. "An American Perspective." Africa Report, 21, 5, September/October 1976, 17-20.

2184 KITCHEN, HELEN, (ed.). "Options for U. S. Policy toward Africa." AEI Foreign Policy and Defense Review, 1, 1, 1979, 1-76.
　　　Presents six options for US policy, each elaborated by statements from diverse viewpoints by such people as Julius Nyerere, Cyrus Vance, and Immanuel Wallerstein.

2185 KORNEGAY, FRANCIS A., JR. "Africa and Presidential Politics." Africa Report, 21, 4, July/August 1976, 7-20.

2186 _____. "Black America and U. S. - Southern African Relatons: An Essay - Bibliographical Survey of Developments During the 1950s, 1960s, and Early 1970s." In: El-Khawas, M. A. and Kornegay, F. A., Jr., (eds.). American-Southern African Relations: Bibliographic Essays. Westport, Conn., 1975, 138-78, bibl.
　　　Surveys literature pertaining to black Americans and Southern Africa, and also concerning black American-African relations in general and the question of an American constituency for Africa.

2187 _____. "Conclusion: American-Southern African Relations at the Crossroads." In: El-Khawas, M. A. and Kornegay, F. A., Jr., (eds.). American-Southern African Relations: Bibliographic Essays. Westport, Conn., 1975, 179-88.
　　　Futuristic appraisal of recent trends in these relations, as viewed in the literature.

2188 _____. "Kissinger and Africa." Africa Report, 20, 6, November/December 1975, 37-40.

2189 KRAMER, JACK. "Our French Connection in Africa." Foreign Policy, 29, Winter 1977/1978, 160-6.
　　　France is taking an increasingly large role in Africa and many believe that France is doing for the USA what the USA does not dare or wish to do for itself. If true, there are dangers for the US in such a relationship.

2190   KUNERT, DIRK. "Carter, the Tradition of American Foreign Policy, and Africa." South Africa International, 8, 2, October 1977, 65-78 and 99-105.
        Carter's emphasis on human rights is naive and dangerous. The Soviets see Africa as a chance to reverse the balance of power and the human rights motif cripples US ability to counteract them.

2191   LAKE, ANTHONY. The "Tar-Baby" Option: American Policy toward Southern Rhodesia. New York: Columbia University for the Carnegie Endowment for International Peace, 1976. 316 p.
        "This account of American policy and the bureaucratic and congressional politics behind it includes . . . information that was not previously on the public record." The author was (and is) a participant in the US foreign policy decision-making process.

2192   LANDIS, ELIZABETH S. "American Responsibilities towards Namibia: Law and Policy." Africa Today, 18, October 1971, 38-48.

2193   LEFEVER, ERNEST W. "The Return of Tshombe and America's Congo Policy, 1964." In: Gappert, G. and Thomas, G., (eds.). The Congo, Africa and America. Syracuse, 1965, 1-12.
        "By and large, American policy toward the Congo has been wise and reasonably effective."

2194   _____. "United States Policy, the United Nations and the Congo." Orbis, 11, 2, Summer 1967, 394-413.
        The UN Operation in the Congo served US interests, although such interests might have been served by better means.

2195   LEMARCHAND, RENE, (ed.). American Foreign Policy in Southern Africa: The Stakes and the Stance. Washington: University Press of America, 1978. 450 p.

2196   _____. "The CIA in Africa: How Central? How Intelligent?" Journal of Modern African Studies, 14, 3, September 1976, 401-26. Also in: Revue Française d'Etudes Politiques Africaines, 136, April 1977, 73-101.
        The author raises several questions about the role of the CIA in African politics and development; questions that are not usually factored in to research designs.

2197   LEMELLE, TILDEN J. "Race, International Relations, United States Foreign Policy, and the African Liberation Struggle." Journal of Black Studies, 3, 1, September 1972, 95-109.
        White-dominated racist systems are an important aspect of international relations.

2198 LEMELLE, TILDEN J. "Whither U. S. African Policy? An Editorial." Africa Today, 25, 2, April-June 1978, 5-6.
    In spite of Carter's "new" policy toward Africa, the Administration reacted with the old duplicity and contradictions to Zaire and Cubans in Angola, seeing internal African conflicts as reflective of the US/USSR conflict.

2199 LOCKWOOD, E. "The Future of the Carter Policy toward Southern Africa." Issue, 7, 4, 1977, 11-5.

2200 LOFCHIE, MICHAEL F. "The British-Rhodesian Agreement and United States Policy." Africa Today, 19, 1, Winter 1972, 44-50.
    Scores the decision by President Nixon to lift the American ban on the importation of Rhodesian chrome and feels this action may indicate a major shift in US policy toward that country.

2201 LOGAN, R. W. "Assessment of Current American Influence in Africa." Annals of the American Academy of Political and Social Sciences, 366, July 1966, 99-107.

2202 MACEBUH, STANLEY. "Misreading Opportunities in Africa." Foreign Policy, 35, Summer 1979, 162-9.
    An editor of the Nigerian Daily Times discusses US policy for Africa.

2203 MARCUM, JOHN A. "Lessons of Angola." Foreign Affairs, 54, 3, April 1976, 407-25.
    A catalog of errors in US policy for Southern Africa.

2204 _____. Politics of Indifference: Portugal and Africa, a Case Study in American Foreign Policy. Syracuse, N. Y.: Program of Eastern African Studies, Syracuse University, 1972. 41 p.
    A lecture, including questions from the audience and answers, presented in 1972.

2205 _____. "Southern Africa and United States Policy: A Consideration of Alternatives." In: Shepherd, G. W., Jr., (ed.). Racial Influences on American Foreign Policy. New York, 1970, 186-219.

2206 _____. "The United States and Portuguese Africa: A Perspective on American Foreign Policy." Africa Today, 18, 4, October 1971, 23-37.

2207 McHENRY, DONALD F. "Captive of No Group." Foreign Policy, 15, Summer 1974, 142-9.
    The State Department has tried without success to develop American blacks as a constituency for US Africa policy.

2208 McKAY, VERNON. "The African Operations of United States Government Agencies." In: Goldschmidt, W., (ed.). The United States and Africa. New York, 1963, 273-95, tabls.
After a brief description of the Africa activities of several branches of the US government, McKay concludes that there is a "vast problem of organization and coordination in the [US] Africa policy machine."

2209 _____. "Southern Africa and Its Implications for American Policy." In: Hance, W. A., (ed.). Southern Africa and the United States. New York, 1968, 1-32, tabls.
US policy must be consonant with its beliefs on government by consent and non-discrimination on the basis of race.

2210 METER, KARL MICHAËL VAN. "Atlantique Sud et Afrique Australe: De Kissinger à Carter." Revue Française d'Etudes Politiques Africaines, 144, December 1977, 59-75.
SATO, a South Atlantic Treaty Organization, is a concept close to Kissinger's heart, but little progress has been made in bringing it into existence.

2211 MILLER, JAKE C. The Black Presence in American Foreign Affairs. Washington, D. C.: University Press of America, 1978. 305 p.
"Analyzes the participation and influence of Black Americans in the foreign policymaking process . . . . Interwoven throughout the book are discussions of foreign policy issues of concern to Black Americans. Primary consideration is given to issues related to Africa and the Third World."

2212 MORRIS, M. D. "Black Americans and the Foreign Policy Process: The Case of Africa." Western Political Quarterly, 25, 3, September 1972, 451-63.
Have black Americans been influential in the making of US African policy? In the past, no, but this is changing today.

2213 MORROW, JOHN H. First American Ambassador to Guinea. New Brunswick, N. J.: Rutgers University, 1968. 291 p., ill., index.
The memoirs of the first US Ambassador, an Afro-American, to Sekou Toure's Guinea. A chapter on Soviet relations with Guinea is included.

2214 MUCHNIK, N. "L'Aide de l'OTAN au Portugal." Temps Modernes, 305, December 1971, 807-31.
For strategic reasons and because of the mineral resources in the Portuguese colonies the NATO states provided secret assistance to Portugal during the liberation wars in those colonies.

2215 NAGORSKI, ANDREW. "U. S. Options vis-à-vis South Africa."
In: Whitaker, J. S., (ed.). Africa and the United States:
Vital Interests. New York, 1978, 187-211.
"This study will make the case that South Africa is
important to the West, but for different reasons than those
presented by the South Africans themselves."

2216 NEILSON, WALDEMAR A. African Battleline: American Policy
Choices in Southern Africa. New York: Harper and Row for
the Council on Foreign Relations, 1965. 155 p., index, map.
The author sees Southern Africa as a major crisis for
US foreign policy, not only US policy for Africa but US
policy for all of the Third World. He describes the situation
and comments on US policy for each territory of the region.

2217 "Ein Neuer Versuch zur Lösung des Rhodesian-Problems. Die
Afrika-Initiative des Amerikanischen Regierung." Europa-
Archiv, 31, 23, 10 December 1976, 619-33.

2218 NEWSOM, DAVID D. "American Interests in Africa and African
Development Needs." Issue, 2, 1, Spring 1972, 44-8.
The Assistant Secretary of State for African Affairs
in the US Government here stresses that economic development
is the major goal of African states, and the US position in
Africa depends on the role the US is willing to play in
that development.

2219 _____. "Southern Africa: Constant Themes in U. S. Policy."
Department of State Bulletin, 67, 1726, July 1972, 119-25.
Although articles in this publication have not been
fully cited in this bibliography, the Department of State
Bulletin is a good source for statements of official US policy.

2220 NYAMEKO, R. S. "U. S. Labour Conspiracy in Africa." African
Communist, 72, 1978, 51-62.
"The African-American Labour Centre now operating in
many African countries is promoting the aims of neo-colonialism
and hampering the fight against international capitalism and
imperialism."

2221 _____. "U. S. Subversion of South African Labour Movement." African Communist, 73, 1978, 74-88.
"The second installment of an exposure of attempts
backed by the CIA and the US State Department to win the
support of the African trade union movement for imperialism."

2222 NYANG, SULAYMAN. "Short Bibliographical Essay on U. S.
Policy toward Southern Rhodesia (Zimbabwe)." In: El-Khawas,
M. A. and Kornegay, F. A., Jr., (eds.). American-Southern
African Relations. Westport, Conn., 1975, 32-46.
Surveys the literature in an effort to show that posi-
tions taken by the US do have serious implications.

2223   NYERERE, JULIUS K. "America and Southern Africa." Foreign
       Affairs, 55, 4, July 1977, 671-84.
            The US could use its economic power to bring about
       rapid political change in Southern Africa. Concrete
       examples are provided.

2224   OBICHERE, BONIFACE I. "American Diplomacy in Africa: Pro-
       blems and Prospects." Pan-African Journal, 7, 1, Spring
       1974, 67-80, tabl.
            A review of America's Africa policies since 1960 with a
       call for a "new African policy . . . based on the needs and
       aspirations of the Africans, as well as on the interests of
       the USA." Author is Nigerian.

2225   OFUATEY-KODJOE, W. B. "Conflicting Political Interests of
       Africa and the United States." In: Arkhurst, F. S., (ed.).
       U.S. Policy towards Africa. New York, 1975, 198-226.
            A summary of US-African relations in recent years.

2226   OKOLO, JULIUS EMEKA. [Book Review] Journal of Modern
       African Studies, 15, 1, March 1977, 142-4.
            Review of US Policy toward Africa edited by Frederick
       S. Arkhurst, and US Neocolonialism in Africa by Stewart
       Smith.

2227   OUDES, BRUCE. "Dialog: Roger's Visit to Africa." Africa
       Report, 15, 4, April 1970, 22-4.

2228   _____. "Evolving American Views of South Africa." In:
       Bissell, R. E. and Crocker, C. A., (eds.). South Africa
       into the 1980s. Boulder, Col., 1979, 159-86.

2229   _____. "New Agenda for Africa Policy." Africa Report,
       20, 5, September/October 1974, 54-6.

2230   _____. "Southern Africa Policy Watershed." Africa
       Report, 19, 6, November/December 1974, 46-9.

2231   _____. "Viewpoint: The Lion of Judah and the Lambs of
       Washington." Africa Report, 16, 5, May 1971, 21-3.

2232   PAHAD, ESSOP. "Washington's 'New African Policy'." World
       Marxist Review, 19, 11, November 1976, 29-35.
            Commentary on the "New African policy" of the US.

2233   PAYNE, RICHARD J. "The Soviet/Cuban Factor in the New United
       States Policy toward Southern Africa." Africa Today, 25, 2,
       April-June 1978, 7-26.
            The major contention is that the shift in US policy is
       a direct response to political realities in Southern Africa
       which have been highlighted by Cuban and Soviet intervention.

2234 PIMENOV, I. and TEPLOV, L. "Imperialist Plans for the South Atlantic." International Affairs, 7, July 1977, 92-100.

2235 POLAN, DIANE. Irony in Chrome: The Byrd Amendment Two Years Later. Washington: Carnegie Endowment for Peace, 1973. 37 p., bibl.

2236 POMEROY, WILLIAM J. Apartheid Axis: The United States and South Africa. New York: International, 1971. 95 p.
  The USA is seen as a major supporter of apartheid as a means of maintaining low wages and high profits for US investors in South Africa.

2237 PRADOS, JOHN. "Sealanes, Western Strategy and Southern Africa." In: Western Mass. Association of Concerned African Scholars, (ed.). U. S. Military Involvement in Southern Africa. Boston, 1978, 58-81.
  The US continues to complain about the Soviet presence in the Indian Ocean, yet the US is the predominant military power in that Ocean.

2238 "Präsident Carter zur Amerikanischen Afrika-Politik. Rede in Lagos (Nigeria) am 1. April 1978." Europa-Archiv, 33, 9, 10 May 1978, D279-83.

2239 PRICE, ROBERT M. U. S. Foreign Policy in Sub-Saharan Africa: National Interest and Global Strategy. Berkeley: Institute of International Studies, University of California, 1979. 69 p. [Policy Papers in International Affairs, No. 8].

2240 PRINSLOO, DAAN. United States Foreign Policy and the Republic of South Africa. Pretoria: Foreign Affairs Association, 1978. 141 p., bibl.
  A South African view of US-South African relations. The major emphasis is on the Carter administration's policies.

2241 QUIGG, PHILIP W. "The Changing American View of Africa." Africa Report, 14, 1, January 1969, 1-10.

2242 RANDOLPH, R. SEAN. "The Byrd Amendment: A Postmortem." World Affairs, 141, 1, Summer 1978, 57-70.

2243 RIVKIN, ARNOLD. "Lost Goals in Africa." Foreign Affairs, 44, 1, October 1965, 111-26.
  The US policy for Africa has lost its credibility and Africa has lost its importance to US policy makers. This must be changed. Aid for African goals rather than the containment of communism must guide US policy.

2244 RIVKIN, A. "Principal Elements of United States Policy towards Underdeveloped Countries." International Affairs, 37, 4, October 1961, 452-64.
For Africa there should be a single policy for all of the free world states.

2245 ROBERTS, K. M. "Sanctions in Southern Africa: United States Policy Dilemma." Genève-Afrique, 9, 1, 1970, 67-87.
The US must support the extension of UN sanctions against Rhodesia to the case of South Africa. Support of the status quo in South Africa is damaging to the US in other parts of the world.

2246 ROBINSON, ALMA. "Africa and Afro-America." Africa Report, 20, 5, September/October 1974, 7-10.

2247 ROSHCHIN, O. "The Economic Underpinning of U. S. Expansion in Africa." International Affairs, 4, April 1978, 43-9.

2248 RUBIO GARCIA, L. "Estados Unidos y Africa. Algunas Premisas e Implicaciones." Revista de Política Internacional, 75, September/October 1964, 123-40.
"The United States and Africa: Some Premises and Implications." A review of the US policies for Africa from 1956 through the Kennedy administration.

2249 RULLI, G. "Negoziati e Guerriglia per la Rhodesia." Civiltà Cattolica, 3037, January 1977, 82-97.
Soviet progress in Southern Africa has caused the US to become interested in this area. Various efforts sponsored by the US and the UK to settle the Rhodesian problem are discussed.

2250 RUSTIN, BAYARD and GERSHMAN, CARL. "Africa, Soviet Imperialism and the Retreat of American Power." Commentary, 64, 4, October 1977, 33-43.
US policy for Africa must be reevaluated. Carter's downplay of the Soviet threat is a severe mistake.

2251 SATTERTHWAITE, J. C. "Our Role in the Quickening Pace towards Independence in Africa." Annals of the American Academy of Political and Social Sciences, 330, July 1960, 37-49.
The author concentrated on the potential role of the US as an aid donor in the African economic development process.

2252 SCHAUFELE, WILLIAM E. "United States Congo Policy and Africa." In: Gappert, G. and Thomas, G., (eds.). The Congo, Africa and America. Syracuse, 1965, 37-45.
The author served with the US Embassy in Leopoldville.

2253 SCHAUFELE, W. E. "US Relations in Southern Africa." Annals of the American Academy of Political and Social Science, 432, July 1977, 110-9.
   An administration statement of US policy for Southern Africa.

2254 SCHLESINGER, ARTHUR M., JR. A Thousand Days: John F. Kennedy in the White House. Boston: Houghton Mifflin, 1965. 1087 p.
   This memoir by a close associate of President Kennedy deals largely with foreign affairs. See Chapter 21,"Africa, the New Adventure," pp. 551-85.

2255 SCHUTZ, BARRY M. "Issues in U. S. Policy toward Africa." Africa Today, 25, 3, July-September 1978, 67-72.
   Review of Africa: From Mystery to Maze, by Helen Kitchen.

2256 SEILER, JOHN. [Book Review] Journal of Modern African Studies, 15, 1, March 1977, 144-7.
   Reviews The Kissinger Study of Southern Africa: National Security Study Memorandum 39 (Secret), edited by Mohamed A. El-Khawas and Barry Cohen, and The 'Tar Baby' Option: American Policy toward Southern Rhodesia, by Anthony Lake.

2257 _____. U. S. Foreign Policy toward Southern Africa: Continuity and Change. Johannesburg: South African Institute of International Affairs, 1973. 15 p., bibl.
   The author is an American scholar.

2258 SHEPHERD, GEORGE W., JR. "The Conflict of Interest in American Policy on Africa." Western Political Quarterly, 12, 4, December 1959, 996-1004.
   Humanitarian, economic, and strategic interests are conflicting in US policy for Africa.

2259 _____. "From Counter-revolution to Majority Rule: Can U. S. Policy Change?" Africa Today, 23, 3, July-September, 1976, 5-16.
   In response to the success of the revolutions in the former Portuguese colonies, and the pro-African constituency of many US non-governmental organizations, Kissinger is attempting to change the image (but not the substance) of US policy toward South Africa.

2260 _____. "The Racial Dimension of United States Intervention in Africa and Asia." In: Shepherd, G. W., Jr., (ed.). Racial Influences on American Foreign Policy. New York, 1970, 220-8.

2261 SHEPHERD, G. W., JR., (ed.). Racial Influence on American Foreign Policy. New York: Basic Books, 1970. 238 p., index.
See references to essays by Shepherd on racism in African foreign relations, Emerson on US Africa policy, Marcum on the US and Southern Africa, and Shepherd's concluding essay on race in US foreign policy.

2262 _____. The United States and Non-aligned Africa. Denver: Graduate School of International Studies, University of Denver, 1969. 76 p., bibl.
US foreign policy and relations with Angola.

2263 SKLAR, RICHARD L. "Dialog: The United States and Biafra." Africa Report, 14, 7, November 1969, 22-3.

2264 SKURNIK, W. A. E. "Recent U. S. Policy in Africa." Current History, 64, 379, March 1973, 97-101+.

2265 _____. "The United States Role in Africa." Current History, 60, 355, March 1971, 129-35+.

2266 SMITH, ROBERT A. The American Foreign Policy in Liberia: 1822-1971. Monrovia: Providence , 1972. 93 p., bibl.

2267 SMITH, ROBERT S. "The Nature of American Interests in Africa." Issue, 2, 2, Summer 1972, 37-44.
The author was US Deputy Assistant Secretary of State for African Affairs.

2268 SMITH, STEWART. United States Neocolonialism in Africa. New York: International Publishers, 1974. 270 p., bibl.
Stuart J. Seborer is another name used by this author. An analysis by an economist of the strategy and tactics of US imperialism in Africa, with some comparison to Soviet policy.

2269 SOUTH AFRICA FOUNDATION. South Africa and United States Policy. Johannesburg: S. A. F., 1966. 27 p.
This pamphlet contains brief notes by Denis Worrall; W. S. Yeowart and Jean Le May; and Anthony B. Davenport and H. Goldberg. US foreign policy, economic sanctions, and American investment in South Africa are the major topics.

2270 SPIERS, RONALD I. "U. S. National Security Policy and the Indian Ocean Area." Department of State Bulletin, 65, 1678, August 1971, 199-203.
The recent build-up of the Soviet naval presence in the Indian Ocean is causing the US to reconsider its traditional policy for that region.

2271 SPIRO, HERBERT J. "The American Response to Africa's Participation in the International System." Issue, 2, 2, Summer 1972, 20-3.
US Africa policy in its global context.

2272 ———. "U. S. Policy: An Official View." In: Arkhurst, F. S., (ed.). U. S. Policy towards Africa. New York, 1975, 56-71.
The author, a long-time student of African affairs, was US Ambassador to the United Republic of Cameroon at the time this was written.

2273 STOCKWELL, JOHN. In Search of Enemies: A CIA Story. New York: Norton, 1978. 285 p., figs., ill., index.
An expose of the CIA role in Angola. The author was an active participant.

2275 TALBOT, STEPHEN. "United States Intervention in Southern Africa: The New Era." Socialist Revolution, 7, 4, July/August 1977, 7-28.
Washington's policy in Southern Africa is predicated upon the need to maintain influence there. This now means fostering the emergence of pro-west black majority rule.

2276 TURKATENKO, N. D. "Rascety i Proscety Vasingtona v Jusnoj Afrike." SSA, 2, February 1977, 24-35.
"Washington's Designs and Miscalculations in Southern Africa." US policy is based on saving Pretoria by sacrificing Rhodesia in order to maintain US imperialist interests.

2277 ULLMAN, RICHARD H. "Human Rights and Economic Power: The United States versus Idi Amin." Foreign Affairs, 56, 3, April 1978, 529-43.
Argument for economic sanctions against the Amin regime.

2278 VENKATARAMANI, M. S. "The Ford-Kissinger Safari in Angola: Ramifications of American Policy." Foreign Affairs Reports, 25, 9/10, September/October 1976, 131-78.
Kissinger deliberately encouraged Soviet-Cuban intervention in Africa in order to make the Soviets appear to be interventionists and to frighten the African governments.

2279 VILLIERS, GAS DE. The African-American Manifesto: A Guideline to Double Standards. Pretoria: Foreign Affairs Association, 1977.
A South African view of American foreign policy.

2280   WALLERSTEIN, IMMANUEL. "Africa, the United States, and the
       World Economy: The Historical Bases of American Policy."
       In: Arkhurst, F. S., (ed.). U. S. Policy towards Africa.
       New York, 1975, 11-37, tabls.
          "If one wants to understand the bases of American for-
       eign policy toward Africa, one has to place both America and
       Africa in their historical relationship to the evolving
       world economy."

2281   WALTERS, RONALD E. "Apartheid and the Atom: The United
       States and South Africa's Military Potential." Africa Today.
       23, 3, July-September 1976, 25-35.
          US trade in nuclear materials with South Africa is
       based on (1) competition for the South African market and
       (2) support of apartheid through concurrent policies of "selec-
       tive relaxation" of economic restriction but "public" oppo-
       sition to racial repression.

2282   WALTERS, RONALD W. "The Global Context of United States
       Policy toward Southern Africa." Africa Today, 19, 3, Summer
       1972, 13-30.
          What are the interests of the global powers in South
       Africa and what is the global significance of bilateral US-
       South African relations?

2283   _____. "U. S. Policy and Nuclear Proliferation in S.
       Africa." In: Western Mass, Association of Concerned Afri-
       can Scholars, (ed.). U. S. Military Involvement in Southern
       Africa. Boston, 1978, 172-97.
          The US has played a major role in the transfer of
       nuclear technology.

2284   WASHINGTON TASK FORCE FOR AFRICA. Impact: U. S. Consti-
       tuency for Africa. Washington, D. C.: African Bibliographic
       Center, 1973. 32 p. [Current Reading List, 10, 4].
          Contains brief essays on the Africa pressure groups in
       the USA, American relations with Portuguese and Southern
       Africa, and other bibliographic material.

2285   WASHINGTON TASK FORCE ON AFRICAN AFFAIRS. Congress and
       Africa. Washington, D. C.: African Bibliographic Center,
       1973. 37 p., tabl. [Current Reading List, 10, 1].
          Contains several brief essays on congressional voting
       patterns and resources for the study of the US Congress and
       Africa.

2286   WEIL, M. "Can the Blacks Do for Africa What the Jews Did
       for Israel?" Foreign Policy, 15, Summer 1974, 109-30.
          If black Americans are to exert pressure on behalf of
       Africa, they must gain electoral power, organize, and relate
       to American ideals.

2287 WEISSMAN, STEPHEN R. *American Foreign Policy in the Congo, 1960-1964*. Ithaca: Cornell University, 1974. 325 p., bibl, index.
"I have sought to determine precisely what US policies were . . . and on what intellectual assumptions they were based." Research is based on a large number of interviews and an extensive survey of documentary and published materials.

2288 _____. "Controlling Our Secret Service." *Africa Today*, 25, 3, July-September 1978, 51-8.
Review of *In Search of Enemies* by John Stockwell about CIA's covert action program in Angola 1975-76.

2289 WESTERN MASSACHUSETTS ASSOCIATION OF CONCERNED AFRICAN SCHOLARS, (ed.). *U. S. Military Involvement in Southern Africa*. Boston: South End, 1978. 276 p., bibl., index.
Major headings are "The Crisis in Southern Africa," "Western Strategy in Southern Africa," and "The U. S. Contribution to the South African Military Build-up." See references to essays by C. Cox, J. Prados, J. Dingeman, C. Enloe, S. Gervasi, M. Klare/E. Prokosch, R. W. Walters, N. Magleta/A. Seidman, and R. Sylvester.

2290 WHITAKER, JENNIFER S., (ed.). *Africa and the United States: Vital Interests*. New York: New York University for the Council on Foreign Relations, 1978. 255 p., maps, tabls.
Contains seven essays dealing with US economic, political, and strategic interests. One essay analyzes Soviet strategic interests.

2291 _____. "Introduction: Africa and U. S. Interests." *In*: Whitaker, J. S., (ed.). *Africa and the United States: Vital Interests*. New York, 1978, 1-20.

2292 _____. "U. S. Policy toward Africa." *In*: Whitaker, J. S., (ed.). *Africa and the United States: Vital Interests*. New York, 1978, 212-44.
"The reiteration here of the need to minimize the military competition in Africa and to pursue political and economic realities rooted in African realities springs, then, from a calculation of real US options."

2293 WILKINS, R. "What Africa Means to Blacks." *Foreign Policy*, 15, Summer 1974, 130-42.
American blacks are only now taking a significant interest in African affairs.

2294 WILLIAMS, FRANKLIN H. "Towards an Africa Policy." *Africa Report*, 21, 4, July/August 1976, 2-6.

2295 WILLIAMS, G. MENNEN. Africa for the Africans. Grand Rapids, Mich.: Eerdmans, 1969. 218 p., index.
See Part 3, "US Policy" with chapters on policy, the Peace Corps, and trade/investment. The author was Assistant Secretary of State for African Affairs in the Kennedy administration.

2296 _____. "Diplomatic Rapport between Africa and the United States." Annals of the American Academy of Political and Social Sciences, 354, July 1964, 54-64.
Ideological similarities between Africans and Americans are only one of several factors leading to US-African rapport. Author was responsible for US Africa policy.

2297 WISEBERG, LAURIE S. and NELSON, GARY F. "Africa's New Island Republics and U. S. Foreign Policy." Africa Today, 24, 1, January-March 1977, 7-30, map.
These 4 new nations (Cape Verde, Sao Tomé and Principe, Comoro Islands, and Seychelles) desperately need aid; it will be unfortunate if the West denies aid because of nationalization, socialism, or strategic bargaining.

2298 WITHERELL, JULIAN W., (comp.). The United States and Africa: Guide to U. S. Official Documents and Government-Sponsored Publications on Africa, 1785-1975. Washington: Library of Congress, 1978. 949 p., index.
This bibliography is organized by time periods, and sub-categorized by country. The very adequate index provides cross-referencing. There are 8,827 items included.

2299 YARBOROUGH, WILLIAM P. Trial in Africa: The Failure of U. S. Policy. Washington, D. C.: Heritage Foundation, 1976. 86 p.
General Yarborough argues that US policy makers do not understand Africa as well as their Soviet counterparts. ECOWAS and the EAC deserve US support and American military forces could play important roles in engineering, medicine, and other activities to ensure stability.

Other States:
Relations with Africa

2300 'ABD AL-RAHMAN, 'AWATIF. Isrā'īl wa-Ifrīqiyā, 1948-1973. Beirut: Palestine Liberation Organization , Research Center, 1974. 139 p. [Silsilat Kutub Filastiniyah, 57].
An historical analysis of relations between Israel and Africa with emphasis on Israeli policy and activity.

2301 'ABD AL-RĀZIG, HUSAYN. "Isrā'īl 'Tutrad' Siyāsīyan min Ifrīqiyā." Al-Kātib, 153, December 1973, 60-72.
Description of the failure of Israel's policy for Africa in 1973.

2302 ABU-LUGHOD, IBRAHIM. "Africa and the Islamic World." In: Paden, J. and Soja, E., (eds.). The African Experience, Vol. 1. Evanston, Ill., 1970, 545-67, maps.
An historical survey of Afro-Arab relations in a textbook designed as an introduction to African studies.

2303 ADAIR, DENNIS G. and ROSENSTOCK, JANET. Canada and Southern Africa. Don Mills, Ontario: Fulcrum, 1973. 38 p.
A statement on present and appropriate relations between Canada/Canadians and South Africa.

2304 ADAM, SAMUEL B. "Israel: Africa's Friend or Foe?" African Communist, 3rd. quarter, 1970, 35-54.

2305 "Erste Afrikanisch-Arabische Gipfelkonferenz in Kairo. Politische Erklärung der Konferenz der Staats-und Regierungschefs der Organisation der Afrikanischen Einheit und der Arabischen Liga." Europa-Archiv, 32, 9, 10 May 1977, 101-201.

2306 "The Afro-Arab Summit." Africa Institute Bulletin, 15, 3/4, 1977, 56-64, ill., maps.

2307 AJAMI, FOUAD and SOURS, MARTIN H. "Israel and Sub-Saharan Africa: A Study in Interaction." African Studies Review, 13, 3, December 1970, 405-13.
Several variables influence African relations with Israel and the Arab states, relations described as "ambivalent." Trade data, diplomatic exchange data, and UN voting patterns are used to describe this ambivalence.

2308 AKINSANYA, A. "On Lagos Decision to Break Diplomatic Relations with Israel." International Problems, 17, 1, Spring 1978, 65-79, bibl.

2309 'ALĪ, 'ALĪ SA'D. "Al-Badīl al-'Arabī lil-Wujūd al-Isrā'īlī fī Ifrīqiyā." Al-Ishtirākī, 5, January 1974, 21-8.
Israel's initial successes in Africa were followed by failure. The Arab states should now take steps to replace the Israelis in Africa.

2310 ALI, S. S. "Continuité des Relations Indo-Africaines." Afrique Contemporaine, 96, March/April 1978, 1-4.

2311 ALUKO, OLAJIDE. "Israel and Nigeria: Continuity and Change in Their Relationship." African Review, 4, 1, 1974, 43-59.
"Elements of change and continuity . . . will be the theme of this article."

2312 ANGLIN, D. G. "Canada and Apartheid." International Journal, 15, 2, Spring 1960, 122-37.
The Canadian people oppose apartheid but government policy fails to reflect this.

2313 _____. "Canada and Southern Africa in the Seventies." Canadian Journal of African Studies, 4, 2, Spring 1970, 261-76.
Lists arguments for and against Canadian involvement in the liberation struggle in South Africa. Favors bolder action by Canada in support of that struggle.

2314 _____. "Towards a Canadian Policy on Africa." International Journal, 15, 4, Autumn 1960, 290-310.
Canadian foreign policy and structures were not adequate to the emergence of African independence. The author pleas for the development of an adequate and coherent policy.

2315 _____; SHAW, T.; and WIDSTRAND, C., (eds.). Canada, Scandinavia and Southern Africa. Uppsala: Scandinavian Institute of African Studies, 1978. 190 p., tabls.
Contains thirteen essays by scholars and politicians. Many of the essays originate from a political economy view point.

2316   ARNOLD, HANS. "Afro-German Cultural Co-operation." Aussenpolitik, 27, 1, 1976, 76-83. In the German edition: "Kulturelle Zusammenarbeit mit Afrika." 27, 1, 1976, 72-9.
        The author is head of the Department of Cultural Affairs in the Foreign Office, Bonn.

2317   ASFAHANY, NABYA. "Afro-Arab Cooperation: Political and Financial Development." Spettatore Internazionale, 12, 1, January-March 1977, 3-54.
        Background on and analysis of current economic and political relations between African and Arab states.

2318   ASTAKHOV, S. "Alliance between Tel Aviv and Pretoria." International Affairs, 8, August 1977, 62-6.

2319   AL-'ATĪYAH, GHASSĀN. "Isrā'īl wa-Ūghandā." Jāmi'at Baghdād. Markaz al-Dirāsāt al-Filastīnīyah. Majallat Markaz al-Dirāsāt al-Filastīnīyah, 2, September 1973, 9-48.
        Israel and Uganda.

2320   AVRIEL, EHUD. "Israel's Beginnings in Africa." In: Curtis, M. and Gitelson, S. A., (eds.). Israel in the Third World. New Brunswick, N. J., 1976, 69-74.
        The author was once Israel's ambassador to Ghana.

2321   'AWDAH, 'ABD AL-MALIK. "Al-Harb wa-al-Tadāmun al-Ifrīqī." Al-Siyāsah al-Duwalīyah, 35, January 1974, 147-52.
        Post-1973 Israeli relations with Africa.

2322   AYNOR, H. S. Notes from Africa. New York: Praeger, 1969. 163 p.
        The author served as an Israeli diplomat in Zaire, Gambia, and Senegal. He uses this novel to describe the problems of diplomacy in Africa.

2323   BASS, ROBERT and BASS, ELIZABETH. "Eastern Europe." In: Brzezinski, Z., (ed.). Africa and the Communist World. Stanford, Cal., 1963, 84-115, tabl.
        Tha authors argue that although the East European states have been active in Africa, "they have not acted simply as Moscow's pawns."

2324   BECK, C. F. "Czechoslovakia's Penetration of Africa, 1955-1962." World Politics, 15, 3, April 1963, 403-16.
        Describes Czechoslovakia's African activities as being tied to Soviet policy, but also as a means for the Czechoslovak government to gain prestige.

2325   BERNER, W. " Kubaner Interventionen in Afrika und Arabien."
       Aussenpolitik, 27, 3, 1976, 325-31.
             The Cubans intervened in Guinea in 1966 and the intervention in Angola was preceeded by several other instances. The relationship between such interventions and Soviet policy is evident.

2326   BIELENSTEIN, DIETER, (ed.). Perspectives in Afro-German Relations. Bonn: Friedrich-Ebert-Stiftung, 1975. 241 p., bibl.
             Report on a conference held in 1975 on relations between Africa and West Germany.

2327   BISSELL, RICHARD E. "Africa and the Nations of the Middle East." Current History, 71, 421, November 1976, 158-60.

2328   BLOUIN, GEORGES. "Canadian Policy toward Southern Africa: The Decision-Making Process." In: Anglin, D.; Shaw T.; and Widstrand, C., (eds.). Canada, Scandinavia and Southern Africa. Uppsala, 1978, 59-63.
             The author is Assistant Under-Secretary, Department of External Affairs, Canada.

2329   BOURGI, A. "Afrique Noire-Monde Arabe: De la Solidarité Politique à la Coopération Institutionnelle." Revue Française d'Etudes Politiques Africaines, 132, December 1976, 22-34.
             An analysis of the increasing strength of the Afro-Arab relationship.

2330   BOUVIER, PAULE. "L'Afrique à l'Heure Algérienne: Les Chances du Rapprochement Arabo-Africain." Studia Diplomatica, 28, 3, 1975, 305-25.

2331   BRASSINE, JACQUES. "Les Relations entre la Belgique et le Congo de 1968 à 1970." Etudes Africaines de CRISP, June 1971, 1-67.

2332   "Brazil's View of Africa." West Africa, 2893, 20 November 1972, 1556-7.
             Review of Brazilian press statements on Brazil's relations with Africa.

2333   BRECHER, M. "Israel and 'Afro-Asia'." International Journal, 16, 2, Spring 1961, 107-37.
             Israel desires to gain Afro-Asian neutrality on the Middle East conflict. Technical aid is an important tool in her policy implementation.

2334 BREYTENBACH, W. J. "Black Africa and the Arabs." Bulletin of the Africa Institute of South Africa, 12, 1974, 432-5.
Author argues that Afro-Arab relations are disintegrating because of the Arab's failure to provide sufficient economic aid.

2335 BRIONNE, B. "Le Déchirement de l'Afrique Portugaise." Défense Nationale, July 1974, 63-83.
Portugal's possible foreign policies for its ex-colonies in Africa.

2336 BROWN, ALEX; BUNTING, PETER; and SANGER, CLYDE. Southern Africa: Some Questions for Canadians. Ottawa: Canadian Council for International Cooperation, 1973. 43 p.
Canadian policy for Southern Africa and suggestions for changes in that policy.

2337 BRUNEAU, THOMAS C. "Out of Africa and into Europe: Towards an Analysis of Portuguese Foreign Policy." International Journal, 32, 3, Spring 1977, 288-314.

2338 BUTENSCHØN, PETER. "Humanitarian Intervention and Humanitarian Foreign Policy." Internasjonal Politikk, 3, 1969, 377-81.

2339 CERVENKA, ZDENEK. "Scandinavia: A Friend Indeed for Africa?" Africa Report, 20, 3, May/June 1974, 39-43.

2340 _____ and ROGERS, BARBARA. The Nuclear Axis: Nuclear Collaboration between West Germany and South Africa. London: Julian Friedmann, 1978. 288 p., index, maps.
West German and US support to South Africa have been important to the development of that state's nuclear capabilities.

2341 CHAZAN, NAOMI. "Israel and Africa - Dynamics of Relationships in the Seventies." Kidma, 1, 2, 1973, 7-12.
Written in an era of good relations between Israel and Africa.

2342 CHHABRA, H. S. "The Competition of Israel and African States." India Quarterly, 31, 4, October-December 1975, 362-70.
Israel's position in African attitudes has varied from aid giving friend to imperialist aggressor. Today the view is mixed.

2343 CHIBWE, E. C. Afro-Arab Relations in the New World Order. London: Julian Friedmann, 1978. 256 p., index.
The author is a Zambian diplomat and former Director of the Bank of Zambia.

2344 CHIBWE, E. C. Arab Dollars for Africa. London: Croom Helm, 1976. 147 p.
Afro-Arab relations and the desire of African states for Arab investments are the major topics. There is also some material on the EEC and Africa.

2345 COLA ALBERICH, JULIO. "España y el Sahara Occidental. Antecendentes de una Descolonization." Revista de Política Internacional, 154, November/December 1977, 9-52.
Analysis of Spain's policy for the Western Sahara.

2346 COLITTI, M. "Rapport fra l'Italia e i Paesi Africani." Affari Esteri, 23, July 1974, 93-109.
"Relations between Italy and the African Countries." Italy intends to cooperate with African states in agricultural and petroleum development.

2347 CONSTANTIN, F. and COULON, C. "Le Développement des Relations entre l'Afrique Noire et le Monde Arabe en 1972." Année Africaine 1972, 1972, 281-306.
Analysis of common interests, points of conflict, and diplomatic relations between the Arab and African states.

2348 CORDERO TORRES, JOSE MARIA. "Los Tratados entre España y Guinea Ecuatorial." Revista de Política Internacional, 119, January/February 1972, 331-5.

2349 CROCKER, CHESTER A. "Comment: Making Africa Safe for the Cubans." Foreign Policy, 31, Summer 1978, 31-3.

2350 CURTIS, MICHAEL and GITELSON, SUSAN A. Israel in the Third World. New Brunswick, N. J.: Transaction Books, Rutgers University, 1976. 410 p., index, tabls.
In addition to general essays specific writings on Israel and Africa are included. See references under Kreinin, Avriel, Decalo, Alpan, Gitelson, and Kochan/Gitelson/Dubek.

2351 DECALO, SAMUEL. "Israel and Africa - A Selected Bibliography." Journal of Modern African Studies, 5, 3, November 1967, 385-99.

2352 _____. "Israeli Foreign Policy and the Third World." Orbis, 11, 3, Fall 1967, 724-45.
Between 1957 and 1962 Israeli Third World activities concentrated on Africa and after that on Latin America in an attempt to gain support for the Israeli position in the Middle East conflict.

2353 DECRAENE, PHILLIPE. "Africa and the Mideast Crisis: Is the Romance with Israel Over?" Africa Report, 18, May/June 1973, 20-4.
A statement of African views on the causes of the cooling of relations betweeen Africa and Israel.

2354 DOUTRELOUX, A. "Mythe et Réalité du Colonialisme." Genéve-Afrique, 4, 1, 1965, 7-23.
Zaire and Belgium as a case study of relations between ex-colonial and ex-colonized states.

2355 DURCH, WILLIAM J. "The Cuban Military in Africa and the Middle East: From Algeria to Angola." Studies in Comparative Communism, 11, 1/2, Spring/Summer 1978, 34-74, figs.
Cuban activity in Angola was neither on orders from the Kremlin nor a unique departure from Cuban foreign policy.

2356 DZIDZIENYO, ANANI. "Brazil's View of Africa." West Africa, 2892, November 1972, p. 1521 and 2893, November 1972, 1556-7.

2357 FABIUS. "Biafra, Israel, Deutschland." Schweizer Monatshefte, 49, 10, January 1970, 985-8.

2358 FALLĀHAH, MAHMŪD. "Isrā'īl wa Tshād: Nahw Siyāsah 'Arabīyah Jadīdah fī Ifrīqīyah." Shu'ūn Filastīnīyah, 18, February 1973, 114-24.
The example of relations between Chad and Israel indicates means whereby Arab states might improve their relations with African states.

2359 FARGUES, G. "Le Portugal Face aux Nationalismes Africains." Afrique et Asie, 65, 1964, 13-25.
International relations and Portugal's attempts to hold on to its African colonies.

2360 FEDERAL REPUBLIC OF GERMANY. Fact v. Fiction: Rebuttal of the Charges of Alleged Co-operation between the Federal Republic of Germany and South Africa in the Nuclear and Military Fields. Bonn: Press and Information Office, 1978. 52 p.

2361 FERREIRA, E. de Sousa. "Portugal and Her Former African Colonies: Prospects for a Neo-Colonial Relationship." Ufahamu, 5, 3, 1975, 159-70, tabl.

2362 FESSARD de FOUCAULT, B. "La Question du Sahara Espagnol." Revue Française d'Etudes Politiques Africaines, 119, November 1975, 74-106 and 120, December 1975, 71-105.
Detailed background on the local and international aspects of Spain's attempt to dispose of its Saharan colony.

2363 FISCHLOWITZ, E. "Subsidios para a 'Doutrina Africana' de Brasil." Revista Brasileira de Política Internacional, 3, 9, March 1960, 82-95.
   Brazil's Africa policy; political and economic interests of Brazil in Africa.

2364 FRANCK, CHRISTIAN. "La Politique Extérieure de la Belgique en 1977." Res Publica, 20, 2, 1978, 357-65.
   Although Belgium's foreign policy is tied to that of the European Community, Africa remains of special importance.

2365 FREEMAN, LINDA. "Canada and the Frontline States." In: Anglin, D.; Shaw, T.; and Widstrand, C., (eds.). Canada, Scandinavia and Southern Africa. Uppsala, 1978, 69-84.
   "This study is an examination of Canadian interests in four of these states - two 'frontier' countries, Tanzania and Zambia, and two states inside the region, Malawi and Botswana - from the early 1960s to the present."

2366 GERSDORFF, R. VON. "Africa, Colonialismo e Assistência Econômica." Revista Brasileira de Política Internacional, 4, 13, March 1961, 92-120.
   Conditions in Africa are appropriate for Soviet penetration, but economic aid and the Eurafrique concept oppose this. Brazil is suited to play an important role in the harmonization of Portuguese interests in her territories.

2367 GINIEWSKI, PAUL. "Israel-Ouganda: Les Conséquences de la Rupture." Rivista di Studi Politici Internazionali, 29, 3, July-September 1972, 439-42.

2368 GITELSON, SUSAN A. "Israel's African Setback in Perspective." In: Curtis, M. and Gitelson, S. A., (eds.). Israel in the Third World. New Brunswick, N. J., 1976, 182-99, tabl. Also in: Israel's African Setback in Perspective. Jerusalem: Hebrew University, 1974. 27 p. [Jerusalem Papers on Peace Problems, 6].
   To understand Israel's setback in Africa in the 1970s we must distinguish between bilateral relations and the multilateral level of international organizations.

2369 GLASGOW, ROY. "Pragmatism and Idealism in Brazilian Foreign Policy in Southern Africa." Munger Africana Library Notes, 4, 23, 1974.

2370 GOLDWORTHY, D. "Australia and Africa: New Relationships?" Australian Quarterly, 45, 4, December 1973, 58-72.
   Traditionally, Australia has been closer to the white governments of Africa but the Whitlam regime moved toward black Africa.

2371 GONZALEZ, EDWARD. "Complexities of Cuban Foreign Policy."
Problems of Communism, 26, 6, November/December 1977, 1-15.
Some explanation of Cuba's active policy in Africa can be found in analysis of elite coalitions in Cuban politics.

2372 GOOD, KENNETH. "The Intimacy of Australia and South Africa."
African Review, 2, 3, 1972, 417-32.
Australia and South Africa have increasingly close relations. Racism and anti-communism are major ties.

2373 GORI, U. "Italy's Attitude to African Problems at the United Nations with Particular Reference to Decolonization." Africa Quarterly, 9, 4, January-March 1970, 374-83.
Italy as an anti-colonial state.

2374 GOUMOIS, MICHEL DE. "Le Canada et la Francophonie." Etudes Internationales, 5, 2, June 1974, 355-66.
Brief descriptions of the various aspects of relations between Canada and other francophone states, with a section on Africa. This is in a special issue, "La Coopération Internationale entre Pays Francophones."

2375 GRIFFITH, WILLIAM E. "Yugoslavia." In; Brzezinski, Z., (ed.). Africa and the Communist World. Stanford, Cal., 1963, 116-41, tabls.
Yugoslavia's Africa policy is described as one aspect of Tito's non-alignment policy.

2376 GUPTA, ANIRUDHA. "A Note on Indian Attitudes to Africa."
African Affairs, 69, 275, April 1970, 170-8.
India's gesture of friendship and solidarity with Africans remains only symbolic, and Indian leaders still labor under stereotypes about Africa which are unhelpful for an objective and proper understanding of the continent.

2377 GUREWITZ, A. "Emidat ha'-Aravim Klafei ha-Israeli le-Africa."
International Problems, 11, 1/2, July 1972, 31-4.
"Arab Attitudes towards Israeli Assistance to Africa."
A brief but general note on Arab-Israeli competition in Africa.

2378 HARNETTY, P. "Canada, South Africa and the Commonwealth, 1960-1961." Journal of Commonwealth Political Studies, 2, 1, November 1963, 33-44.
Canada supported African opposition to South Africa's Commonwealth membership.

2379 HEATHCOTE, N. "Ireland and the United Nations Operations in the Congo." International Relations, 3, 11, May 1971, 880-927.
The effects within Ireland of her participation in the UN Operation in the Congo.

2380 HIGGOTT, R. A. "Rhetoric and Reality: Australia's African Relations under Labor." Journal of Commonwealth and Comparative Politics, 14, 2, July 1976, 1158-76, tabls.
Labour Government policy for Africa (1972-75) is examined. In spite of much rhetoric, economic factors were more important than ideological.

2381 IQBAL, MEHRUNNISA H. "Pakistan's Relations with Africa and Latin America." Pakistan Horizon, 27, 2, 1974, 57-61.

2382 "Israel's Africa Wounds." Bulletin of the Africa Institute of South Africa, 11, 1973, 363-73.

2383 JAMIESON, DONALD C. "Canada's Attitude toward Southern Africa." In: Anglin, D.; Shaw, T; and Widstrand, C., (eds.). Canada, Scandinavia and Southern Africa. Uppsala, 1978, 153-8.
The author is Secretary of State for External Affairs, Canada.

2384 JAWWĀD, SA'ĪD. "Siyāsat Isrā'īl fī Ifrīqiyā wa-Hisāduhā fī Harb Tishrīn." Shu'ūn Filastīnīyah, 28, December 1973, 155-62.
Israel's policy and goals in Africa.

2385 JOSHUA, W. "Belgium's Role in the United Nations Peacekeeping Operations in the Congo." Orbis, 11, 2, Summer 1967, 414-38.
Analysis of Belgium's various policies for the Congo and the UN Operations in the Congo.

2386 KANAFĀNĪ, MUHAMMAD NU'MĀN. "Isrā'īl wa-Ūghandah." Shu'ūn Filastīnīyah, 18, February 1973, 101-13.
Israeli aid to Uganda indicates the dominance of political factors in Israeli aid policy.

2387 KEREKES, TIBOR, (ed.). The Arab Middle East and Muslim Africa. New York: Praeger, 1961. 126 p., index.
Although each of the seven essays in this issue is relevant in some degree to the study of sub-Saharan Africa, only that of William H. Lewis is directly relevant.

2388 KHAN, R. "India and the Decolonization of Africa." Africa Quarterly, 8, 3, October-December 1968, 238-46.
Although not alone in the struggle, India has played an important role in the process of decolonization.

2389 KIM, CHONGHAN. "Korea's Diplomacy toward Africa." Orbis, 11, 3, Fall 1967, 885-96.
Competition with North Korea and a desire for African support at the UN are major motivations for South Korean policy.

2390 KRASILNIKOV, A. "West Germany-South Africa: Nuclear Alliance." International Affairs, 8, August 1969, p. 111.

2391 KREININ, MORDECHAI E. "Israel and Africa: The Early Years." In; Curtis, M. and Gitelson, S. A., (eds.). Israel in the Third World. New Brunswick, N. J., 1976, 54-68.
A survey of "Israel's technical assistance to Africa during the late 1950s and early 1960s, the years in which the program was formed and took on its special character."

2392 LAUFER, LEOPOLD. Israel and the Developing Countries: New Approaches to Cooperation. New York: Twentieth Century Fund, 1967. 298 p., figs., index, tabls.
"The present work examines in depth Israel's experiences in international technical cooperation and also reviews briefly similar programs undertaken by other countries."

2393 _____. "Israel and the Third World." Political Science Quarterly, 87, 4, December 1972, 615-30.
Aid, trade, and political relations are examined. The author sees the Third World as playing at least a limited role in resolution of the Middle East conflict.

2394 LEVESQUE, JACQUES. "La Guerre d'Angola et le Rôle de Cuba en Afrique." Etudes Internationales, 9, 3, 1979, 429-37.

2395 LEVIN, N. "Israel in Africa." Reconstructionist, 39, March 1973, 11-8.

2396 LE VINE, VICTOR T. and LUKE, TIMOTHY W. The Arab-African Connection: Political and Economic Realities. Boulder, Col.: Westview, 1979. 130 p., index.
Analysis of the dynamics of Afro-Arab relations.

2397 LIBA, M. "Perek Hadash be-Yehasei Israel-Afrika." International Problems, 13, 4, September 1974, 20-7.
Israel had positive relations with most African states until the 1973 war, after which most African countries broke relations. Those states are rethinking their position as Arab aid promises have not been kept.

2398 LINHARES LEITE, M. Y. "Brazilian Foreign Policy and Africa." World Today, 18, 12, December 1962, 532-40.
There are various tendencies in Brazil with respect to Africa. Africa is seen as a potential market and there is a romantic trend in thoughts of cooperation with Portuguese-speaking states.

2399 LLOYD, W. B., JR. "Solidarity and Autonomy: Africa and the Swiss Example." Genève-Afrique, 5, 2, 1966, 179-88.

2400 LORCH, N. "Gormei Yessod Bi-Yahassei Yisrael-Afrika." Hamizrah Hebadash, 12, 1/2, 1962, 1-18.
"Basic Factors of Israeli-African Relations." Various aspects of Israel's history and economic development give a sense of similarity with the new states of Africa, many of which are turning to Israel for aid in spite of the opposition of the USSR and Egypt.

2401 _____. "Israel and Africa." World Today, 19, 8, August 1963, 358-68.
A general survey of various aspects, economic and political, of Israel's relations with Africa.

2402 LOTTEM, E. "Yehassi Israel-Afrika be-r'ai ha-'itonut ha-Israelit." International Problems, 14, 3/4, Fall 1975, 25-44.
The Israeli press has made Israel's relations with Africa more difficult.

2403 LÖWIS of MENAR, H. VON. "Das Engagement der DDR im Portugiesischen Afrika." Deutschland Archiv, 10, 1, January 1977, 32-42.

2404 MARIÑAS OTERO, LUIS. "La Conferencia Ministerial Afro-Arabe de Dakar." Revista de Política Internacional, 148, November/December 1976, 121-9.
A review of the positive development of Afro-Arab relations since 1973 with emphasis on the 1976 ministerial conference.

2405 _____. "La Cumbre Afro-Arabe de El Cairo." Revista de Política Internacional, 153, September/October 1977, 15-30.
A review of the discussions and agreements reached at the March 1977 Afro-Arab summit meeting in Cairo.

2406 MARSH, WILLIAM W. "East Germany and Africa." Africa Report, 14, 3/4, March/April 1969, 59-62.

2407 MATTHEWS, R. O. "Africa in Canadian Affairs." International Journal, 26, 1, Winter 1970/1971, 120-50.
The general aspects of Canadian policy are discussed and then considered in respect of several cases. The Commonwealth and francophonie are important influences on that policy.

2408 _____. "L'Afrique Noire dans la Politique Etrangère du Canada." Etudes Internationales, 1, 4, December 1970, 59-72.
A critical appraisal of the context and effects of ten years of Canadian-African relations.

2409 MATTHEWS, R. O. "Canada's Relations with Africa." International Journal, 30, 3, Summer 1975, 536-68.
   After describing the history of Canada's Africa policies since 1945, the author concludes that such policies vary constantly, are complex, and that at times there are actually several policies in effect.

2410 _____ and PRATT, CRANFORD. "Canadian Policy toward Southern Africa." In: Anglin, D.; Shaw, T.; and Widstrand, C., (eds.). Canada, Scandinavia and Southern Africa. Uppsala, 1978, 164-80.
   "What we are seeking is an explanation of the timid policies followed by Canada."

2411 MAURY, MICHAEL. "Taiwan et l'Afrique Noire." Revue de Défense Nationale, 25, July 1969, 1213-9.

2412 MAZRUI, ALI A. "Afro-Arab Relations and the Role of the Gulf States of Eastern Arabia." Pan-Africanist, 6, June 1975, 30-2.

2413 MEDZINI, MERAN. "Israel and Africa: What Went Wrong? Midstream: A Monthly Jewish Review, December 1972, 25-34.
   What were the underlying causes of Israel's diplomatic setbacks in Africa during 1972?

2414 MENDES VIANA, A. "O Mundo Afro-Asiático: Sua Significaça para o Brasil." Revista Brasileira de Política Internacional, 2, 8, December 1959, 5-23.
   Although Brazil has not pursued its relations with the Afro-Asian states aggressively, there are large areas of agreement between Brazil and these states, as is shown by an examination of UN activities. Issues concerning Portuguese colonialism are a major exception.

2415 MEULEN, J. VAN DER. "Israel's Relations with Africa." Kroniek van Afrika, 2, 1974, 166-84.

2416 MIQUEZ, ALBERTO. "Le Sahara Occidental et la Politique Maghrébine de l'Espagne." Politique Etrangère, 43, 2, 1978, 173-80.
   Spain's policy on the Western Sahara is a major factor leading to the continued unsettled conditions there.

2417 MILENKY, EDWARD S. "Lateinamerika und die Dritte Welt." Europa-Archiv, 32, 14, July 1977, 441-52.
   "Latin America and the Third World." There is some discussion of links betweeen African and Latin American states.

2418 MILLER, J. C. "African-Israeli Relations: Impact on Continental Unity." Middle East Journal, 29, 4, Autumn 1975, 393-408.
A survey of African-Israeli relations, 1957-1975.

2419 MONETA, C. J. "Argentina y Africa en el Contexto de los Países No-alineados." Revista de Derecho Internacional y Ciencias Diplomáticas, 41/42, 1972, 124-54.
"Argentina and Black Africa in the Context of the Non-aligned Countries." Mainly a consideration of the possibilities of expanding trade between Africa and Argentina.

2420 MORENO, A. "Sahara Español: Una Descolonización Controvertida." Revista de Política Internacional, 149, 1977, 147-73.
A Spanish view of the problems faced by Spain in trying to end its colonial rule in Spanish Sahara.

2421 MOUVEMENT ANTI-APARTHEID DE GENEVE. L'Afrique Sud et Nous. Neuchâtel: A La Baconnière, 1971. 203 p., ill., map, tabls.
The first two sections are devoted to analysis of apartheid. The third section contains foreign policy suggestions for opposition to apartheid.

2422 MUHLEMANN, CHRISTOPHER. "Kuba, die Sowjetunion und Afrika." Schweizer Monatshefte, 56, 2, May 1976, 91-4.

2423 MWINYI, ABOUD JUMBE. "India and Tanzania." India Quarterly, 29, 1, January-March 1973, 1-8.

2424 NABIL, MOSTAFA. "Future of Arab-African Cooperation: Interview with Arab League Secretary-General." Africa Newsletter, 15 May 1974, 8-15.
The interview is mainly concerned with the problem of meeting Africa's petroleum needs.

2425 NAHUMI, M. "New Directions in Israeli-African Relations." New Outlook; Middle East Monthly, 16, September 1973, 14-24.

2426 NARAYANAN, R. "The Role of Cuba in Africa." Foreign Affairs Reports, 27, 5, May 1978, 80-93.

2427 NICOLAS, GUY. "L'Expansion de l'Influence Arabe en Afrique Sudsaharienne." L'Afrique et l'Asie Modernes, 117, 1978, 23-46.

2428 ODED, A. "Yehasei Afrika 'im Medinot Arav ve-Israel le' or Haknus ha-Kol Afrikanei be-Kampala." International Problems, 15, 1/2, Spring 1976, 8-15.
"Afro-Arab Relations and Israel in the Light of the OAU Conference in Kampala." There is still a split between the Arab and African states but Israel is not yet able to reestablish its position in Africa. Furthur settlement of the Middle East crisis must occur first.

2429 OUDES, B. J. "Portogallo e Afrika: Il Momento della Trattiva." Affari Esteri, 23, July 1974, 56-69.
   What will be the effects on African international relations, especially in Southern Africa, of the revolution in Portugal?

2430 PALMBERG, MAI. "Present Imperialist Policies in Southern Africa: The Case for Scandinavian Disassociation." In: Anglin, D.; Shaw, T.; and Widstrand, C., (eds.). Canada, Scandinavia and Southern Africa. Uppsala, 1978, 124-52.
   Argues that US policy is in favor of the maintenance of the status quo and therefore that Scandinavia should not be associated with US policy.

2431 PARK, R. L. "Indian-African Relations." Asian Survey, 5, 7, July 1965, 350-8.
   A review of India's policies for and relations with African states. Trade, aid, technical assistance, and the position of Indian citizens and descendants living in Africa are included.

2432 PARK, SANG-SEEK. "Africa and the Two Koreas: A Study of African Non-alignment." African Studies Review, 21, 1, April 1978, 73-88.

2433 PRAAG, NICHOLAS VON. "European Political Cooperation and Southern Africa." Spettatore Internazionale, 12, 1, January-March 1977, 67-81.
   Europe, i. e. the European Community, has not developed a common policy for Southern Africa after the revolution in Portugal. Consensus on such a policy is necessary.

2434 PREISWERK, A. ROY. "Neokolonialismus oder Selbstkoloniierung? Die Kulterbegegnug in den Europäisch-Afrikanischen Beziehungen." Europa-Archiv, 28, 4, December 1973, 845-53.

2435 AL-QAR'Ī, AHMAD YŪSUF. "Al-Jadīd fi al-'illāqāt al-'Arabīyah al-Ifriqīyah." Al-Siyāsah al-Duwalīyah, 41, July 1975, 166-71.
   Description of increasing Arab political, economic, and social relations with Africa.

2436 RANCIC, DRAGOSLAV. "Return of the Germans." Atlas, 14, 1, July 1967, 50-1.
   This note on German activities in Africa is reprinted from the March 7, 1967 issue of Borba, a Yugoslav journal.

2437 RIPKEN, PETER. "West German Options in Southern Africa." Journal of Southern African Affairs, 7, 4, October 1978, 133-52.

2438   RODRIGUES, JOSE HONORIO. Brazil and Africa. Los Angeles: University of California, 1965. 382 p., index, map, tabl.
       Most of the volume is devoted to relations between Brazil and Africa from 1500 to 1960, but there are several chapters directly relevant to current relations.

2439   _____. "La Política Internacional del Brasil y Africa." Foro Internacional, 4, 3, January-March 1964, 313-46.
       Policies for Africa of recent Brazilian administrations are compared. The problems of close relations with colonialist Portugal and with independent African states have been significant for Brazil.

2440   _____. "O Presente e o Futuro das Relaceõs Africano Brazileiras." Revista Brasileira de Política Internacional, 5, 18, June 1962, 263-84 and 5, 19, September 1962, 501-16.
       The first part of the essay is a description of relations between African states and between Africa and Europe. The second part examines Brazil's policies over recent administrations. Until recently, Brazil's policy has generally been aligned with that of the colonial powers.

2441   RONFELDT, DAVID F. "Superclients and Superpowers: Cuba-Soviet Union/Iran-United States." Conflict, 1, 4, 1979, 273-302.
       Provides some explanation (and prediction) of Cuba's role in African affairs.

2442   ROPP, KLAUS VON DER. "German Attitudes to South Africa." South Africa International, 2, 4, April 1972, 218-27.
       A major factor in overall German policies for Africa is the division of Germany into two competing parts. German attitudes on South Africa are, accordingly, different in each case.

2443   RUBINSTEIN, ALVIN Z. Yugoslavia and the Nonaligned World. Princeton: Princeton University, 1970. 353 p., bibl. index.
       Although largely of general interest there are chapters on relations with Ethiopia and Yugoslav Congo policy.

2444   _____. "Yugoslavia and Africa." Africa Report, 15, 8, November 1970, 14-7.

2445   _____. "Yugoslavia's Nonaligned Role in Africa." Africa Report, 15, 8, November 1970, 14-7, ill, map, tabls.
       Tito offers aid, trade, and political support without strings, as part of his efforts to woo Africa and the Third World for political and economic reasons.

2446   RUBIO GARCIA, L. "Africa Negra y Estados Arabes: Petroléo y Mundo Pobre." Revista de Política Internacional, 145, May/June 1976, 127-54.
After 1973 most African states turned from close relations with Israel to improved relations with the Arab states. The Africans needed Arab oil but they were also reacting against Israel's ties with South Africa.

2447   RÜHMLAND, U. "Sowjetzonale Aktivität in den Staaten Afrikas." Aussenpolitik, 11, 3, March 1960, 182-9.
"East German Activities in the African States." A West German view of East German policies for and activities in Africa. East German goals are international recognition and the discrediting of West Germany.

2448   SABOURIN, LOUIS. "Quebec and Africa: Language and Politics." Africa Report, 15, 4, April 1970, 16-7.

2449   SALPETER, E. "Israel and Africa - A Reappraisal." The American Zionist, 63, March/April 1973, 9-12.

2450   SANGER, CLYDE. "Canada and Africa: Aid and Politics." Africa Report, 15, 4, April 1970, 12-5.

2451   SANKARI, FAROUK A. "The Cost and Gains of Israeli Influence in Africa." Africa Quarterly, 14, 1/2, 1974, 5-19, tabls.
"The purpose of this essay is to demonstrate the dynamics and workings of Israeli aid to Africa with reference to three areas of Israel's foreign policy objectives: (1) containment of Arab influence; (2) strengthening economic ties; and (3) gaining political support against the Arabs within the United Nations and the Organization of African Unity."

2452   SANNESS, JOHN. "Nigeria, Biafra and Norway." Internasjonal Politikk, 3, 1979, 306-9.

2453   SARAJCIC, I. "Yugoslavia and Africa." Africa Quarterly, 10, 4, January-March 1971, 375-81.
A positive assessment of relations between Yugoslavia and the states of Africa.

2454   SCHENCK, D. VON. "Das Problem der Beteiligung der Bundesrepublik Deutschland an Sanktionen der Vereinten Nationen besonders im Falle Rhodesians." Zeitschift für Ausländisches Öffentliches Recht und Völkerrecht, 29, 2, May 1969, 257-315.
"The Problem of the Participation of the Federal Republic of Germany in United Nations Sanctions with Special Regard for the Rhodesian Case." Although Germany is not a member of the UN and is not required to obey the sanctions decision, the German government has agreed to institute sanctions.

2455 SCHLEGEL, JOHN P. *The Deceptive Ash: Bilingualism and Canadian Policy in Africa: 1957-1971.* Washington: University Press of America, 1978. 463 p., bibl., figs.
   This study includes chapters on relations with Nigeria, Ghana, Tanzania, francophone Africa, and Southern Africa. The author presents "the genesis and roots of Ottawa's policy in Sub-Saharan Africa."

2456 SELCHER, W. A. *The Afro-Asian Dimension of Brazilian Foreign Policy, 1956-1972.* Gainesville: University of Florida, 1974. 252 p., bibl., tabls. [Latin American Monographs, Second Series, 13].
   Bibliography contains many references not referenced here.

2457 _____. "Brazilian Relations with Portuguese Africa in the Context of the Elusive 'Luso-Brazilian Community'." *Journal of Interamerican Studies and World Affairs*, 18,1, February 1976, 25-58.
   Brazil is developing closer economic and political ties with black Africa, but the continuation of relations with South Africa poses problems for Brazil.

2458 SENGHOR, LEOPOLD A. "Africa, the Middle East and South Africa." *Africa Report*, 20, September/October 1975, 18-20.
   Senegal, the Middle East and South Africa.

2459 SHA'LĀN, HUSAYN. "Mawqi' Ifrīqīya min al-Sirā al-'Arabī al-Isrā'īlī ba'd Uktūbar: Da'wah lil-Hiwār." *Al-Talīah*, 12, December 1973, 59-63.
   Arab and Israeli relations with Africa and the significance of the 1973 war in the Middle East.

2460 SHAW, TIMOTHY M. "Scandinavia and Canadian Policy Issues." In: Anglin, D.; Shaw, T.; and Widstrand, C., (eds.). *Canada, Scandinavia and Southern Africa.* Uppsala, 1978, 181-90.

2461 SHIMONI, YAACOV. "Israel, the Arabs and Africa." *Africa Report*, 21, 4, July/August 1976, 55-9.

2462 SIMONET, HENRI. "La Politique Etrangère de la Belgique: Hasard ou Nécessité?" *Studia Diplomatica*, 30, 5, 1977, 463-82.
   African relations in Belgian foreign policy.

2463 SKINNER, E. P. "African States and Israel: Uneasy Relations in a World of Crises." *Journal of African Studies*, 2, 1, Spring 1975, 1-23.
   Claims that the break in relations between Israel and some African states at the time of the 1973 Mid-East war resulted from long-standing illusions on the part of both sides.

2464   SOUKUP, J. R. "Japanese-African Relations: Problems and
        Prospects." Asian Survey, 5, 7, July 1965, 333-40.
             Japan could play an important role as a bridge between
        the western democracies and the African states, but first
        Japan must settle several problems with and inconsistencies
        in her Africa policies. Trade imbalances, and relations
        with South Africa are the most important of these.

2465   STEINBACH, UDO. "Arabische Politik rund um das Horn von
        Afrika." Aussenpolitik, 28, 3, 1977, 300-11.
             "Arab Policy around the Horn of Africa." The Arab
        states, especially the conservatives under Saudi Arabia's
        leadership, wish to isolate the Horn from Great Power in-
        fluence.

2466   STERPELLONE, A. "Il Primo 'Vertice' Afro-Arabo." Affari
        Esteri, 34, April 1977, 301-7.
             A review of the discussions and agreements reached at
        the March 1977 Afro-Arab summit meeting in Cairo.

2467   STEVENS, RICHARD P. and ELMESSIRI, ABDELWAHAB M. Israel and
        South Africa: The Progression of a Relationship, rev. ed.
        New Brunswick, N. J.: North American, 1977. 228 p., index.
             Foreign relations of Israel and South Africa and the
        Jewish population of South Africa. A large collection of
        readings, largely newspaper articles, is included and there
        is a chapter on the UN role.

2468   STOKKE, OLAV. "Foreign Policy Effects of Governmental
        Humanitarian Intervention in Nigeria-Biafra." Internasjonal
        Politikk, 3, 1969, 415-35.

2469   _____. "Norsk Politikk Overfor det Sørlige Afrika."
        Internasjonal Politikk, 3, July-September 1978, 381-426.
             "The Norwegian Policy towards Southern Africa."

2470   STOLTENBERG, THORVALD. "Nordic Opportunities and Responsi-
        bilities in Southern Africa." In: Anglin, D.; Shaw, T.;
        and Widstrand, C., (eds.). Canada, Scandinavia and Southern
        Africa. Uppsala, 1978, 105-10.
             The author is the Under-Secretary of State, Royal Mini-
        stry of Foreign Affairs, Oslo.

2471   THOMPSON, RICHARD. Retreat from Apartheid: New Zealand's
        Sporting Contacts with South Africa. Wellington, N.Z.:
        Oxford University, 1975. 102 p., bibl.
             New Zealand's rugby and cricket matches have been the
        cause of serious protest by African governments.

2472 THUNBORG, ANDERS. "Nordic Policy Trends towards Southern Africa." In: Anglin, D.; Shaw, T.; and Widstrand, C., (eds.). Canada, Scandinavia and Southern Africa. Uppsala, 1978, 111-23.
The author is the Swedish Ambassador to the UN.

2473 TIMMLER, M. "Getäuschte oder Enttäuschete Afrikaner." Aussenpolitik, 25, 3, 1974, 329-40.
"Disappointed or Misled Africans." Discusses relations between Arab and African states in respect of the 1974 Conference of African Heads of State.

2474 TOMEH, GEORGE S. "The Unholy Alliance: Israel and South Africa." Mazungumzo, 2, 3, 1972, 7-16.

2475 TOUMA, E. "Deteriorating Israeli Position in Africa." African Communist, 54, 3, 1973, 75-81.

2476 TREVERTON, GREGORY F. "Kuba nach der Intervention in Angola." Europa-Archiv, 32, 1, 10 January 1977, 18-27.

2477 AL-'UWAYNĪ, MUHAMMAD 'ALĪ. "Al-'Ilāqāt al-'Arabīyah al-Ifrīqīyah fī a'Qāb al-Harb al-'Arabīyah-al-Isrā'īlīyah al-Rābi'ah." Al-Shu'ūn al-Filastīnīyah, 44, April 1975, 77-84.
Arab-African relations and steps to improve these.

2478 VAUGEOIS, DENIS. "La Coopération du Québec avec l'Extérieur." Etudes Internationales, 5, 2, June 1974, 376-87, tabl.
Description of Quebec's relations - political and economic - with other francophone states with a section on Africa. The article is contained in a special issue, "La Coopération Internationale entre Pays Francophones."

2479 VEDOVATO, GIUSEPPE. "L'Africa e il Conflitto Arabo-Israliano." Rivista di Studi Politici Internazionali, 39, 2, April-June 1972, 301-2.

2480 VENKATASUBBIAH, H. "India and the Apartheid Question." India Quarterly, 33, 1, January-March 1977, 62-70.
A brief history of India's role since 1946 in the international campaign against apartheid.

2481 WELLINGTON, G. "Israel, an Agent of Social and Economic Change in Africa." New Outlook; Middle East Monthly, 16, July/August 1973, 17-27.

2482 WISHNEWSKI, H. J. "La Politica della RFT nei Confronti dei Paesi Africani." Affari Esteri, 23, July 1974, 80-92.
"The FRG's Policy towards the Conflicts between African Countries." Germany opposes racism but cannot interfere in the domestic affairs of other states. The development of commerce is the best way to eliminate tension.

2483 ZAMPGLIONE, GERADO. "Europeismo, Africa Nera e Mondo Arabo." La Comunità Internasjonale, 28, 2/3, 1973, 376-94.

2484 ZOGHBY, SAMIR M. Arab-African Relations, 1973-1975. Washington: Library of Congress, 1976. 26 p. [Maktaba Afrikana Series].
An annotated bibliography containing references to numerous Arabic sources, many of which are included herein.

Economic Factors in African International Relations

2485 ADELMAN, KENNETH L. "Energy Crisis Brightens Zaire's Future!' Africa Today, 22, 4, October-December 1975, 49-55.
Zaire is potentially an energy supplier for Central and Southern Africa.

2486 ADELSON, CHARLES E. "Western Tourism and African Socialism!' Africa Report, 21, 5, September/October 1976, 43-55.

2487 "L'Afrique et la Crise de l'Energie." Revue Française d'Etudes Politiques Africaines, 102, June 1974, 35-88.
Oil price increases have had serious repercussions for Africa.

2488 "Agreement Establishing an Association between the European Economic Community and the East African Community." In: Tandon, Y., (ed.). Readings in African International Relations, Vol. I. Nairobi, 1972, 367-80.
Text of the Agreement signed at Arusha on September 24, 1969.

2489 AKINSANYA, A. "The European Common Market and Africa." International Problems, 16, 1/2, Spring 1977, 99-117. Also in: Pakistan Horizon, 28, 3, 1975, 38-55. Also in: West African Journal of Sociology and Political Science, 1, 2, January 1976, 147-63.
The Lomé Convention provides for a reliable supply of raw materials to the industrial states of Europe. It does not so clearly benefit the African states.

2490 ALBRECHT, F.-W. "Das Neue Assoziierungsabkommen von Jaunde." Verfassung und Recht Übersee, 3, 2, 1970, 201-19.
A description of the Yaounde Agreement of 1969 between the EEC and 18 African states.

2491 ALIBONI, ROBERTO. "Renewal of the Yaounde Convention." Africa Quarterly, 9, 2, July-September 1969, 95-100. See also: "An Evalution of the Yaounde Convention Association." In: Tandon, Y., (ed.). Readings in African International Relations, Vol. I. Nairobi, 1972, 337-44.
    Reappraises the convention and asserts that the benefits to the African states have been meagre.

2492 ALKAZAZ, AZIZ. "Die Multilaterale Arabische Entwicklungshilfe und die Afrikanischen Staaten. Die Institutionen und ihre Leistungen." Orient, 18, 3, September 1977, 115-58.
    Reviews changes in Arab aid policies for Africa since 1973. "Multilateral Arab Development Aid to the African States: The Institutions and Their Achievements."

2493 ALLEN, PHILLIP M. "The Technical Assistance Industry in Africa: A Case for Nationalization." International Development Review, 12, 3, 1970, 8-15.
    "African nations must develop the means to bargain effectively for foreign aid."

2494 ALPAN, MOSHE. "Israeli Trade and Economic Relations with Africa." In: Curtis, M. and Gitelson, S. A., (eds.). Israel in the Third World. New Brunswick, N. J., 1976, 100-10, tabls.
    A review of the characteristics and history of Israel's economic relations with independent Africa.

2495 ALTING VON GEUSAU, F. A. M., (ed.). The Lomé Convention and a New International Economic Order. Leyden: Sijthoff, 1977. 249 p., tabls.

2496 AMIN, S. "After Nairobi: Preparing the Non-aligned Summit in Colombo: An Appraisal of UNCTAD IV." Journal of Contemporary Asia, 6, 3, 1976, 309-13.
    An analysis of the refusal of the developed states to meet Third World demands at the UNCTAD Conference in Nairobi. Recommends formation of Third World producer cartels.

2497 AMIR, SHIMEON. Israel's Development Cooperation with Africa, Asia, and Latin America. New York: Praeger, 1974. 133 p., bibl., index, tabls.
    Agriculture, youth programs, cooperatives and trade unions, and other aspects of Israel's technical aid program are considered.

2498 AMOA, R. K. "Euro-African Association and the Economic Development of Africa South of the Sahara." Politica Internazionale, 7/8, July/August 1974, 27-52.
    Economic aspects of the Yaounde Convention are considered as well as the possible effects of the EurAfrica relationship on African economic development.

2499 ANGLIN, DOUGLAS G. The International Arms Traffic in Sub-Saharan Africa. Ottawa: Norman Patterson School, Carleton University, 1971. 37 p. [Occasional Paper, 12].

2500 "The Arab African Bank." Middle East News: Economic Weekly, 14, 12 April 1975, 15-6.

2501 ARKADIE, B. VAN. "Central Banking in an East-African Federation." In: Leys, C. and Robson, P., (eds.). Federation in East Africa. Nairobi, 1965, 145-57.
"This chapter attempts a straightforward account of some of the problems that are likely to arise in the administration of a Central Bank in an East African federation."

2502 ARNOLD, GUY. Aid in Africa. New York: Nichols, 1979. 250 p.
A detailed look at who is giving what to whom, and the consequences of such aid for all the parties concerned. Some of the questions considered are: who are the donors and why are they providing aid?, how is aid divided between loans and grants and technical assistance?, what is the effect of aid on recipient economies?, how much development results?, what is the effect of aid on self-reliance?, etc.

2503 AVRI-SEGRE, DAN. "Macro and Micro-Cooperation among Nations." In: Gardiner, R. K. A., (ed.). Africa and the World. Addis Ababa, 1970, 211-9.
Looks at why international aid to less-advanced countries has gone wrong in so many instances.

2504 BADOUIN, ROBERT. "Régime Foncier et Développement Economique en Afrique Intertropicale." Civilisations, 20, 1, 1970, 50-65.

2505 BAILEY, MARTIN. "Chinese Aid in Action: Building the Tanzania-Zambia Railway." World Development, 3, 7/8, July/September 1975, 587-94.

2506 BAKER, JONATHAN. "Oil and African Development." Journal of Modern African Studies, 15, 2, June 1977, 175-212, maps, tabls.
Outlines the current state of the African oil industry and highlights some of the repercussions of increased prices.

2507 BARTKE, WOLFGANG. China's Economic Aid. Translated by Waldtraut Jarke. London: C. Hurst for the Institute of Asian Affairs, Hamburg, 1975. 215 p., figs., tabls.
A global perspective is presented in the introductory essay. Forty-six country studies follow. Twenty-nine of these are African states.

2508 BASILE, N. D. "Il Fondo Europeo di Sviluppo, un Importante Strumento di Cooperazione." Politica Internazionale, 1, January 1977, 38-44.
"The European Development Fund: An Important Cooperation Instrument." A review and analysis of the activities and role of the FED in Third World economic development. FED activities have been concentrated in Africa.

2509 BASSO, JACQUES. "La Politique Française de Coopération Internationale (Plus Particulièrement avec les Pays en Voie de Développement)." Etudes Internationales, 5, 2, June 1974, 342-54, tabls.
In a special issue, "La Coopération Internationale entre Pays Francophones."

2510 _____ and SPINDER, JACQUES. "Quelques Jalons pour l'Analyse Financière de la Coopération Bilatérale Française avec les Pays en Voie de Développement." Etudes Internationales, 5, 2, June 1974, 269-301, tabls.
In a special issue, "La Coopération Internationale entre Pays Francophones."

2511 BAURET, H.; FORLACROIS, C.; and GILLET, M. "Assistance et Coopération en Afrique." Année Afrique, 1969, 1969, 86-113.
In general, this considers French aid, although the US, Germany, Japan, Britain, and the World Bank are also included.

2512 BELFIGLIO, VALENTINE J. "Israeli Foreign Aid Programs to Africa." International Problems, 15, 3/4, Fall 1976, 132-44.
Israel has proven its friendship to Africa, but the Africans have been swayed by Arab oil money. African states should reconsider and rescind the 1975 anti-Zionist vote in the UN General Assembly.

2513 _____. "United States Economic Relations with the Republic of South Africa." Africa Today, 25, 2, April-June 1978, 57-68, figs., tabl.
US has continued normal (and close) commercial relations with S. Africa in spite of US disapproval of apartheid.

2514 BERARD, JEAN-PIERRE. "La Coopération Française avec les Etats Africains et Malgache: Etat Actuel, Perspectives d'Avenir." Développement et Civilisation, 13, March 1963, 79-88.

2515 BERGE, ELIAS. "The Norwegian Church Relief and Humanitarian Relief Efforts in Nigeria-Biafra." Internasjonal Politikk, 3, 1969, 452-7.

2516 BERNSTEIN, S. J. and ALPERT, E. J. "Foreign Aid and Voting Behavior in the United Nations: The Admission of Communist China." Orbis, 15, 3, Fall 1971, 963-77.
Indication that US aid advances US political interests.

2517 BERTOLIN, GORDON. "U. S. Economic Interests in Africa: Investment, Trade, and Raw Materials." In: Whitaker, J. S., (ed.). Africa and the United States. New York, 1978, 21-59.
"Greater global interdependence and the increasing reliance of the United States on foreign sources of oil and other raw materials have enhanced the economic importance of Africa to the United States."

2518 BETTS, T. F. "Development Aid from Voluntary Agencies to the Least Developed Countries." Africa Today, 25, 4, October-December 1978, 49-68.
Traces the development of voluntary agencies (ICVA, Oxfam, ACORD) and contrasts their funding and approach to those of UN specialized agencies (FFHC, UNHCR, FAO).

2519 BETZ, F. H. Entwicklungshilfe an Afrika. Munich: Weltforum Verlag for the Institut für Wirtschaftforschung, 1970. 120 p.
A collection of statistics on African development and Euro-African relations. It is particularly informative on aid, especially German aid.

2520 BEZY, F. "Vers des Mutations Structurelles dans l'Intégration Eurafricaine?" Politica Internazionale, 7/8, July/August 1974, 87-97.
Economic relations between Europe and Africa are changing; Europe is becoming dependent upon Africa's high priced primary products. Europe's industry thus cannot compete with US industry and even with some African industry.

2521 BIENEFELD, MANFRED. "Special Gains from Trade with Socialist Countries: The Case of Tanzania." In: Nayyar, D., (ed.). Economic Relations between Socialist Countries and the Third World. Montclair, N. J.: 1978, 18-52, figs., tabls.
"Tanzania has not derived any special benefits from its trade with socialist countries other than gaining access to some additional export markets."

2522 BLACK, L. and CONROY, F. D. "United States Economic Aid to Africa." African Studies Bulletin, 7, 4, March 1964, 1-11.

2523 BOGAERT, E. VAN. "De Associatieverdragen van Jaoende en Lome." Studia Diplomatica, 29, 1, 1976, 41-64.

2524 BOSELLO, FRANCO. "La Cooperazione Finanziaria tra la CEE egli Stati Africani e Malgascio Associati." Il Politico, 25, 1, March 1970, 106-32.

2525 BOSTOCK, MARK and HARVEY, CHARLES, (eds.). Economic Independence and Zambian Copper: A Case Study of Foreign Investment. New York: Praeger, 1972. 274 p., ill., maps, tabls.
Essays by economists, a geologist, a lawyer, and a student.

2526   BOTCHWAY, K. "The ACP-EEC Convention: New Order, or Old Order with a New Face." University of Ghana Law Journal, 13, 1/2, 1976, 133-9, tabls.

2527   BOURGE, Y. "La Coopération Franco-Africaine et Malgache." Revue de Défense Nationale, 26, May 1970, 709-22.

2528   BOUTROS-GHALI, BOUTROS. "Les Fonds Arabes pour le Développement Economique." Annuaire Français de Droit International, 21, 1975, 65-72.
        Brief description of institutions involved in, and the effects of, Arab economic aid to Africa.

2529   BRIGGS, W. "Negotiations between the Enlarged European Economic Community and the African, Caribbean and Pacific Countries." Nigerian Journal of International Affairs, 1, 1, July 1975, 12-32.

2530   BROADBENT, KIERAN. "Chinese Aid and Trade in African Countries." Contemporary Review, 216, 1248, January 1970, 19-22.

2531   BROWN, ROBERT W. "African Development: Locked on Oil." In: Brown, R. W., et al. Africa and International Crises. Syracuse, 1976, 1-34, maps, tabls.
        Africa's position as a consumer and supplier of petroleum products and the relationships to economic development are examined.

2532   BUSCH, GARY K. "The Transnational Relations of African Trade Unions." Africa Today, 19, 2, Spring 1972, 22-32, bibl.
        The role and importance of external assistance and intervention in the African trade union movement.

2533   "La CEE et le Tiers Monde." Annuaire du Tiers Monde, 3, 1976/1977, 17-643.
        A collection of essays presented at a conference held in September 1977 under the auspices of the French Association for the Study of the Third World.

2534   CERVENKA, Z. "Africa and the New International Order." Verfassung und Recht in Übersee, 9, 2, 1976, 187-99.
        Economic cooperation in Africa will not succeed without political unity. OAU and ECA compete rather than cooperate. The negotiations for a New International Economic Order are between the oil-producing and the industrialized states; Africa is a by-stander.

2535   CHAPEL, YVES. "Aspects de la Coopération Technique entre la Belgique et les Pays Francophone d'Afrique." Revue Juridique et Politique, 23, 4, April-June 1969, 225-52.

2536   CHEYSSON, CLAUDE. "Europe and the Third World after Lomé."
       World Today, 31, 6, June 1975, 232-9.
           The Lomé Convention and its consequences for the EEC
       and the associated members.

2537   CHICAGO COMMITTEE FOR A FREE AFRICA. Sell the Stock: The
       Divestiture Struggle at Northwestern University and Building
       the Anti-Imperialist Movement. Chicago: Peoples College,
       1978. 146 p.

2538   CHINA, REPUBLIC OF. SINO-AFRICAN COOPERATION COMMITTEE.
       Technical Missions of the Republic of China in Africa and
       Other Areas. Taiwan: Republic of China, 1970. 118 p., tabls.
           Description of twenty-seven projects in Africa.

2539   CONDE, ALPHA. Guinée: Albanie d'Afrique ou Néo-colonie
       Américaine? Paris: Editions Gît-le-Coeur, 1972. 270 p.,
       bibl., ill.
           Guinea's foreign economic relations and the domestic
       political/economic scene. The US role in the economy is
       considered in some detail.

2540   CONSTANTIN, FRANÇOIS and COULON, CHRISTIAN. "Islam, Pétrole
       et Dépendance: Un Nouvel Enjeu Africain." Revue Française
       d'Etudes Politiques Africaines, 113, May 1975, 28-53.
           The Arabs wil provide the capital and the Europeans
       the technology for African development.

2541   COSGROVE, C. A. "The European Economic Community and Its
       Yaounde Associates: A Model for Development." International
       Relations, 4, 2, November 1972, 142-55.
           The Yaounde agreement as a model for relations between
       developed and non-developed states.

2542   _____ and TWITCHETT, K.J. "The Second Yaoundé Convention
       in Perspective: An Examination of the Association of Eigh-
       teen African and Malagasy States with the European Economic
       Community." International Relations, 3, 9, May 1970, 679-89.
           Poverty dictated that the African states sign the
       Yaounde agreement in 1969.

2543   COUSIN, MARIE-ELISABETH. "Quelques Aspects Formels des Con-
       ventions de Coopération entre Pays Francophones." Etudes
       Internationales, 5, 2, June 1974, 326-41.
           In special issue, "La Coopération Internationale entre
       Pays Francophones."

2544   "La Crise de la Coopération Franco-Africaine." Revue Fran-
       çaise d'Etudes Politiques Africaines, 8, 90, June 1973,
       95-115.

2545   CROCKER, CHESTER A. "External Military Assistance to Sub-Saharan Africa." Africa Today, 15, 2, April/May 1968, 15-20.

2546   CRONJE, SUZANNE; LING, MARGARET; and CRONJE, GILLIAN. The Lonrho Connections: A Multinational and Its Politics in Africa. Los Angeles: Bellwether. London: Friedmann, 1976. 316 p.
       A thorough description of the activities of a major, Africa-oriented MNC.

2547   CURRY, ROBERT L., JR. "Global Market Forces and the Nationalization of Foreign-Based Export Companies." Journal of Modern African Studies, 14, 1, March 1976, 137-43, tabls.
       There are ideological, political, and economic incentives for nationalizing foreign-based companies which operate in the export sectors of African economies.

2548   _____ and ROTHCHILD, DONALD. "On Economic Bargaining between African Governments and Multinational Companies." Journal of Modern African Studies, 12, 2, June 1974, 173-90.
       Analysis of the variables affecting the bargaining outcome and suggestions for African governments on ways to improve the outcome.

2549   CURTIN, TIMOTHY. "Africa and the EEC: The Lomé Convention." In: Africa South of the Sahara 1978-79. London: Europa, 1978. 56-66, tabls.
       A brief description of the Lomé Convention and the Stabex fund is followed by several tables on trade and aid.

2550   CURZON, G. and CURZON, V. "Neo-Colonialism and the European Economic Community." Yearbook of World Affairs, 25, 1971, 118-41.
       If the ties between Europe and Africa established at Yaounde survive, then the existing trade links must become stronger. Whether or not this is neo-colonialism depends upon one's view.

2551   DALE, TORSTEIN. "The Norwegian Red Cross and Humanitarian Aid in Nigeria-Biafra." Internasjonal Politikk, 3, 1969, 449-51.

2552   DAMODARAN, K. C. E. E. C.-Third World: Benefits of the Generalized System of Preferences of the European Community to Developing Countries. Bonn/Bad-Godesberg: Friedrich-Ebert-Stiftung, 1976. 49 p.

2553   DAVIS, MORRIS. "Audits of International Relief in the Nigerian Civil War." International Organization, 29, 2, Spring 1975, 501-12.
       Looks at the sources, timeliness, scope, and effectiveness of the relief programs from a political perspective.

2554 DAVIS, M, (ed.). Civil Wars and the Politics of International Relief: Africa, South Asia, and the Caribbean. New York: Praeger, 1975. 111 p., index.
Contains essays by Warren Weinstein on Burundi and Alvin G. Edgell on Nigeria/Biafra.

2555 _____. "The Politics of International Relief Processes in Large Civil Wars: An Editorial Comment." Journal of Developing Areas, 6, 4, July 1972, 487-92.
The experiences of international relief efforts during the Nigerian Civil War are a lesson in the political problems involved in such operations.

2556 DECALO, SAMUEL. "Afro-Israeli Technical Cooperation: Patterns of Setbacks and Successes." In: Curtis, M. and Gitelson, S. A., (eds.). Israel in the Third World. New Brunswick, N. J., 1976, 1-99, tabls.
"The composite picture that emerges of fifteen years of Afro-Israeli cooperation is an intricate mosiac of advances and reverses."

2557 DELORME, NICOLE. L'Association des Etats Africains et Malgache à la Communaute Economique Européene. Paris: R. Pichon et R. Durand-Auzias, 1972. 372 p., bibl., index, tabls.
Evolution, functions, and problems of the association to the EEC of African states. The author argues that for such association to be of real value to Africa, the African states must regroup on a regional basis much as Europe did after World War II. That is, Europe may now play the stimulus role for Africa that the USA played for Europe.

2558 DIABATE, MOUSTAPHA. "Le Marché Commun Européen et l'Afrique (ou Bilan Décennal d'une Association)." In: Reflections on the First Decade of Negro-African Independence. Paris: Présence Africaine, 1971, 155-69, tabls.
The relationship between African states and the Common Market is essentially colonial in character. Recommendations are made for improving the African position in that relationship.

2559 DIALLO, SIRADIOU. "La Zone Franc et les Etats Africains après Dix Ans d'Indépendance." In: Reflections on the First Decade of Negro-African Independence. Paris: Presence Africaine, 1971, 137-54, tabls.
Background and mechanisms of the franc zone in Africa.

2560 DIAMOND, ROBERT A. and FOUQUET, D. "American Military Aid to Ethiopia - and Eritrean Insurgency." Africa Today, 19, 1, Winter 1972, 37-43.

2561 DIAWARA, MOHAMED T. "Coopération Eurafricaine et Nouvelle Division Internationale du Travail Industriel." Chronique de Politique Etrangère, 27, 2, March 1974, 181-91.
A lecture presented by the Minister of Planning of the Ivory Coast to the Royal Institute of International Relations in 1973.

2562 DINWIDDY, BRUCE, (ed.). European Development Policies: The United Kingdom, Sweden, France, EEC, and Multilateral Organizations, New York: Praeger, 1973. 118 p., tabls.
This volume provides information on the aid policies of individual states, the EEC, and other organizations.

2563 DIXON-FYLE, S. R. "Monetary Dependence in Africa: The Case of Sierra Leone." Journal of Modern African Studies, 16, 2, June 1978, 273-94.
Examines the nature of African dependence on foreign currencies and the extent to which this determines their reactions to changes in international monetary conditions, using Sierra Leone as an example.

2564 DJAMSON, ERIC O. The Dynamics of Euro-African Co-operation: Being an Analysis and Exposition of Institutional, Legal and Socio-Economic Aspects of Association/Co-operation with the European Economic Community. The Hague: Martinus Nijhoff, 1976. 370 p., bibl., index.
"The object of this study is to provide a guide to an understanding of the relationship between the EEC and the Developing Countries." History of relations, institutions and treaty analysis, and political/economic factors are discussed.

2565 DRUCKER, P. F. "Multinationals and Developing Countries: Myths and Realities." Foreign Affairs, 53, 1, October 1974, 121-34.
The MNCs are portrayed as positive factors in Third World economic development.

2566 DUBEY, S. "Africa's Role at UNCTAD." Africa Quarterly, 8, 3, October-December 1968, 263-68.
Africa's main interests at UNCTAD concerned primary commodities and special preferences.

2567 DUPONT, PIERRE. "La Somalie et la C. E. E." Revue Française d'Etudes Politiques Africaines, 115, July 1975, 41-62.

2568 E. E. C. INFORMATION SERVICE. "In Defence of the Yaounde Convention." In: Tandon, Y., (ed.). Readings in African International Relations, Vol. I. Nairobi, 1972, 345-58, tabls.
Attempts to answer some of the criticism aimed at the Convention, and gives some of the basic facts and figures about the association.

2569 EDGELL, ALVIN G. "Nigeria/Biafra." In: Davis, M., (ed.). Civil Wars and the Politics of International Relief. New York, 1975, 50-73.
A descripton of the aid program and its relation to Nigerian military policy.

2570 EDMONDS, MARTIN. "Civil War and Arms Sales: The Nigerian-Biafran War and Other Cases." In: Higham, R. D. S., (ed.). Civil Wars of the Twentieth Century. Lexington, Kentucky, 1972, 203-16.
Arms sales to Nigeria and to Biafra by governments and private dealers was a major international issue.

2571 EDOZIEN, E. C. "Some Aspects of the United Nations Technical Cooperation with Africa." Nigerian Journal of Economic and Social Studies, 12, 1, March 1970, 29-44, tabls.
The UNDP has made significant contributions to African development in several ways.

2572 EHRHARDT, C. A. "Die EG-Assoziierung der AKP-Länder." Aussenpolitik, 25, 4, 1974, 384-99.
Description of the increasing number of associated states and analysis of the Yaounde II agreement.

2573 EINBECK, EBERHARD. "Moscow's Military Aid to the Third World." Aussenpolitik, 4th Quarter, 1971, 460-74.
The author, a West German colonel, describes Soviet military aid to Egypt, Sudan, Libya, and Ghana. Military aid is related to Soviet expansionist policy.

2574 EISEMANN, PIERRE MICHEL. "L'Accord International sur le Cacao." Annuaire Français de Droit International, 21, 1975, 738-66.
The problems and disappointments in the attempts to develop cooperation among cocoa producing states are examined.

2575 EL-KHAWAS, MOHAMED A. "Foreign Economic Involvement in Angola and Mozambique." African Review, 4, 2, 1974, 299-314, tabls.
Scope and influence of foreign investment during the wars for liberation.

2576 _____. "A Reassessment of International Relief Programs." In: Glantz, M. H., (ed.). The Politics of Natural Disaster: The Case of the Sahel Drought. New York, 1976, 77-100, tabls.
There are two tasks for aid, food relief in the short-term and technical assistance and capital in the long-term.

2577 ELLIMAH, R. "Financial and Technical Co-operation under the ACP-EEC Convention of Lomé." University of Ghana Law Journal, 13, 1/2, 1976, 1041-32, tabls.

2578 ELLIS, FRANK; MARSH, JOHN; and RITON, CHRISTOPHER. Farmers and Foreigners: Impact of the Common Agricultural Policy on the Associates and Associables. London: Overseas Development Institute, 1973. 86 p.
  The authors describe the concept of association and the Common Agricultural Policy. There are analyses of the effects of the policy upon individual commodities and the trade situation of each associated state.

2579 EL MALLAKH, RAGAEI and MUKERJEE, T. "The Education Dimension of United States Aid to Africa." Africa Today, 17, 3, May/June 1970, 22-4.
  "The United States policy toward Africa remains long on platitudes and short on constructive commitment."

2580 EMEMBOLU, G. E. and PANNU, S. S. "Africa: Oil and Development." Africa Today, 22, 4, October-December 1975, 39-47, figs., tabls.
  Both oil-rich and oil-poor African nations have felt consequences of the rise of oil prices in 1973.

2581 ERB, GUY F. "Africa and the International Economy: A U. S. Response." In: Whitaker, J. S., (ed.). Africa and the United States: Vital Interests. New York, 1978, 60-86.
  A general review of Africa's role in the international economy, with some emphasis on the potential for US-African economic relations.

2582 _____. "Research on Foreign Investment in Africa." Current Bibliography on African Affairs, 6, 3, Summer 1973, 345-54.
  Research suggestions and a list of information sources.

2583 ERLICH, ALEXANDER and SONNE, CHRISTIAN R. "The Soviet Union: Economic Activity." In: Brzezinski, Z., (ed.). Africa and the Communist World. Stanford, Cal., 1963, 49-83, tabls.
  The authors analyze "changes in the 'Marxist-Leninist' attitude to the problem of economic underdevelopment" and provide data on trade, credit, and technical assistance relationships.

2584 ESSEKS, JOHN D. "Humanitarian Assistance in Africa: Some Case Studies." Africa Today, 24, 3, July-September 1977, 73-8.
  Review of three books: (1) The Politics of Natural Disaster: The Case of the Sahel Drought, ed. by Michael H. Glantz, (2) The Politics of Starvation, by Jack Shepherd, and (3) Civil Wars and the Politics of International Relief: Africa, South Asia, and the Caribbean, ed. by Morris Davis.

2585 ESSEKS, J. D. "Soviet Economic Aid to Africa: 1959-72."
In: Weinstein, Warren, (ed.). Chinese and Soviet Aid to
Africa. New York, 1975, 83-119, tabls.
Relative size of aid flows, hypotheses as to why some
states do and some do not receive Soviet aid, benefits to
the USSR from its aid program, and speculation about the
future.

2586 EWING, ARTHUR E. "Industrial Development in Africa: The
Respective Roles of African Countries and External Assistance." In: Gardiner, R.K. A., (ed.). Africa and the World.
Addis Ababa, 1970, 110-21.
Emphasizes three problems: financing, the prerequi-
sites of industrial development, and the instruments of
industrial development. While external assistance is essential, much more can be done by the African countries themselves.

2587 FAJANA, OLUFEMI. "Trends and Prospects of Nigerian-Japanese
Trade." Journal of Modern African Studies, 14, 1, March
1976, 127-36.
Both Japan and Nigeria depend on international trade
for their survival. Assuming that Japan is not seen to be
supporting the continued existence of colonialism in Africa,
a substantial increase in the volume of Nigerian-Japanese
trade can be predicted.

2588 FEUER, G. "La Révision des Accords de Coopération Franco-
Africains et Franco-Malgache." Annuaire Français de Droit
International, 19, 1973, 720-39.
Monetary arrangements between France and African states
have undergone important changes.

2589 FEUSTEL, SANDY. "Commodities: African Minerals and American
Foreign Policy." Africa Report, 23, 5, September/October
1978, 12-8.

2590 FRANCOLINI, B. "Il Problema Politico dell' Africa Nera."
Studi Politici, 8, 1, January-March 1961, 27-46.
African economic development requires outside assistance,
but this has dangers of a new form of colonialism.

2591 FRIDERICHS, H. "Nairobi und die Folgen." Europa-Archiv,
31, 16, August 1976, 517-26.
Description of the debates at the 1976 meeting in
Nairobi of UNCTAD.

2592 GALONI, P. "Sino-Soviet Commercial and Aid Agreements to
Africa: Assistance or Clientship?" Pan-African Journal,
6, 3, Autumn 1973, 349-68, tabls.
The argument over the relationships between aid programs and dependence.

2593   GALTUNG, JOHAN. "La Convention de Lomé et le Néo-capitalisme." Etudes Internationales, 9, 1, March 1978, 75-86.
        The Lomé Convention will provide for the rapid expansion of local capitalism in the the Third World.

2594   _____. "The Lomé Convention and Neo-capitalism." African Review, 6, 1, 1976, 33-42.
        Signing the Convention has certain benefits for the ACP states, but the cost is the maintenance of the international division of labor.

2595   GANN, LEWIS W. "Neo-colonialism, Imperialism and the 'New Class'." The Intercollegiate Review, Winter 1973/1974, 13-27. An earlier version appeared in Survey, 19, 1, Winter 1973, 165-83.
        A refutation of the neo-colonialism concept.

2596   GARRITY, MONIQUE. "Africa and the European Economic Community." Review of Black Political Economy, 2, 1, 1971, 95-109.

2597   GEISLER, WOLFF. "Die Militärische Zusammenarbeit zwischen BRD und RSA im Atomaren und Konventionellem Bereich." Blätter für Deutsche und International Politik, 23, 2, February 1978, 166-85.
        "Military Cooperation between the Federal Republic of Germany and the Republic of South Africa in Atomic and Conventional Fields." Analysis and description of a wide range of military contacts between the two states and some assessment of their importance.

2598   GERGEN, K. J. and GERGEN, M. "Understanding Foreign Assistance through Public Opinion." Yearbook of World Affairs, 28, 1974, 125-40.
        Analysis of survey data from 12 states including Kenya, Nigeria, Senegal, and African students in Germany suggests that "reactions to aid appeared to be influenced by the image of the donor."

2599   GERSCH, GABRIEL. "Israel's Aid to Africa." Commonweal, 85, 8, November 1966, 226-8.
        Israel has learned the development lessons that Africa should be taught and is ready to assist it with practical know-how. Existing aid programs are highly successful.

2600   GIBERT, S. P. "Soviet-American Military Aid Competition in the Third World." Orbis, 13, 4, Winter 1970, 1117-37.
        Analysis of a twelve-year period shows striking similarity in Soviet and US military aid diplomacy.

2601   GINGYERA-PINCYWA, A. G. G. and MAZRUI, ALI A. "Regional Development and Regional Disarmament: Some African Perspectives." In: Arkhurst, F. S., (ed.). Arms and African Development. New York, 1972, 31-54, tabl.
   The South African problem and the failures of Pan-Africanism are factors involved in African disarmament questions.

2602   GITELSON, SUSAN A. "How Are Development Projects Selected? The Case of the UNDP in Uganda and Tanzania." African Review, 2, 2, 1972, 365-79, tabls.
   Concludes that greater efforts are needed to find the proper balance between rational economic thought and responsiveness to political demands.

2603   GORDENKER, LEON. International Aid and National Decisions: Development Programs in Malawi, Tanzania, and Zambia. Princeton, N. J.: Princeton University, 1976. 190 p., index, tabls.
   "This study deals with efforts carried on through international institutions to solve the problem of uneven economic development and to meet pressing demands from the governments of some of the least developed areas in the world for rapid change.

2604   _____. "The 'U. N. Maze' and African Development: A Comment." International Organization, 32, 2, Spring 1978, 562-68.
   Looks at the limitations of the United Nations system of developmental aid, and questions two of its underlying assumptions: that resources brought from the outside to the least developed countries in Africa probably can have useful developmental effects and that such effects should be sought primarily within national states. Also see essay by I. V. Gruhn.

2605   GREAT BRITAIN. CENTRAL OFFICE OF INFORMATION. Britain and the Developing Countries: Overseas Aid, a Brief Survey. London, 1972. 23 p., tabls.

2606   GREEN, R. H. "The Lomé Convention: Updated Dependence or Departure toward Collective Self-Reliance?" African Review, 6, 1, 1976, 43-54.
   The Lomé Convention is dependence-oriented but the unity the ACP states showed in the negotiations is a step toward independence.

2607   _____. "Petroleum Prices and African Development: Retrenchment or Reassessment?" International Journal, 30, 3, Summer 1975, 391-405.
   The situations of the countries which gained and lost the most from OPEC oil price rises are described and broader implications are raised.

2608 GREEN, R. H. "UNCTAD and After: Anatomy of a Failure."
 Journal of Modern African Studies, 5, 2, September 1967,
 243-67.
  Examines the results of the 1964 Geneva conference of
 UNCTAD vis-à-vis Africa, and the outlook for the next meeting
 in New Delhi. Calls for world monetary reform, including
 transfer of GNP from developed to developing nations, an end
 to tied aid, liquidity reform, and new sources of finance.

2609 GRENIER, R. "La 'Crise Globale' et la Coopération Canadienne
 au Développement de l'Afrique Francophone." Etudes Internationales, 5, 2, June 1974, 367-75.
  The author suggests changes in Canadian aid to Africa.

2610 GROHS, GERHARD. "Difficulties of Cultural Emancipation in
 Africa." Journal of Modern African Studies, 14, 1, March
 1976, 65-78.
  Tanzania, Senegal, and South Africa are examined as
 three different approaches to the problem of being in the
 position of having attained political but not economic
 emancipation.

2611 GRUHN, ISEBILL V. British Arms Sales to South Africa: The
 Limits of African Diplomacy. Denver: University of Denver,
 1972. 30 p.
  Military assistance to South Africa and the South
 African defense plans.

2612 _____. "The Lomé Convention: Inching towards Interdependence." International Organization, 30, 2, Spring 1976,
 241-62.
  The negotiation for the Lomé Convention and the agreements made provide important understanding on North-South
 relations.

2613 _____. "The UN Maze Confounds African Development."
 International Organization, 32, 2, Spring 1978, 547-61.
  The least developed countries are dependent upon the UN
 system, and this retards their development. Also see essay
 by L. Gordenker.

2614 GUTTERIDGE, WILLIAM F. "Foreign Military Assistance and
 Political Attitudes in Developing African Countries."
 Bulletin (Institute of Development Studies, University of
 Sussex), 4, 4, September 1972, 24-32.

2615 _____. The Military in African Politics. London:
 Methuen, 1969. 166 p., bibl., maps.
  See "Foreign Military Assistance and the Political Role
 of African Armed Forces," pp. 126-40.

2616 GUTTERIDGE, W. F. "The Political Role of African Armed Forces: The Impact of Foreign Military Assistance." African Affairs, 66, April 1967, 93-103.

2617 GWYER, G. D. "Three International Commodity Agreements: The Experience of East Africa." Economic Development and Cultural Change, 21, 3, April 1973, 465-77.

2618 HALL, RICHARD and PEYMAN, HUGH. The Great Uhuru Railway: China's Showpiece in Africa. London: Gollancz, 1976. 208 p., bibl., ill., index, maps.

2619 AL-HAY'AH AL-'ĀMMAH LIL-ISTI'LĀMĀT. Misr wa-Ifrīqiyā 'alā Tarīq al-Ta'awun al-Mushtarak. Cairo: Wizarat al-I'lām, 1974. 27 p.
Cooperation between Egypt and black Africa in a variety of technical and economic fields.

2620 HAYTER, TERESA. French Aid. London: Overseas Development Institute, 1966. 230 p., index, tabls.
Contains general chapters on the history, economics, and administration of French aid as well as chapters on aid to specific areas of the world. "This study is primarily concerned with the present administration, mechanisms and policies of official French aid . . . . The study concentrates on aid to the African and Malagasy states . . . ."

2621 _____. "French Aid to Africa - Its Scope and Achievements." International Affairs, 41, 2, April 1965, 236-51. Also in: Tandon, Y., (ed.). Readings in African International Relations, Vol. I. Nairobi, 1972, 289-300.
French aid is not aimed at promoting economic independence for African states.

2622 HEBGA, JEAN. "La Place de l'Afrique dans la Politique des Investissements Privés Allemands à l'Etranger." Revue Française d'Etudes Politiques Africaines, 64, April 1971, 36-65.

2623 HEDRICH, MANFRED and ROPP, KLAUS VON DER. "Lomé II im Licht der Erfahrungen mit Lomé I." Aussenpolitik, 29, 3, 1978, 297-312.
"Lomé II in the Light of the Experiences of Lomé I." The differences between the rich and the poor states will make renegotiation of Lomé II very difficult.

2624 HELLEINER, G. K. "New Forms of Foreign Investment in Africa." Journal of Modern African Studies, 6, 1, May 1968, 17-27.
Additional means must be found for the private sector to contribute to African development.

2625 HENGSBACH, F. Die Assoziierung Afrikanisher Staaten an die Europäischen Gemeinschaften: Eine Politik Raumwirtschaftlicher Integration? Baden-Baden: Nomos, 1977. 238 p., bibl., tabls.

2626 HENRY, P. M. "The United Nations and the Problem of African Development." International Organization, 16, 2, Spring 1962, 362-74. Also in: Padelford, N. and Emerson, R., (eds.). Africa and World Order. New York, 1963, 94-106. Also in: Quigg, P. Africa. New York, 1964, 161-74.
    The UN must play an important role in African development. The outlines of the role are appearing.

2627 HENTSCH, T. "Humanitarian Scramble in West Africa: Looking Back to a Spectacular Relief Operation." Genève-Afrique, 15, 1, 1976, 5-14.

2628 HODGKIN, THOMAS. "Some African and Third World Theories of Imperialism." In: Owen, R. and Sutcliffe, B., (eds.). Studies in the Theory of Imperialism. London, 1972, 93-116.
    "My purpose . . . is to discuss a small sample of the theories of imperialism that have been developed by its consumers, or victims, in Africa primarily." Among writers included are Edward Blyden, Marcus Garvey, Aimé Césaire, Frantz Fanon, and Lamine Senghor.

2629 HOLLOWAY, ANNE, F. "Developing a Multinational Case: LONRHO in Africa." Current Bibliography on African Affairs, 8, 3, 1975, 220-31, bibl.
    Multinational corporation behavior and African economic self-determination.

2630 HOOGVELT, ANKIE M. M. and TINKER, ANTHONY M. "The Role of Colonial and Post-Colonial States in Imperialism: A Case-Study of the Sierra Leone Development Company." Journal of Modern African Studies, 16, 1, March 1978, 67-79, figs., tabls.
    After independence, the state becomes a hireling of the multinational corporations. Political sovereignty, however, permits the state to demand an increasingly higher price for its comprador role.

2631 HOUTART, F. "La Conférence Internationale de Khartoum et les Mouvements Révolutionnaires en Afrique (18-20 Janvier 1969)." Cultures et Développement, 1, 3, 1968, 619-48.
    Neo-colonialism, geopolitics, liberation in Southern Africa, and economic development are all linked.

2632 HUGON, PHILIPPE. "Vers une Théorie Economique de la Coopération entre les Pays Francophones." Etudes Internationales, 5, 2, June 1974, 252-68.
In a special issue, "La Coopération Internationale entre Pays Francophones."

2633 HUTCHINSON, ALAN. "Chinese Aid: On the Right Lines." African Development, August 1972, 10-3, map, tabl.
A description of China's aid efforts and goals.

2634 HUTCHINSON, E. C. "American Aid to Africa." Annals of the American Academy of Political and Social Sciences, 354, July 1964, 65-74.
US aid has concentrated on institution-building and infrastructure, but many states are unable to absorb all of this. There is a need to rethink aid policy and to turn to increasing domestic income and different types of infrastructure development.

2635 HUTTON, N. "Africa's Changing Relationship with the EEC." World Today, 30, 10, October 1974, 426-35.
Relationships with the EEC may be loosened as African states diversify their trade relationships.

2636 _____. "Sources of Strain in the Eurafrican Association." International Relations, 4, 3, May 1973, 288-94.
The renegotiation of the agreements and the entrance of Commonwealth members as sources of strain.

2637 HVEEM, HELGE. "Relationship of Underdevelopment of African Land-locked Countries with the General Problem of Economic Development." In: Cervenka, Z., (ed.). Land-locked Countries of Africa. Uppsala, 1973, 278-87, tabls.
Being land-locked is only a part of the world system of dominance and exploitation.

2638 _____ and HOLTHE, O. K. "European Economic Community and the Third World." Instant Research on Peace and Violence, 2, 1972, 73-85.
The EEC-Third World agreements are the basis of continuing European neocolonial exploitation of Africa.

2639 IFFLAND, C. "La Suisse et l'Afrique." Revue Juridique et Politique, 24, 1, January-March 1970, 17-34.
Swiss investment in Africa is very small, but most Swiss aid goes to Africa. The author suggests that Swiss aid might be more effective if it were part of a common European aid program.

2640   INSTITUT D'ETUDES EUROPEENNES. L'Association à la Communauté Economique Européenne: Aspects Juridiques. Brussels: Universitaires de Bruxelles, 1970. 369 p.
        A collection of essays on the legal aspects of association with the EEC.

2641   "International Technical Co-operation." Journal of Local Administration Overseas, 1, 1, January 1962, 47-56 and 1, 2, April 1962, 112-23.
        The article contains descriptions of the structural organization of intergovernmental organizations and technical cooperation organizations.

2642   "Israel in Africa." African Development, December 1972, 65-70.
        Several short articles describing Israeli economic and technical aid to Africa, with emphasis on Zambia and Ivory Coast. Authors agree it has been effective.

2643   JACKSON, BARBARA W. "Free Africa and the Common Market." Foreign Affairs, 40, 3, April 1962, 419-30.
        An early and positive assessment of the role the Common Market could play in African affairs.

2644   JACOB, ABEL. "Foreign Aid in Agriculture: Introducing Israel's Land Settlement Scheme to Tanzania." African Affairs, 71, April 1972, 186-94.
        Israeli aid programs in Africa are generally thought to be technically successful.

2645   _____. "Israel's Military Aid to Africa, 1960-1966." Journal of Modern African Studies, 9, 2, August 1971, 165-87.
        What are the political motives behind Israeli military aid and how does this aid influence the actions of the receivers?

2646   JAYARAMAN, J. K. "Economic Co-operation among Afro-Asian Countries." India Quarterly, 33, 2, April-June 1977, 198-215.
        Long-term economic gains are expected from increased co-operation between African and Asian states.

2647   JEANTELOT, CHARLES B. "Relations Commerciales Franco-Nigériennes." Nigeria Demain, 52, May/June 1978, 5-8.
        This journal is published by the French government.

2648   JESKE, J. "The Association of the BLS Countries with the EEC: General Questions and Particular Problems." South African Journal of African Affairs, 5, 2, 1975, 19-33, tabls.

2649   JOHNSON, WILLARD R. "The Politics of Foreign Investment in Africa." Black World, 24, 2, December 1974, 14-21, ill.
        MNCs are oppressive and prevent economic development.

2650 JONES, DAVID. Europe's Chosen Few: Policy and Practice of the E. E. C. Aid Programme. London: Overseas Development Institute, 1973. 100 p.
"To provide objective and detailed information on the aid given by the European Economic Community to its associated states."

2651 JOSHUA, WYNFRED and GIBERT, STEPHEN P. Arms for the Third World: Soviet Military Aid Diplomacy. Baltimore: Johns Hopkins, 1969. 169 p., bibl., index, tabls.
See chapter 3, "Sub-Saharan Africa," pp. 31-52. Author argues that "Moscow's arms diplomacy in Africa is an integral part of its overall policy of undermining western influence." So far, "the Soviet Union has not realized any dramatic results."

2652 JOUTSAMO, K. "The Lomé Convention: The Sun Never Sets." Instant Research on Peace and Violence, 5, 3, 1975, 138-49.
The Lomé Convention provides for an improved African trade situation but it leaves the international division of labor unchanged.

2653 KAMARA, LATYR. "Aide Internationale à l'Afrique pour le Développement et Détériotation des Termes de l'Echange." Présence Africaine, 77, 1971, 52-87.
The author analyzes numerous problems with foreign aid and concludes that if Africa is to develop, the African states must work together and rely on their own resources.

2654 KAMARCK, ANDREW M. "The African Economy and International Trade." In: Goldschmidt, W., (ed.). The United States and Africa. New York, 1963, 156-84, tabls.
Largely an analysis of US economic relations with and interests in Africa.

2655 KANAREK, JEHUDI J. Israeli Technical Assistance to African Countries. Geneva: Geneva-Africa Institute, 1969. 115 p.
A brief survey of Israeli aid.

2656 KANE, CHEIKH HAMIDOU. "Exposé sur l'Aide Suisse au Tiers-Monde et à l'Afrique en Particulier." Genève-Afrique, 12, 2, 1973, 43-61.

2657 KANZA, THOMAS NSENGA. "Chinese and Soviet Aid to Africa: An African View." In: Weinstein, Warren, (ed.). Chinese and Soviet Aid to Africa. New York, 1975, 222-37.
The author had held positions in the government of Congo prior to taking up a scholarly post at the University of Massachusetts.

2658 KATOND, DIUR. "La Comunidad Económica de Estados de Africa, el Caribe y el Pacifico (ACP)." Revista de Política Internacional, 144, March/April 1976, 283-300.

2659 KAVIRAJ, SADIPATA. "Multi-National Firms in Africa." Africa Quarterly, 16, 3, January 1977, 54-66.

2660 KAYA, Y. K. "Volta Dam: An Example of International Cooperation." Turkish Yearbook of International Relations, 11, 1971, 102-18.
   The construction of the dam required aid from 16 countries.

2661 KEMP, A. G., (ed.). Africa and the E. E. C. in the Aftermath of Lomé and UNCTAD IV. Aberdeen: Aberdeen University African Studies Group, 1977. 101 p.

2662 KLEEMEIER, LIZZ LYLE. "Empirical Tests of Dependency Theory: A Second Critique of Methodology." Journal of Modern African Studies, 16, 4, 1978, 701-4.
   Furthur discussion of Patrick McGowan's essay in the same journal in March 1976.

2663 KLINGHOFFER, ARTHUR J. "The Strategy of Aid." Africa Report, 17, 4, April 1972, 12-4.

2664 KOPPENFELS, G. VON. "Die Bedeutung des Abkommens von Lome für die Entwicklung der Beziehungen zwischen Europa und der Dritten Welt." Europa-Archiv, 31, 1, January 1976, 10-8.
   "The Significance of the Lomé Agreement for the Evolution of the Relations between Europe and the Third World." Although the Lomé agreements are not perfect, they are a model for relations between industrialized and developing states.

2665 KREININ, MORDECAI E. Israel and Africa: A Study in Technical Cooperation. New York: Praeger, 1964. 206 p.
   An economic analysis of Israeli aid to Africa.

2666 KROHN, H.-B. "Das Abkommen von Lomé zwischen den EG und der AKP-Staaten: Eine Neue Phase der EG-Entwicklungshilfepolitik." Europa-Archiv, 30, 6, March 1975, 177-88.
   "The Lomé Convention between the EEC and the ACP Countries: A New Phase in the EEC's Development Aid." The author views the Lomé Convention as a development contract.

2667 LAMBERT, J. R. "The European Economic Community and the Associated African States: Partnership in the Making." World Today, 17, 8, August 1961, 344-55.
   Examination of the problems caused by the existence at that time of two economic groups in Europe and the split between French- and English-speaking states in Africa.

2668 LANGDON, S. "The Poltical Economy of Dependence: Note toward Analysis of Multinational Corporations in Kenya." Journal of East African Research and Development, 4, 2, 1974, 123-59.

2669 LANGHAMMER, ROLF J. "Die Wirtschaftsgemeinschaft Westafrikanischer Staaten (ECOWAS): Ein Neuer Integrationsversuch." Europa-Archiv, 31, 5, March 1976, 163-8.

2670 LARKIN, BRUCE D. "Chinese Aid in Political Context: 1971-73." In: Weinstein, Warren, (ed.). Chinese and Soviet Aid to Africa. New York, 1975, 1-28, tabls.
"The object of this paper is to show the extent of commitments since 1970 and something of the policy context from which they have sprung."

2671 LEE, J. J. VAN DER. "Association between the European Economic Community and African States." African Affairs, 66, 264, July 1967, 197-212.
Outlines the origins and development of the association, and some of its strengths and weaknesses. Feels the system of aid to the developing countries has worked well and should continue.

2672 _____. "The European Common Market and Africa." World Today, 16, 9, September 1960, 370-76.
Reviews Common Market arrangements for relations with Africa and recommends that western states should increasingly coordinate their Africa policies.

2673 LEISTNER, G. M. E. Aid to Africa. Pretoria: Africa Institute, 1966. 22 p., bibl., figs., map, tabl.
Motivations for giving aid, analysis of programs of major donors, and results of aid programs in Africa are briefly considered.

2674 LEYMARIE, P. "Les Accords de Coopération Franco-Malgaches." Revue Française d'Etudes Politiques Africaines, 78, June 1972, 55-60.

2675 _____. "L'Agence de Coopération Culturelle et Technique ou la Francophonie Institutionelle." Revue Française d'Etudes Politiques Africaines, 122, February 1976, 13-24.

2676 LI, K. T. "Republic of China's Aid to Developing Nations." Pacific Community, 1, 4, July 1970, 664-71.
Taiwan's aid is highly successful because of the greater relevance of aid from one developing state to another.

2677   LIBBY, R. T. "External Co-optation of a Less Developed Country's Policy Making: The Case of Ghana, 1969-1972." World Politics, 29, 1, October 1976, 67-89.
The policies of external creditors forced the Ghanaian government to follow unpopular domestic policies and this led to a coup d'état.

2678   LIGOT, M. "La Coopération entre la France et le Ruanda." Revue Juridique et Politique, 18, 1, January-March 1964, 107-20.
Although it had been a Belgian colony, Rwanda signed aid agreements with France. It was the first non-ex-French state in Africa to do so.

2679   _____. "La Coopération Militaire dans les Accords Passés entre la France et les Etats Africains et Malgache d'Expression Française." Revue Juridique et Politique d'Outre-Mer, 17, 4, October-December 1963, 517-72.
The agreements provide for French military aid in terms of materials and troops.

2680   _____. "Vue Générale sur les Accords de Coopération." Revue Juridique et Politique d'Outre-Mer, 16, 1, January-March 1962, 3-20.
An analysis of the original cooperation agreements signed by France and the ex-French colonies in Africa.

2681   LITTLE, I. M. D. Aid to Africa: An Appraisal of U. K. Policy for Aid to Africa South of the Sahara. Oxford: Pergamon. New York: Macmillan, 1964. 76 p.
"The countries which can, and do, use [aid] best should get most."

2682   LOEHR, WILLIAM and RAICHUR, SATISH. "A Decade of United States Investment Activity in Africa." Africa Today, 20, 1, Winter 1973, 45-58, tabls.
Examines some of the trends in foreign private investment - specifically that of US businessmen, discusses the pros and cons of such investment, and offers some suggestions aimed at preventing the "tragic situation" in which many developing countries find themselves."

2683   LUCRON, CLAUDE. "La Convention de Lomé, Exemple de Coopération Réussie." Studia Diplomatica, 30, 1/2, 1977, 7-99.
A comparison of key features of the Yaoundé and Lomé Conventions.

2684   MANSUR, ANTWĀN. "Al-'Ilāqāt al-Iqtisādīyah Bayn Isra'il wa-Ifriqiya." Shu'ūn Filastīnīyah, 29, January 1974, 79-104.
Israel's economic relations with Africa, an example of attempts to establish dependency relationships.

2685 MANSŪR, SĀMĪ. "Al-Taʻāwun al-Arabī al-Ifrīqī wa-Qadiyyat Filastīn." Shuʼūn Filastīnīyah, 39, November 1974, 60-5.
Arab aid programs in Africa need to be increased in size and scope.

2686 MARIÑAS OTERO, L. "El Acuerdo de Lomé." Revista de Política Internacional, 139, May/June 1975, 53-71.
The Lome Agreement between the EEC and the Associated States alters the political economy of Afro-European relations, but conflicts between Britain and France make the agreement lack a coherent policy.

2687 _____. "La IV Conferencia de la UNCTAD en Nairobi (Mayo de 1976)." Revista de Política Internacional, 151, May/June 1977, 145-56.

2688 MAY. R. "A Country-by-Country Analysis of Aid under the Terms of the Lomé Convention: African Countries." In: May, R. Convention of Lomé: EEC Financial Aid to the ACP States. Brussels, 1977, 19-191, tabls.

2689 MAYNARD, GEOFRREY. "The Economic Irrelevance of Monetary Independence: The Case of Liberia." Journal of Development Studies, 6, 2, January 1970, 111-32.

2690 MAZRUI, ALI A. "African Attitudes to the European Common Market." International Affairs, 39, 1, January 1963, 24-36.
The author notes several factors that may underlie Africa's negative views of the EEC.

2691 McGOWAN, PATRICK J. "Economic Dependence and Economic Performance in Black Africa." Journal of Modern African Studies, 14, 1, March 1976, 25-40, tabls.
Correlates 3 measures of economic dependence with 23 indicators of economic performance. Finds little support for dependency theory, and concludes that more testing is necessary.

2692 MCLAUGHLIN, RUSSELL U. Foreign Investment and Development in Liberia. New York: Praeger, 1965. 217 p., bibl., figs., tabls.
"This study attempts to define and delineate the role which foreign and private investment have played in Liberian economic development." Foreign investment in the private sector and foreign aid, mainly US, in the public sector are the central features.

2693 MEHDEN, F. R. VON DER. "South East Asian Relations with Africa." Asian Survey, 5, 7, July 1965, 341-9.
The author provides information on trade and political relations between the two areas.

2694 MEHMET, OZAY. "Effectiveness of Foreign Aid: The Case of Somalia." Journal of Modern African Studies, 9, 1, May 1971, 31-47.

"Somalia presents a unique opportunity for a case study of the effectiveness of foreign aid to a country at an early stage of development." It is unique because of the state's unusually heavy dependence upon foreign aid for development expenditure.

2695 MELLAH, M. F. "La Convention de Lomé: Nouveau Type de Relations entre les Pays Sous-développés et les Pays Développés?" Notes Africaines, 152, October 1976, 102-11.

2696 "Memorandum: Norway and Relief Activity in Nigeria-Biafra." Internasjonal Politikk, 3, 1969, 382-8.

2697 MERTENS, P. "Les Modalités de l'Intervention du Comité International de la Croix-Rouge dans le Conflit du Nigéria." Annuaire Français de Droit International, 15, 1969, 183-209.

The civil war in Nigeria had many lessons for international relief programs.

2698 MINTER, WILLIAM. "Imperial Network and External Dependency: Implications for the Angolan Liberation Struggle." Africa Today, 21, 1, Winter 1974, 25-39, bibl.

Angola is a dependency of Portugal but is involved overall in a network of dependency involving several capitalistic powers.

2699 MOHAMMED, DURI. "Notes on the Common Market and Africa." In: Gardiner, R. K. A., (ed.). Africa and the World. Addis Ababa, 1970, 122-7, figs.

Assesses some of the economic implications of the European Economic Community for the development problems of African countries.

2700 MONTGOMERY, JOHN D. "The Infrastructure of Technical Assistance: American Aid Experience in Africa." In: Karp, M., (ed.). African Dimensions. Boston, 1975, 137-53.

A typology of technical assistance projects and evaluation of such projects are the major topics.

2701 MORRIS, ROBERT C. Overseas Volunteer Programs: Their Evolution and the Role of Governments in Their Support. Lexington, Mass.: Lexington Books, 1973. 352 p., bibl., figs., index, maps.

Descriptions and analyses of each major overseas volunteer program, except the US Peace Corps, provides information on a major source of aid for African development.

2702 MORRISON, T. K. "Africa and the Common Fund: UNCTAD's Integrated Program for Commodities." Africa Today, 24, 3, July-September 1977, 61-7, tabls.
The Fund may discriminate against the poorest countries of Africa in terms of benefit distribution.

2703 MORTON, KATHRYN. Aid and Dependence: British Aid to Malawi. London: Croom Helm, 1975. 189 p., index, map, tabls.
This volume is one of a series on British aid published by the Overseas Development Institute. "How far can Malawi's performance be attributed to its aid receipts? Did aid help or . . . hinder the development process?"

2704 MÜLLER, KURT. "Soviet and Chinese Programmes of Technical Aid to African Countries." In: Hamrell, S. and Widstrand, C. G., (eds.). The Soviet Bloc, China and Africa. Uppsala, 1964, 101-30, figs., tabls.
Soviet and Chinese aid programs are considered together. Author concludes that these programs have had no success in converting Africa to communism.

2705 MUTHARIKA, B. W. T. "The Trade and Economic Implications of Africa's Association with the Enlarged E. E. C." Economic Bulletin for Africa, 10, 2, 1974, 40-86, tabls.

2706 MWANGO, G. G. Foreign Aid and Tanzania Development Strategy. Dar es Salaam: University of Dar es Salaam, 1972. 17 p.

2707 MYTELKA, LYNN K. "The Lomé Convention and a New International Division of Labour." Journal of European Integration, 1, 1, September 1977, 63-76.
The Lomé Convention is a device aimed at stabilizing a new international division of labor.

2708 NADUBERE, D. W. "Stablisation of Export Earnings ("Stabex") in the Lomé Convention." University of Ghana Law Journal, 13, 1/2, 1976, 159-79.

2709 NATIONAL CHENGCHI UNIVERSITY. PROGRAM OF AFRICAN STUDIES. Agreements on Technical Co-operation between the Republic of China and African States. Mushan, Taipei: National Chengchi University, 1974. 269 p.

2710 NAYYAR, DEEPAK, (ed.). Economic Relations between Socialist Countries and the Third World. Montclair, N. J.: Allanheld, Osmun, 1978. 265 p., index, figs., tabls.
Several of the general essays in this volume are indirectly relevant to Africa affairs. See the references under Bienefeld and Stevens for those with more direct relevance.

2711 NDEGWA, PHILIP. The Common Market and Development in East Africa, 2nd. edition. Nairobi: East African Publishing, 1968. 228 p.
A cost-benefit analysis from the points of view of each state in the EAC.

2712 NDONGKO, WILFRED A. "The Economic Implications of Multi-membership in Regional Groupings - The Case of Cameroon and Nigeria." Afrika Spectrum, 11, 3, 1976, 319-23.

2713 _____. "The Economic Origins of the Association of Some African States with the European Economic Community." African Studies Review, 16, 2, September 1973, 219-32, tabls.
French desires to gain associate membership in the EEC for ex-French colonies is seen as a continuation of the French colonial economic policy of Jean-Baptiste Colbert, of the late 1600s.

2714 _____. "The External Trade Pattern of Cameroon, 1957-72." Africa Quarterly, 16, 1, July 1976, 76-87.

2715 _____. "From Economic Domination to Association: Africa in the EEC." Présence Africaine, 99/100, 1976, 181-95.
The author, a Cameroon scholar, argues that association with the EEC is of benefit to the African states.

2716 NEGRE, LOUIS-PASCALE. "La Banque Africaine de Développement: Solidarité et Coopération au Service de l'Unité." Revue Française d'Etudes Politiques Africaines, 128, August 1976, 27-42, tabls.

2717 _____. "Les Dimensions de la Coopération Afro-Arabe." Jeune Afrique, 740, March 1975, 36-40.
Author was Minister of Finance and Commerce in Mali and Vice President of the ADB. He recommends more thorough planning for Afro-Arab economic cooperation.

2718 NKRUMAH, KWAME. Challenge of the Congo: A Case Study of Foreign Pressures in an Independent State. New York: International Publishers. London: Thomas Nelson, 1967. London: Panaf, 1969. 304 p., bibl., index.
Foreign imperialists are attracted to Congo (Zaire) because of its mineral wealth. These imperialists manipulate events in Congo through economic means in an effort to maintain political control of the state.

2719 NORBERG, V. H. Swedes in Haile Selassie's Ethiopia, 1924-1952: A Study in Early Development Co-operation. Uppsala: Scandinavian Institute of African Studies, 1977. 320 p., bibl., maps, tabls.

2720 O'BRIEN, RITA C. "Colonization to Co-operation? French Technical Assistance in Senegal." Journal of Development Studies, 8, 1, October 1972, 45-58. Also in: Bernstein, H., (ed.). Underdevelopment and Development. Harmondsworth, 1973, 323-40.
   This case study of French technical assistance in Senegal suggests important limitations upon the effectiveness of foreign technical personnel in African bureaucracies.

2721 "Objectif de la Coopération Française: Former des Cadres Adaptés aux Réalités Africaines." France Eurafrique, 24, 235, 1972, 2-4.
   Brief description of French technical assistance projects.

2722 OKIGBO, PIUS N. C. Africa and the Common Market. Evanston: Northwestern University, 1967. 183 p., index, tabls.
   The development of relationships between African states and the EEC with particular emphasis on Nigeria. There is a chapter titled "The Possibilities of an African Common Market." The author has been an economic advisor to the Nigerian team in negotiations with the EEC.

2723 OLOFIN, S. "Ultra-import-biased Taste in Nigeria's External Trade Relations." In: Akinyemi, A. B., (ed.). Nigeria and the World. Ibadan, 1978, 32-44.
   The economic benefits to Nigeria of the oil boom may be restricted by "the country's ultra-import-biased taste."

2724 OSTRANDER, F. TAYLOR and ARMSTRONG, WINIFRED. "US Private Investment in Africa." Africa Report, 14, 1, January 1969, 38-41.

2725 PACKENHAM, R. A. "Political Development Doctrines in the American Foreign Aid Program." World Politics, 18, 2, January 1966, 194-235.
   Interviews with US AID officials indicate that political development is not a factor considered by those who administer US aid programs.

2726 PAGAMONCI, A. "Francia e Terzo Mondo." Rivista di Studi Politici Internazionali, 37, 4, October-December 1970, 545-81.
   French aid is an extension and revision of French colonial policies. This essay includes discussion of relations between France and Africa.

2727 PALOT, LALE. "L'Assistance Technique Française en Matière Administrative en Faveur des Etats Africains et Malgache." Revue Juridique et Politique, 23, 2, April/June 1969, 253-74.

2728 PATAL, RAPHAEL. "Africa and Israel: Quiet Partners in the Third World." Tuesday Magazine [Chicago], November 1972, 13-37.
    A summary of Israeli economic and technical assistance programs in Africa.

2729 PELEG, ILAN. "Arms Supply to the Third World: Models and Explanations." Journal of Modern African Studies, 15, 1, March 1977, 91-103, fig., tabl.
    Points out the increasing importance of the arms trade, reviews scholarly works on the topic, and suggests a theoretical framework and model for understanding and analyzing arms supply as an international political phenomena.

2730 PETER, J. E. "Le Fonds de Solidarité Africain." Revue Juridique et Politique. Indépendance et Coopération, 31, 1, January-March 1977, 30-42.
    The FSA originates from an agreement between France and fifteen francophone African states.

2731 PETROVIC, NEGOSAVA. "Relationships between the EEC and the 18 Associated African Countries." Medunarodni Problemi, 22, 2, 1970, 61-76.

2732 PLATE, BERNARD VON. "DDR-Aussenpolitik Richtung Afrika Araber." Aussenpolitik, 29, 1, 1978, 73-83. In English edition: 29, 1, 1978, 75-86.
    "The GDR's Foreign Policy toward Africa and the Arabs." International recognition is the primary objective of the GDR'S Africa policy.

2733 PLESSIS, J. A. DU. "Russian Aid to Africa." Bulletin of the Africa Institute of South Africa, 12, 7, 1974, 277-87, tabls.

2734 POLIT-ECON SERVICES, INC. Problems of Voluntary Agencies in African Development. Washington, D. C., 1973. 25 p.
    Analysis of the problems of voluntary agency work in African development.

2735 POPOVIC, VOJISLAV. Tourism in Eastern Africa. München: Weltforum Verlag, 1972. 208 p., bibl., map.

2736 PRAIN, RONALD. "Metals and Africa: Economic Power in an International Setting." African Affairs, 77, 307, April 1978, 236-46.
    "First, an analysis of Africa's mineral wealth . . . . Secondly, an assessment of the importance of Africa to the rest of the world as a supplier of minerals and metals . . . . Thirdly, Africa's potential for political-economic influence."

2737 PROKOPCZUK, JERZY. "Poland's Relations with Asian, African and Latin America Countries." Studies on the Developing Countries, 1, 1972, 9-26, tabls.
Primarily an analysis of economic relations between Poland and the Third World.

2738 RADETZKI, MARIAN. Aid and Development: A Handbook for Small Donors. New York: Praeger, 1972. 323 p.
This is based on a case study of Swedish aid to Kenya and Tanzania.

2739 REICH, B. "Israel's Policy in Africa." Middle East Journal, 18, 1, Winter 1964, 14-26.
Israeli education programs, technical assistance, joint economic ventures, loans, and trade with Africa are discussed. The author argues that African states see close relations with Israel as compatible with neutralism.

2740 RENDELL, WILLIAM. "Commonwealth Development Corporation Experience with Joint Ventures." In: Ady, P., (ed.). Private Foreign Investment and the Developing World. New York, 1971, 243-68.
The CDC, a British organization, is a major source of investment for development projects in African Commonwealth states.

2741 RIVKIN, ARNOLD. Africa and the European Common Market, 2nd. edition. Denver: University of Denver, 1966. 67 p., bibl., tabls.
"The European Common Market . . . seems a fruitful arrangement for the African Associates and those of their African neighbors who may wish to join them."

2742 ROBERTS, GEORGE O. "The Sierra Leone Experience with Foreign Assistance." Journal of African Studies, 3, 1, Spring 1976, 83-100.
The role of foreign aid in the development of a concept of Sierra Leone nationalism.

2743 ROBERTS, GLYN. Volunteers and Neo-colonialism: An Inquiry into the Role of Foreign Volunteers in the Third World. Manchester, England: the author, 1968. 44 p., bibl.
A thought-provoking examination with some information on Swedish technical assistance.

2744 RODNEY, WALTER. How Europe Underdeveloped Africa. Dar es Salaam: Tanzania Publishing. London: Bogle l'Ouverture, 1972. 316 p.
An important statement on the economic and political effects of colonial rule.

2745   ROGERS, BARBARA. White Wealth and Black Poverty: American Investments in Southern Africa. Westport, Conn.: Greenwood, 1976. 331 p., index. [Center on International Race Relations, University of Denver, Studies in Human Rights, No. 2].
   An examination of US investments, their effects on South African society, and the effects of sanctions and other forms of economic pressure.

2746   RONDOT, PIERRE. "Etats Arabes: Prémices d'une Coopération avec l'Afrique." Revue Française d'Etudes Politiques Africaines, 104, August 1974, 15-8.
   Effect of the oil embargo on Fourth World countries.

2747   ROOD, LESLIE L. "Foreign Investment in African Development." Journal of African Studies, 5, 1, Spring 1978, 18-33.
   "Discusses whether private foreign direct investment is beneficial to the black African countries."

2748   _____. "Foreign Investment in African Manufacturing." Journal of Modern African Studies, 13, 1, March 1975, 19-34, tabls.
   An assessment of the possibilities for and risks of major investments in Africa by multinational corporations.

2749   _____. "Nationalism and Indigenisation in Africa." Journal of Modern African Studies, 14, 3, September 1976, 427-47.
   Description, analysis, and commentary on the nationalization of foreign investments by African governments.

2750   ROWE, E. T. "Aid and Coups d'Etat: Aspects of the Impact of American Military Assistance Programs in the Less Developed Countries." International Studies Quarterly, 18, 2, June 1974, 239-55.
   US military aid increases the possibility of coups d'etat, instability, and the long-term survival of military rule.

2751   RUBINSTEIN, G. "Aspects of Soviet-African Economic Relations." Journal of Modern African Studies, 8, 3, October 1970, 389-404, maps, tabls.
   "The last decade has witnessed a rapid development of economic relations between the Soviet Union and independent African countries . . . . The principles of Soviet foreign economic policy are equality, non-interference, mutual advantage, and assistance to the developing countries in building their national economies."

2752 RUBIO GARCIA, L. "Hacia la Asociación entre Europa y Africa Negra.' Revista de Política Internacional, 60, March-April 1962, 69-84.
"Towards Association between Europe and Black Africa." Political independence has not led to the solution of African economic problems and so a new relationship with Europe has been formed. Association with the EEC is an example.

2753 RWEYEMAMU, J. F. "International Trade and the Developing Countries." Journal of Modern African Studies, 7, 2, July 1969, 203-20.

2754 SABOURIN, L. "Les Programmes Canadiens de Coopération avec les Etats de l'Afrique, Particulièrement avec l'Afrique Francophone." Etudes Internationales, 1, 4, December 1970, 73-87.
Canadian aid has passed through three major phases. Programs with anglophone Africa have gone smoothly, but certain difficulties exist in programs with the francophone states.

2755 SAMUELS, J. W. "Humanitarian Relief in Man-Made Disasters: The International Red Cross and the Nigerian Experience." Behind the Headlines, 34, 3, 1975, 1-44.

2756 SAUL, JOHN S. "Canadian Bank Loans to South Africa." In: Anglin, D.; Shaw, T.; and Widstrand, C., (eds.). Canada, Scandinavia and Southern Africa. Uppsala, 1978, 28-36.
Canadian policy for South Africa is analyzed, with stress on the differences between "words and deeds."

2757 SAWYER, CAROLE A. Communist Trade with Developing Countries 1955-1965. New York: Praeger, 1966. 126 p., bibl., tabls.
"The major purpose of this study is to evaluate communist claims about their trade with developing economies and to analyze the most important facets of this trade . . . . An attempt is made to seek an economic explanation." USSR and Eastern Europe receive the majority of the author's attention.

2758 SAWYERR, A. "Industrial Co-operation under the ACP-EEC Convention of Lomé." University of Ghana Law Journal, 13, 1/2, 1976, 93-103.

2759 SCHAAR, STUART H. "Patterns of Israeli Aid and Trade in East Africa." American Universities Fieldstaff Report, East Africa Series, 7, 1 and 2, 1968, 13 + 15 p.
Part I is "Israel's African Experience and Its Shifts Away from Joint-Company Partnerships." Part II is "East Africans in Israel and Israelis in East Africa."

2760 SCHATZ, SAYRE P. "Crude Private Neo-Imperialism: A New Pattern in Africa." Journal of Modern African Studies, 7, 4, December 1969, 677-88.
"The term ["crude private neo-imperialism"] refers to unprincipled exploitation by some foreign firms of African government-controlled, directly productive enterprises in an effort to accelerate economic development."

2761 SCHEEL, W. "Neue Beziehungen der EWG-Gruppe zu Afrika." Aussenpolitik, 11, 6, June 1960, 379-87.
"New Relationships of the Common Market with Africa." Emphasis is on Germany's role in economic relations between African states and the EEC.

2762 _____. "Weltpolitische Perspektiven der Europäischafrikanischen Zusammenarbeit. Die Ergebnisse der Strassburger Beratungen über die Künftigen Beziehungen zwischen der EWG und den Assoziierten Ländern." Europa-Archiv, 16, 20, October 1961, 555-62.
"Political Prospects of European-African Cooperation: The Achievements of the Strasbourg Conference on the Future Relations between the EEC and the Associate States."

2763 SCHIFFLER, G. "Das Abkommen von Lomé zwischen der Europäischen Wirtschaftsgemeinschaft und 46 Staaten Afrikas, des Karabischen und Pazifischen Raums." Jahrbuch für Internationales Recht, 18, 1975, 320-39.
Analysis and description of the 1975 Lomé Agreement between the EEC and the asssociated states.

2764 SCHRÖDER, DIETER. "Europa und die Assoziierten in Übersee - zum Völkerrechtlichen Begriff der Assoziation." Verfassung und Recht in Übersee, 10, 1, 1977, 125-31.
"Europe and the Overseas Associates: On the Concept of Association in International Law." Description of the changing status of the relationship between the EEC and the various categories of associated members.

2765 SECCHI, C. "L'Associazione tra CEE e SAMA e i suoi Efetti sul Processo di Integrazione Economica in Africa." Politico, 37, 4, December 1972, 731-58.
"The Association between the EEC and the AASM and its Effects on African Economic Integration." Cooperation between the EEC has been beneficial to integration in Central Africa, harmful in East Africa, and negligible in West Africa.

2766 SEGAL, AARON. "Africa Newly Divided?" The Journal of Modern African Studies, 2, 1, 1964, 73-90. Also in: Tandon, Y., (ed.). Readings in African International Relations, Vol. I. Nairobi, 1972, 321-36.
The division between African states over the EEC.

2767 SHAW, J. A. "German Church Aid to Africa." Bulletin of the Africa Institute of South Africa, October 1972, 383-6.
Protestant and Catholic Church contributions to German development aid effort in Africa.

2768 SHAW, T. M. "The Political Economy of African International Relations." Issue, 5, 4, Winter 1975, 29-38.
A note on the effects of dependency status on African states' social, political, and economic conditions. Shaw develops a tentative typology of states based on political economy concepts.

2769 _____ and GRIEVE, M. J. "The Political Economy of Resources: Africa's Future in the Global Environment." Journal of Modern African Studies, 16, 1, March 1978, 1-32.
"In this article we analyze Africa's place in the global economy - paying particular attention to the impact of the environment on its development prospects - and we examine different projections of its future problems and opportunities."

2770 SHEETS, HAL and MORRIS, ROGER. "Disaster in the Desert." In: Glantz, M. H., (ed.). The Politics of Natural Disasters. New York, 1976, 25-76.
The ineffective international response to the drought raises serious questions about international disaster assistance programs.

2771 SHIVJI, ISSA, (ed.). Tourism and Socialist Development. Dar es Salaam: Tanzania Publishing, 1973. 97 p.
Tourism is a major source of international transactions for many African states. These authors consider the appropriate role - if any - for tourism in a developing socialist state.

2772 SHUTE, J. C. M. "Notes and Documents: Canadian University Technical Assistance Programs in Africa." Canadian Journal of African Studies, 6, 3, 1972, 491-500.
Outlines some problems which must be dealt with in the course of carrying out technical assistance contracts, based on experiences which seem to crop up when a university deals triangularly with itself, an agency (like CIDA), and a cooperating university or ministry abroad.

2773 SINGH, M. "Regional Development Banks." International Conciliation, 576, January 1970, 1-84.
The ADB is one of the examples cited by the author in this study of the origins, role, and financial requirements of regional development banks.

2774 SKLAR, RICHARD L. Corporate Power in an African State: The Political Impact of Multinational Mining Companies in Zambia. Berkeley: University of California, 1975. 245 p., bibl., index, maps, tabls.

"The present study suggests a conceptual approach to the political analysis of multinational business enterprise on the basis of a single intensively researched case." Author provides information on the effects of MNCs on governmental policies and the effects of governments on MNC policies.

2775 SKORODUMOV, A. "Soviet-African Trade." International Affairs (USSR), 5, May 1977, 117-9.

2776 SMITH, S. "U. S. Capital in Africa." International Affairs (USSR), 4, April 1974, 52-6.

2777 SMITH, SHEILA M. "Economic Dependence and Economic Empiricism in Black Africa." Journal of Modern African Studies, 15, 1, March 1977, 116-8.

Critiques McGowan's article in this journal.

2778 SMITH, TIMOTHY. "U. S. Firms and Apartheid: Belated Steps Analyzed." Africa Today, 24, 2, April-June 1977, 29-33.

Churches oppose US investment in South Africa as a form of support for apartheid.

2779 SMOCK, AUDREY. "The Politics of Relief." Africa Report, 4, 8, December 1969, 24-6.

2780 SOBQUI, GABRIEL. Fonds Européen de Développement et Economie des Etats de UDEAC. Paris: CAPU, 1969. 98 p., maps, tabls.

General description of UDEAC, analysis of FED activities in UDEAC, and recommendations.

2781 SOMMER, J. G. Beyond Charity: U. S. Voluntary Aid to the Third World: What Is Its Future? Washington: Overseas Development Council, 1975. 65 p.

2782 SOPER, TOM. "The European Economic Community and Aid to Africa." International Affairs, 41, 3, July 1965, 463-77.

Examines the nature of the association from its origins and its importance to Africa. Assesses the results of the aid programs, and looks at continuing problems.

2783 _____. "A Note on European Trade with Africa." African Affairs, 67, 267, April 1968, 144-51.

First describes the specific trade arrangements and some of the problems involved, and then looks at more fundamental issues that affect the philosophy and organization of world trade.

2784 SOUBEYROL, J. and HOLTHE, O. K. "Les Difficiles Négociations de 'Yaounde II'." Année Africaine 1969, 1969, 53-85.
   The Yaounde II agreement between the EEC and the African associates is the result of compromises between several sets of interests.

2785 STANDARD BANK, ECONOMIC-DEPARTMENT. "Commonwealth Africa and the Enlarged European Community: Notes on the Possible Effects of Britain's Entry." African Affairs, 71, 285, October 1972, 427-36, tabls.
   Assesses the effects of UK entry into the EEC on African Commonwealth countries, and pinpoints the safeguards which these territories will need to have included in any agreement with the enlarged community.

2786 STEEL, K. Considerations Critiques sur la Convention de Lomé. Bruxelles: Centre Etudes et Documentation Africaine, 1976. 42 p., bibl., map, tabls.

2787 STENT, ANGELA. "Soviet Aid to Guinea and Nigeria: From Politics to Profit." In: Weinstein, Warren, (ed.). Chinese and Soviet Aid to Africa. New York, 1975, 142-82, tabls.
   These two case studies highlight changes in Soviet motivations in African policy.

2788 STEVENS, CHRISTOPHER. "Entente Commerciale: The Soviet Union and West Africa." In: Nayyar, D., (ed.). Economic Relations between Socialist Countries and the Third World. Montclair, N. J., 1978, 78-104, tabls.
   A comparison of Soviet-Ghanaian economic relations from 1960 through 1972. Trade and aid are considered.

2789 _____. "In Search of the Economic Kingdom: The Development of Economic Relations between Ghana and the USSR." Journal of Developing Areas, 9, 1, October 1974, 3-26, tabls.
   A positive utilization of foreign aid requires good planning within an overall development plan with long-term considerations.

2790 STOCKHOLM INTERNATIONAL PEACE RESEARCH INSTITUTE. Arms Trade Registers: The Arms Trade with the Third World. Stockholm: Almquist and Wiksell. Cambridge: MIT, 1975. 176 p., figs., tabls.
   This annual publication lists sales of arms by source and recipient, number and type of item, year of transfer, and other information.

2791 STOCKHOLM INTERNATIONAL PEACE RESEARCH INSTITUTE. The Arms Trade with the Third World. Harmondsworth: Penguin, 1975. 362 p., bibl., index, tabls.
There are chapters on each of the major suppliers of arms as well as on the recipients. See "Sub-Saharan Africa and South Africa," pp. 230-258. There is some data on the effects of the arms embargo on South Africa. One such effect has been the growth of the Republic's indigenous arms industry.

2792 STOKKE, BAARD RICHARD. Soviet and Eastern European Trade and Aid in Africa. New York: Praeger, 1967. 326 p., bibl., index, tabls.
"This study presents an analysis of the economic relations which have developed between Africa and the centrally planned economies since the mid-1950s to [1967]." Case studies of Egypt, Algeria, Tunisia, Ghana, Guinea, Mali, The Sudan, Ethiopia, Somalia, and Kenya-Tanzania-Uganda are included.

2793 "The Strange Case of Lonrho." Africa Report, 19, 2, March/April 1974, 40-5.

2794 STREETEN, PAUL. Aid to Africa: A Policy Outline for the 1970s. New York: Praeger, 1972. 169 p., figs., tabls.
This study was conducted for the ECA as an assessment of the quantity and quality of aid flows, a statement of Africa's aid needs from an African perspective, and a statement of support for the concept of an African Development Fund.

2795 _____ and SUTCH, HELEN. Capital for Africa: The British Contribution. London: Africa Publications Trust, 1971. 36 p., bibl.
British aid and investment potentials for African development.

2796 TATON, ROBERT. "La Coopération Arabo-Africaine: La Banque Arabe pour le Développement Economique en Afrique a Accordé Ses Premiers Prêts." Europe Outremer, 550, November 1975, 35-6, 47.

2797 TETZLAFF, R. "Das Abkommen von Lomé und das 'Rapprochement' Europa-Afrika: Fakten, Argumente und eine Zwischenbilanz." Afrika Spectrum, 11, 2, 1976, 157-72.

2798 TORELLI, MAURICE. "L'Influence des Accords d'Association de la Communauté Economique Européenne sur les Relations Internationales des Etats d'Afrique Noire." Etudes Internationales, 2, 1, March 1971, 182-230.

2799 TROCLET, LEON E. L'Association du Marché Commun et de Vingt-quatre Etats Africains. Brussels: Presses Universitaries de Bruxelles, 1971. 116 p., bibl., tabls.

2800 TUNTENG, P-KIVEN. "External Influences and Subimperialism in Francophone West Africa." In: Gutkind, P. C. W. and Wallerstein, I., (eds.). The Political Economy of Contemporary Africa. Beverly Hills, 1976, 212-31, tabl.
   A racial analysis of colonial and neo-colonial relationships is inadequate. "The pattern of economic interests and the resulting linkages - internal and external - must be analyzed in order to derive more fruitful generalizations."

2801 TURNER, LOUIS. Multinational Companies and the Third World. New York: Hill and Wang, 1973. 294 p., bibl., index.
   This volume deals in general with Africa, but see especially Chapter 9,"Free, White, and Beleaguered - Corporations and Southern Africa."

2802 TWITCHETT, KENNETH J. "Yaounde Association and the Enlarged Community." World Today, 30, 2, February 1974, 51-63.
   Bargaining between the EEC and the African states may lead to a split between anglophone and francophone states and between Africa and the rest of the Third World.

2803 UKPONG, IGNATIUS I. "The Impact of Foreign Aid on Electricity Development in Nigeria." Journal of African Studies, 2, 2, Summer 1975, 275-86, tabls.
   Analyzes power development, power consumption, and the impact of foreign aid on power production and distribution.

2804 UNGER, KARL. "Die EG und die Entwicklungsländer. Das Abkommen von Lomé als Grundstein einer Nuen Weltwirtschaftsordnug?" Blätter für Deutsche und Internationale Politik, 22, 3, March 1977, 301-17.
   "The Economic Community and the Developing Countries: The Lomé Agreement as a Foundation for a New World Economic Order?"

2805 UNITED STATES. CENTRAL INTELLIGENCE AGENCY. Communist Aid to the Less Developed Countries of the Free World. Washington: C. I. A., 1977. 33 p., figs., tabls.
   Data and analysis of Soviet, Cuban, Chinese, and Eastern European programs are presented.

2806 VALLEE, C. "Regards sur la Convention de Lomé." Revue Iranienne des Relations Internationales, 4, Autumn 1975, 173-257.
   The Lomé Convention includes many more associated states than previous agreements, but this Convention also has much greater objectives. An English summary is provided.

2807 VAN CHIEN, NGUYEN. "L'Afrique et la Marché Commun." Cahiers Zairois d'Etudes Politiques et Sociales, 1, April 1973, 51-66.

2808 VEDOVATO, GUISEPPE. "La Convention de Lomé: Promesses d'un Vrai Dialogue." Revue Roumaine d'Etudes Internationales, 10, 31, 1976, 85-98.
   A positive interpretation of the significance of the Lomé Convention.

2809 _____. "La Convenzione CEE-ACP di Lomé: Promesse di un Vero Dialogo." Rivista di Studi Politici Internazionali, 42, 3, July-September 1975, 359-77.
   Although full of promises for the developing states, the Lomé Convention is a very fragile agreement.

2810 VENGROFF, RICHARD. "Dependency and Underdevelopment in Black Africa: An Empirical Test." Journal of Modern African Studies, 15, 4, December 1977, 613-30, tabls.

2811 VOLKOVA, I. "Japanese Monopolies in Africa." International Affairs (USSR), 9, September 1973, p. 114.

2812 WALLERSTEIN, IMMANUEL. "Dependence in an Interdependent World: The Limited Possibilities of Transformation within the Capitalist World Economy." African Studies Review, 17, 1, April 1974, 1-26.
   An important statement of the history of Africa's involvement in the economic system of the West and the position Africa holds in that system.

2813 _____. "The Three Stages of African Involvement in the World-Economy." In: Gutkind, P. C. W. and Wallerstein, I., (eds.). The Political Economy of Contemporary Africa. Beverly Hills, 1976, 30-57.
   An introduction to the dependency view of Africa's position in the international economic system.

2814 WEINSTEIN, WARREN. "Burundi." In: Davis, M., (ed.). Civil Wars and the Politics of International Relief. New York, 1975, 5-24.
   "This chapter deals with the humanitarian aid effort directed toward Burundi between April 1972, when its civil strife began, and early 1973."

2815 _____. "China's Aid to Africa." In: Weinstein, W., (ed.). Chinese and Soviet Aid to Africa. New York, 1975, 275-82, tabls.

2816 WEINSTEIN, W., (ed.). Chinese and Soviet Aid to Africa. New York: Praeger, 1975. 290 p., tabls.
　　　See references to essays by Larkin, Yu, Weinstein, Esseks, V. P. Bennett, Stent, Desfosses, Glantz/El-Khawas, and Kanza. Appendixes contain statistics on communist aid and trade with Africa.

2817 _____. "Communist States and Developing Countries: Aid and Trade in 1972." In: Weinstein, W., (ed.). Chinese and Soviet Aid to Africa. New York, 1975, 238-74, figs., tabls.
　　　This is a report prepared by the State Department's Bureau of Intelligence and Research. It covers through 1972.

2818 _____. "The Limits of Military Dependency: The Case of Belgian Military Aid to Burundi, 1961-1973." Journal of African Studies, 2, 3, Fall 1975, 419-31, tabls.

2819 WEPSIEC, JAN. Serial Publications on the Foreign Trade of the Countries of Africa South-of-the-Sahara. Waltham, Mass.: African Studies Association, 1971. 40 p.
　　　This bibliography lists the relevant publications by country. It includes a list of international statistical publications.

2820 WEST, ELEANORA and WEST, ROBERT L. "Conflicting Economic Interests of Africa and the United States." In: Arkhurst, F. S., (ed.). U. S. Policy towards Africa. New York, 1975, 153-84, figs., tabls.
　　　A profile of economic relations is followed by a discussion of potential sources of US-African conflicts.

2821 WIDSTRAND, CARL G., (ed.). Multinational Firms in Africa. Uppsala: Scandinavian Institute of African Studies, 1975. New York: Africana, 1976. 425 p., tabls.
　　　The proceedings of a conference held in Dakar in 1974.

2822 _____ and CERVENKA, ZDENEK. Scandinavian Developments with African Countries. Uppsala: Scandinavian Institute of African Studies, 1971. 74 p., fig., tabl.
　　　Summary, analysis, and recommendations for Scandinavian aid programs in Africa.

2823 WILSON, DICK. "China's Economic Relations with Africa." Race, 5, 4, April 1964, 61-71.
　　　In a special issue containing several articles on China and Africa.

2824 WILSON, ERNEST J., III. "Energy, Africa and World Politics." Review of Black Political Economy, 3, 4, 1973, 27-41.

2825 WILSON, E. J., III. "The Energy Crisis and African Underdevelopment." Africa Today, 22, 4, October-December 1975, 11-37.
Although African nations broke off relations with Israel, Arab aid has not been forthcoming. Africa should try to develop a measure of autonomy and self-reliance.

2826 WIRSING, ERICH. "Der Gemeinsame Markt und die Entwicklungsländer. Zum Beginn der Verhandlungen um eine Neue Periode der Assoziierung der Afrikanischen Länder und Madagaskars." Europa-Archiv, 24, 3, January 1969, 89-100.

2827 WISEBERG, LAURIE S. "Christian Churches and the Nigerian Civil War." Journal of African Studies, 2, 3, Fall 1975, 297-331, tabl.
Examines the role of the churches from a detached and analytical social science perspective, for the sake of a more accurate understanding of the war, and in order to address the theoretical question of assessing the influence that the churches - as nongovernmental or transnational actors - exerted in the crisis.

2828 _____. "Humanitarian Intervention: Lessons from the Nigerian Civil War." Human Rights Journal, 7, 1, 1974, 61-98.

2829 _____. "An International Perspective on the African Famines." In: Glantz, M. H., (ed.). The Politics of Natural Disaster. New York, 1976, 101-27.
Drought is a natural phenomenon, but the suffering that accompanies drought only in the Third World is a man-made phenomenon resulting from the structure of international economic and political relations.

2830 _____. "An International Perspective on the African Famines." Revue Canadienne des Etudes Africaines, 9, 2, 1975, 293-314.

2831 WODDIS, JACK. "Neo-colonialism in Africa: Schemes and Failures." World Marxist Review, 14, 6, June 1971, 103-12.
The majority of African states are still under imperialist domination.

2832 WRIGHLY, C. C. "Empire and Commerce in Africa." Journal of Commonwealth Political Studies, 7, 3, November 1969, 246-57.

2833 YAKEMTCHOUK, ROMAIN. Assitance Economique et Pénétration Industrielle des Pays de l'Est en Afrique. Leopoldville: Institut de Recherches Economiques et Sociales, 1966. 104 p., tabls.
Soviet aid to Africa with strong warnings from the author of the threat of communist influence through aid and trade relationships.

2834 YAKEMTCHOUK, R. La Convention de Lomé: Nouvelles Formes de la Coopération entre la C. E. E. et les Etats d'Afrique, des Caraïbes et du Pacifique. Brussels: Académie Royale des Sciences d'Outre-Mer, 1977. 181 p.

2835 YU, GEORGE T. "Chinese Aid to Africa: The Tanzania-Zambia Railway." In: Weinstein, W., (ed.). Chinese and Soviet Aid to Africa. New York, 1975, 29-55, tabl.
    A revised version of Yu's 1971 essay in Asian Survey.

2836 ZARTMAN, I. WILLIAM. "Europe and Africa: Decolonization or Dependency?" Foreign Affairs, 54, 2, January 1976, 325-43.
    The author argues that African independence is being gained gradually through decolonization, a process still on-going.

2837 _____. "The EEC's New Deal with Africa." Africa Report, 15, 2, February 1970, 28-31.

2838 _____. The Politics of Trade Negotiations between Africa and the European Economic Community: The Weak Confront the Strong. Princeton: Princeton University, 1971. 243 p., index, tabls.
    This analysis of the negotiations between the EEC and the African states is a study in the manner and process by which the weak states negotiate with the strong, the nature of North-South diplomacy.

2839 _____. "Les Transferts d'Armements en Afrique." Etudes Internationales, 8, 3, September 1977, 478-86.

2840 ZEYLSTRA, WILLEM G. Aid or Development: The Relevance of Development Aid to Problems of Developing Countries. Leyden: A.W. Sijthoff, 1975. 269 p., bibl.
    Aid is not development-oriented; it is for political purposes. Aid serves to recolonize the recipient state. The diagnosis of underdevelopment is wrong and therefore the theory of aid is wrong.

Subject Index

AAPAM, 1091
AAPC, 1032
AAPSO, 1955
Accra Conference (1965), 454, 479
ACORD, 2518
ACP. See EEC
ADB, 675, 885, 944, 1038, 2716, 2773
Addis Ababa Agreement (1972), 453, 937
Aden, 639
Afars and Issas, 487, 602, 638-9, 1851, 1923, 1979
Africa: place of, in world politics, 19, 35, 76, 81, 89, 94, 249, 299, 326-7, 1237, 2581, 2769, 2812-3
Africa Institute (Moscow), 546
African-American Labour Centre, 2220
African Common Market, 2722
African Development Fund, 2794
African Diaspora, 37, 1044, 1880
African Personality, 932
African Socialism, 1895, 1965, 2157, 2486, 2771
African unity, 23, 36, 62, 91, 292, 305, 365, 391, 426, 460, 627, 780, 806, 821, 914, 936, 964, 971, 979, 1040, 1066, 1286; definition of, 1068, 1125; goals of, 931, 1042, 1063, 1088, 1095; history of, 865, 910, 941, 970, 996, 998, 1024, 1046, 1049, 1056, 1081-2, 1085, 1092, 1117, 1122, 1132, 1135
Africanism. See Pan-Africanism
Afrikaaners, 1496
Afrique Occidental Française. See French West Africa
Afro-Americans, 37, 347, 917, 1009, 2104-5, 2120, 2135, 2145, 2165, 2173, 2175, 2186, 2207, 2211-2, 2246, 2286, 2293
Afro-Arab Ministers Conference, Dakar (1976), 2404
Afro-Arab Summit Conference, Cairo (1977), 2305-6, 2405, 2466
Afro-Asian Conference, Second, 357, 409, 446
Afro-Asian Solidarity Conference, 454, 479
Afro-Asian states, 1, 55, 66, 99, 139, 184, 248, 282, 295, 357-8, 409, 446, 454, 479, 949, 1141, 1191, 1195, 1253, 1257
Agency for International Development. See USA, aid programs of
Ahidjo, Ahmadou, 113, 157
Air Afrique, 675, 692, 902
Akinyemi, A. B.: works of, reviewed, 119

Akumu, Dennis, 555
Algeria: foreign relations of, 127, 155-6, 168, 182, 353, 368, 423, 2038, 2330, 2355, 2792; relations of, with Morocco, 507, 593, 649, 657-8, 661
Algiers Conference (1965), 357, 409, 446
Aliens Compliance Order, Ghana, 491
Aluko, O.: works of, reviewed, 627
American Committee on Africa, 2169
Amin, Idi, 1127; foreign policy of, 241, 440, 2277
Anglo-Nigerian Defence Agreement, 278
Anglophone states, 563, 867, 2754; relations of, with francophone states, 535, 659, 2667, 2802
Angola, 21, 995, 1006, 1147, 1342, 1775, 2027, 2457, 2575; crisis and war in, 57, 309, 315, 390, 1160, 1444, 1499, 1538, 1694, 1747, 1758, 1825, 1827, 1860, 1935, 2002, 2009; in Southern Africa, 1407, 1520, 1563, 1629, 1630, 1659; relations of, with Cuba, 2037, 2088, 2198, 2278, 2325, 2355, 2394, 2476; relations of, with Portugal, 2335, 2359, 2361, 2698; relations of, with South Africa, 1489, 1545, 1641, 1643, 1723, 1863; relations of, with USA, 57, 1576, 1827, 1863, 1917, 1935, 2027, 2048, 2088-9, 2095, 2115, 2121, 2125, 2129, 2132, 2158, 2163, 2198, 2203, 2262, 2273, 2278; relations of, with USSR, 1747, 1827, 1841, 1863, 1902, 1916, 1917, 1935, 1967, 2002, 2019, 2037, 2048, 2158, 2199, 2278, 2288

Anti-Apartheid Movement, 1577
Anti-communism, 2371
Anti-imperialism: in African foreign policies, 223, 366, 1136
Anti-Semitism, 1691
AOAS, 1091
Arab Bank for African Economic Development, 2500, 2796
Arab League, 611, 2305, 2424
Arab states, 620, 1213-4, 1252, 2325, 2424, 2446, 2465, 2732, 2746; aid programs of, 2334, 2397, 2492, 2528, 2685, 2825; and the OAU, 974, 984-5, 1026, relations of, with Africa, 22, 106, 111, 114, 120, 152, 155, 157, 159, 167-70, 175-6, 186, 214, 218-20, 228, 244, 250, 329, 339, 424, 440, 445, 453, 471, 509, 545, 557, 559, 570, 724, 908, 914, 957, 984-5, 1175, 1213-4, 1651, 2302, 2305-7, 2309, 2317, 2329-30, 2334, 2343-4, 2347, 2358, 2377, 2387, 2396-7, 2404-5, 2412, 2424, 2427-8, 2435, 2446, 2451, 2459, 2461, 2466, 2473, 2477, 2479, 2483-4, 2500, 2512, 2540, 2717
Arabism, 111
Argentina, 1848; relations of, with Africa, 2419
Arkhurst, F.: works of, reviewed, 2226
Arms, 11, 1460, 1519, 2570, 2651, 2729, 2790-1; sales of, to Africa, 30, 1984, 2499, 2839; sales of, to South Africa, 245, 911, 1460, 1482, 1519, 1577, 1606, 1703, 1719, 1787, 1804, 1923, 2147, 2611
Arms control, 11, 92, 2601
Arusha Declaration, 407
Asia, 32, 994, 1193, 1317, 1738, 1783, 1887, 1889, 1993, 2061, 2070, 2333, 2414, 2456, 2497, 2554, 2646, 2693, 2737; and Africa at the UN, 1141,

Asia (continued), 1191,
1195, 1253, 1281, 1789
Asians, 356, 1299, 1607, 1853,
2431
Australia: relations of, with
Africa, 2370, 2380; relations of, with South
Africa, 1455, 1599, 2372

Bailey, M.: works of, reviewed, 2008
Balewa, Abubakr, 119, 153,
310-1
Balkanization, 73, 1037, 1748
Banda, Hastings, 267
Bantustan, 261, 1494
Basin. See River basin
Belgium, 2178; aid programs
of, 2535, 2818; relations
of, with Africa, 2364,
2462; relations of, with
Congo/Zaire, 1257, 2178,
2331, 2354, 2385
Benin, 483, 596, 696, 726,
765-6, 864, 1227. Also
see Conseil de l'Entente
Bibliography, 28-9, 33-4, 48,
79, 85, 88, 91, 93, 124,
572, 579, 688, 717, 762,
795, 1013, 1195, 1781,
1784, 1920, 1922, 2351,
2456, 2484, 2819; of international relations in
Southern Africa, 1430,
1433, 1501, 1520, 1654,
2134, 2284; of Pan-Africanism, 91, 813, 819, 825,
883, 972, 981, 1009, 1055,
1060, 1072; of US relations
with Africa, 1009, 1809,
2078, 2129, 2134, 2165,
2186, 2222, 2284, 2298;
Blyden, Edward: views of,
on imperialism, 2628
Bophuthatswana, 1391
Border conflicts, 507-8, 510,
522-3, 546, 554, 560, 634,
643, 646, 652, 655, 657-8,
664, 666, 916. Also see
specific cases

Borders; African, 508, 512,
529, 543, 550, 588, 605,
608, 610, 614, 616-7, 633-4,
643, 645, 648, 656, 667,
1261, 1425, 1585; as cause
of conflict, 489, 507, 510,
519, 546, 560, 634, 643,
645, 652, 655, 666, 1505;
economic aspects of, 633,
655, 702, 733; legal aspects
of, 495-6, 512, 664, 1261,
1353; OAU on, 522-3, 628,
646, 662
Botswana, 240, 257, 532, 609,
1147; aid programs in,
1514, 1545, 1547; domestic
politics in, 200, 261-2,
532; foreign relations of,
160, 165, 199-203, 240,
256-7, 261-2, 286, 307,
425, 450-1, 476, 551, 1560,
1752, 2015, 2182, 2365,
2648; relations of, with
Southern African, 517,
1407, 1413, 1419, 1512,
1524, 1551, 1563, 1641,
1643, 2118
Brazil, 1848, 2414, 2456;
relations of, with Africa,
2332, 2356, 2363, 2366,
2398, 2414, 2438-40, 2456-7;
relations of, with Portuguese-speaking states,
2366, 2398, 2414, 2439,
2457; relations of, with
Southern Africa, 2369,
2457
Brezhnev Doctrine, 1974
British South Africa Company,
227
Brzezinski, Z.: works of,
reviewed, 1944
Burundi, 588, 623, 693, 724,
818, 1821, 2051, 2096,
2554, 2814, 2818

Cabora Bassa, 1572
CAFRAD, 1091
Cairo Conference (1957), 248

Cameroon, 589, 653, 739, 790, 900, 1200; foreign relations of, 113, 121, 157, 301, 381, 2712, 2714
Canada: aid programs of, 1523, 2450, 2609, 2754, 2772; and the Commonwealth, 2378, 2407; relations of, with Africa, 2314, 2407-9, 2448, 2450, 2455, 2754; relations of, with francophonie, 2374, 2407, 2455, 2478, 2609, 2754; relations of, with Southern Africa, 1523, 1528, 1577, 2303, 2312-3, 2315, 2328, 2336, 2365, 2377, 2383, 2410, 2455, 2460, 2756,
Canary Islands, 967
CAP. See EEC
Cape route: and military strategy, 1468, 1623; USSR threat to, 1428, 1517
Cape Verde, 2297
Capitalism, 75; state, 97
Caribbean, 2554
Carnegie Foundation, 64
Cartels, 2496, 2574
Carter, J.: Africa policies of, 2074, 2082, 2110, 2116, 2122, 2190, 2198-9, 2240, 2250; visit to Nigeria of, 2238
Casablanca group, 659
CCTA, 1064, 1223, 1354
CEAO, 741, 751, 757, 869; bibliography of, 688
Center-periphery relations, 13, 97
Central Africa, 633, 680, 1299, 1754, 2051, 2125, 2485; regionalism in, 693, 801-2, 818, 832, 871, 902, 2765
Central African Empire. See Central African Republic
Central African Federation, 1151
Central African Republic, 192, 623, 697
Césaire, Aimé, 2628
CFAO, 1988

Chad, 558, 565, 589, 623, 697, 809, 897, 2358
Chad Basin Commission, 837
China. See People's Republic of China; Republic of China
Churches: and apartheid, 2778; relief and aid programs of, 2515, 2767, 2827. Also see Biafra, relief operations in; Nigeria, relief operations in; Voluntary organizations
CIA. See USA, covert operations of
CIDA, 1523, 2772
CILSS, 857
Class conflict, 97, 1056
Cobalt, 2152
Cocoa, 765, 2574
Cohen. B.: works of. reviewed, 2256
Cohn, H. D.: works of, reviewed, 1883, 1901
Colbert, Jean-Baptiste, 2713
Cold War, 15, 18, 253, 405, 613, 1482, 1794, 1813, 2001, 2005
Collective security: and the OAU, 1025, 1052
Colonialism, 96, 429, 670, 980, 1322, 2159, 2174, 2354; effects of, 1, 8, 12, 26, 74, 76, 445, 500, 546, 643, 712, 718, 873, 932, 1007, 2178, 2744; UN and, 1188, 1342, 1346. Also see UN, and decolonization
Comintern, 1880
Commission of Mediation, Conciliation and Arbitration. See OAU, institutions of
Commodities, 1576, 2566, 2589, 2617, 2702. Also see Natural resources
Commonwealth, 78, 235, 359, 391, 1523, 1853, 1872, 1934, 2054, 2378, 2407, 2740; and Africa, 53, 444, 535, 931, 1047, 1864, 2058; and the EEC, 1840, 1918, 1934, 2636, 2758; and the

Commonwealth (continued),
Rhodesian crisis, 138, 359,
1504, 1578, 1757, 1934;
and the South African situation, 1415, 1482, 1509,
1573, 1717, 1858, 2378
Commonwealth Development
Corporation, 2740
Communism, 14, 36, 79, 145,
1031, 1359, 1486, 1620,
1717, 1862, 1896, 1941,
1943-4, 2243, 2704, 2833
Communist states, 1141, 1875,
2710, 2757, 2817; aid and
economic relations of,
with Africa, 1532, 2521,
2757, 2805, 2817; relations of, with Africa,
90-1, 99, 195, 342-3, 444,
464, 482, 974, 1770-1,
1788, 1793-4, 1811, 1847,
1877-8, 1953, 1975
Comoro Islands, 2297
Comprador, 2630
Conflict, 16, 18, 25, 59, 67,
104, 189, 526, 530, 533, 536,
558, 569, 578-80, 585,
590, 596, 600, 629, 879,
1154, 1187, 1813. Also
see Border conflicts
Conflict resolution, 16
Congress of Racial Equality,
2169
Conseil de l'Entente, 799,
804, 864, 867, 902, 904
Content analysis, 1740
Copper, 2525, 2774
Core state. See Middle powers
Coup d'etat, 10, 597, 635,
922, 1127, 1372, 1937,
2677, 2750
Crabb, Cecil: works of, reviewed, 172
Cuba, 1999, 2325, 2355, 2371;
relations of, with African
states, 525, 1847, 1849,
2325, 2349, 2355, 2422,
2426, 2805; relations of,
with Angola, 2037, 2088,
2198, 2278, 2325, 2355,
2394, 2476; relations of,
with Southern Africa, 1381,
1645, 2233; relations of,
with the USSR, 2325, 2355,
2441
Czechoslovakia, 2324

Dahomey. See Benin
Davis, M.: works of, reviewed,
2584
Decision-making, 484, 665,
1644, 2328
Decolonization, 15, 56, 222,
343, 408, 487, 506, 544,
577, 602, 872, 1423, 1618,
1745, 1810, 1939, 1950,
1995, 2090, 2095, 2345,
2373, 2388, 2420, 2836; UN
role in, 544, 1158, 1173,
1178, 1190-1, 1196, 1251,
1254, 1275, 1319, 1335,
1337, 2373
De Gaulle, Charles: Africa
policy of, 353, 1748, 1778,
1807, 2038, 2052
Deng, Francis, 229
Denmark, 1516
Dependence, 4, 7-8, 10, 13,
26, 47, 50, 61, 65, 75,
86-9, 97, 100-1, 144, 221,
286, 298, 371-2, 390, 413,
433-4, 439-40, 452, 551,
739, 1409, 1478-9, 1652,
1655, 2540, 2606, 2610,
2662, 2668, 2684, 2691,
2698, 2703, 2768, 2812-3,
2818, 2836; cultural, 47,
390, 2108, 2610; economic,
47, 65, 165, 344, 917, 1003,
2563, 2592, 2610; quantitative analysis of, 65,
2662, 2691, 2777, 2810
Détente, 641, 1765, 1812-3,
1833, 1902, 1968, 2048
Deutsch, Karl, 716
Diaspora, 37, 1044, 1880
Diori, H., 225
Diplomatic corps, 3, 70, 128,
563, 599, 2307
Disputes. See Conflict
Divestiture, 2537
Djibouti. See Affars and
Issas

Domestic politics, 3, 10, 17,
23, 34, 38, 45, 53, 59, 70,
83, 85, 90, 93-5, 197, 474,
537, 587, 918, 920, 1441, 1828;
and international conflict,
189, 526, 578, 1827; and
the OAU, 269, 923, 955,
958, 962, 983, 1073, 1098,
1124, 1128, 1345
Drought: in Sahel, 42, 2770;
and famine, 2829-30
Dugard, J.: works of, reviewed, 1423

EAC, 584, 675, 685, 707, 721, 723,
747, 755, 757, 763, 787, 789,
791, 798, 844, 858, 867, 878,
882, 903-4, 2299, 2711; and
the EEC, 758, 2488; bibliography of, 762; break-up of,
761, 775, 807, 820, 852
EACM, 690, 721, 745, 829
EACSO, 686, 687, 748, 886, 1032;
federal university in, 886
EAEC, 677, 916
East Africa, 84, 162, 191,
205, 320, 571, 588, 621,
633, 679, 685, 738, 778,
814, 881, 1299, 1549,
1985; economic regionalism in, 683, 722, 725,
728, 754-5, 787, 800,
815-6, 831, 833, 836, 861,
876-7, 738, 753, 2501,
2765; economic relations
of, 690, 2617, 2735, 2759;
federation in, 686, 694,
746, 770, 793, 800, 814,
833, 849, 861, 870-2, 884,
2501; political regionalism in, 679, 685, 706,
711, 722, 737, 779, 787,
802, 833, 874, 876-7;
regional organizations
in, 885; regionalism in,
62, 151, 699, 737-9, 769,
792, 794, 834-5, 853, 873;
relations of, with the
USSR, 1549, 1663. Also
see names of specific
regional organizations

East African Development Bank,
675
East Germany: relations of,
with Africa, 2406, 2442,
2447
Eastern Europe: aid programs
of, 2792, 2805; relations
of, with Africa, 1770,
2323, 2757, 2792. Also see
names of specific countries
ECA, 675, 681, 689, 926, 1038,
1091, 1155, 1172, 1201,
1220, 1225, 1258, 1266-7,
1300, 1354, 2534, 2794
Economic Community of the
Great Lakes States, 801
ECOWAS, 675, 729, 751, 782, 785,
808, 828, 841, 856, 905-7,
2299, 2669
Education, 886, 908, 1288,
2579, 2739, 2772
EEC, 2533, 2552, 2578, 2638,
2764; aid program of, 2508,
2549, 2562, 2650, 2666,
2688, 2780, 2782; and
Africa, 53, 133, 730, 758,
885, 1116, 1357, 1628,
1840, 1918, 1934, 2344,
2433, 2488, 2524, 2526,
2529, 2550, 2557-8, 2564,
2567, 2596, 2625, 2635-6,
2638, 2640, 2643, 2648,
2658, 2661, 2667, 2671-2,
2690, 2699, 2705, 2713,
2715, 2722, 2731, 2741,
2752, 2761-2, 2765-6, 2785,
2798-9. 2807, 2826, 2837-8;
and the Lomé Convention,
881, 2489, 2495, 2523,
2526, 2536, 2549, 2577, 2593-
4, 2606, 2612, 2623, 2652,
2661, 2664, 2666, 2683,
2686, 2688, 2695, 2707,
2708, 2758, 2763, 2786,
2797, 2804, 2806, 2808-9,
2834; and the Yaoundé
Convention, 312, 2490-1,
2498, 2523, 2541-2, 2550,
2568, 2572, 2683, 2784,
2802
Egypt, 127, 166, 424, 985,
1984, 2573, 2792; relations

Egypt (continued), of, with Africa, 62, 149, 166, 219, 230, 325, 349, 352, 416, 424, 486, 505, 545, 639, 1731, 2619
Ekangaki, Nzo, 1133
Elites, 851
El-Khawas, M. A.: works of, reviewed, 2256
Energy production, 979, 1651, 2487, 2824-5; in Zaire, 2485
Entebbe: Israeli raid on, 120, 1153, 1316, 1327
Environmental problems, 1155
Equatorial Africa, 825, 859
Equatorial Guinea, 572, 1147, 2348; relations of, with Nigeria, 121, 493, 604
Eritrea, 509, 513, 601, 983, 1759; war in, 574, 598, 1731, 2560
d'Estaing, Giscard, 2043
Ethiopia, 482, 509, 539, 545, 588, 602, 639, 1304, 1819, 1823, 1832, 1846, 1851, 1897, 2111, 2231, 2443, 2560, 2719; domestic politics in, 509, 1897, 1998, 2005; foreign policy of, 127, 150, 260, 281, 314; Eritrean crisis in, 509, 513, 545, 574, 598; relations of, with Somolia, 158, 269, 489, 507, 519, 524, 536, 545, 558, 576, 661, 1854; relations of, with USSR, 55, 1823, 1832, 1908-9, 2055-6, 2792
Eurafrique, 66, 213, 1096, 1835, 1838, 1920-1, 1963, 1987, 2366, 2498, 2561, 2636
Europe, 91, 349, 383, 670, 1746, 1755, 1804, 1867, 1953, 1969, 1992, 2026, 2434, 2440, 2483, 2519-20, 2540, 2639, 2783, 2836. Also see names of specific states
European Free Trade Association, 2667

Ewe, 900
Exiles, 57, 316

Factor analysis, 1281
Famines, 2829-30
Fanon, Frantz: views of, on imperialism, 2628
FAO, 1150, 1330, 2518
Fascism, 1658
FED. See EEC
Federal University of East Africa, 886
Federation of Mali, 740, 783, 916
Fermeda Workshop, 536
Fernando Poo. See Equatorial Guinea
FFHC, 2518
Filesi, T: works of, reviewed, 1879
Fonds de Solidarité Africain, 2730
Food, 1288, 2576
Ford, G., 2133, 2278
Foreign aid, 36, 316, 824, 878, 1003, 1144-5, 1198, 1276, 1470, 1523, 1532, 1697, 1734, 1943, 2297, 2316, 2366, 2431, 2445, 2493, 2502-3, 2511, 2516, 2519, 2532, 2535, 2538, 2586, 2590, 2598, 2603, 2608, 2639, 2641, 2653, 2656, 2660, 2663, 2673, 2676, 2694, 2706, 2709, 2742, 2792, 2794, 2803, 2805, 2817, 2840; voluntary organizations in; 2518, 2701, 2734, 2743, 2767, 2781. Also see specific donor countries
Foreign investment: in Africa, 1192, 1784, 1959, 1988, 2076, 2295, 2344, 2517, 2525, 2575, 2582, 2622, 2624, 2639, 2649, 2682, 2692, 2724, 2747-9, 2776, 2795, 2811; in Southern Africa, 1412, 1445, 1449, 1491, 1501, 1528, 1548, 1550, 1552-4, 1556, 1638

Foreign policy: African, 1, 5, 15, 36, 43, 67, 81, 91, 106-484, 510, 530, 569, 640, 642, 778, 1205. Also see names of specific states
Fourth World, 2746
Franc zone, 2559, 2588
France, 565, 975, 1257, 1801, 1838, 1851, 1958, 2189; cultural programs of, 1748, 1819, 1970, 1988, 2675; and decolonization, 1745, 1939, 1995; economic aid programs of, 52, 1748, 1766, 1797, 1837, 1892, 1959, 2032, 2509-11, 2514, 2527, 2562, 2588, 2620-1, 2632, 2674-5, 2678, 2680, 2720-1, 2726-7; economic relations of, with Africa, 1748, 1797, 1928, 1959, 1988, 2713; military assistance program of, 1745, 1748, 1797, 2679; relations of, with Africa, 53, 319, 353, 1755, 1766, 1797-8, 1801, 1807-8, 1817, 1819-20, 1836, 1838, 1843, 1915, 1971-2, 1988, 2021, 2041, 2043, 2049, 2052-3, 2072, 2119, 2189, 2647, 2686, 2726; relations of, with francophone Africa, 124, 179, 305, 354, 379, 456, 795, 969, 1745, 1748, 1755, 1763, 1766, 1776-8, 1810, 1822, 1842, 1852, 1892, 1926-7, 1939-40, 1952, 1959, 1973, 1980, 1995-6, 2004, 2006-7, 2032, 2034, 2047, 2713, 2730; relations of, with Southern Africa, 1492, 1516, 1597, 1685, 1712, 1798, 2027; military concerns of, in Africa, 1745, 1800, 1915, 1928, 1988
Francophonie, 235, 369, 456, 623, 692, 724, 803, 846, 1744, 1958, 2050, 2142, 2535, 2543, 2632, 2754, 2800; bibliography of, 124, 1781; and French language, 1744, 2050; and regional integration, 724, 773, 760, 803, 825, 846, 867, 885; relations of, with anglophone states, 535, 659, 2667, 2802; relations of, with Canada, 2374, 2407, 2455, 2478, 2609; relations of, with France, 124, 1743, 1748, 1763, 1810, 1816, 1822, 1842, 1927-8, 1939-40, 1973, 1980, 1995-6, 2004, 2007, 2032, 2047, 2675, 2679-80, 2713, 2730, 2800
FRELIMO, 427
French Association for the Study of the Third World, 2533
French Community, 354, 535, 1939-50, 2004, 2007
French West Africa, 483, 494, 1763
Front-line states, 988, 1407, 2365

Gabon, 208, 572, 727, 2119
Gambia, 535, 1147, 2322; relations of, with Senegal, 369, 651, 811, 850, 862-3, 900
Garba, J. W., 274
Garvey, Marcus: views of, on imperialism, 2628
Genocide, 2096
Geopolitics, 239-40, 320, 836, 1361, 1468, 1990, 2631. Also see Land-locked states
Ghana, 36, 589, 2660; domestic politics in, 129, 246, 490, 491; foreign African migrants in, 490, 491, 632; foreign relations of, 63, 112, 117, 127, 129-30, 138, 161, 209-10, 222, 238, 246, 252, 303, 367, 383-4, 418, 462-4, 469, 503, 629, 726, 839, 849, 1099, 1151, 2001, 2111, 2125, 2455, 2677;

Ghana (continued), relations of, with Nigeria, 121, 154, 497, 627; relations of, with USSR, 1762, 1824, 1913, 2001, 2018, 2020, 2573, 2788-9, 2792
Ghana-Guinea-Mali Union. See Union of African States
Glantz, M. H.: works of, reviewed, 2584
Gold, 1673
Gowan, General Y., 396
Great Britain, 565, 871, 1342, 1384, 1853, 2046, 2155; aid programs of, 1514, 2015, 2511, 2562, 2605, 2681, 2703, 2795; arms sales to South Africa by, 245, 1787, 1904, 1923, 2611; relations of, with Africa, 161, 459, 969, 975, 1257, 1754-5, 1810, 1828, 1840, 1843, 1853, 1865, 1893, 1906, 1934, 1938, 1970, 1982, 2059, 2072, 2611, 2686, 2795; relations of, with the Commonwealth, 1757, 1853, 1918, 1934; relations of, with the EEC, 1840, 1918, 1934, 2785; relations of, with Nigeria, 132, 161, 278, 397, 401, 455, 1740-1, 1801; relations of, with Rhodesia, 138, 414, 1268, 1426, 1438, 1516, 1578, 1602, 1637, 1660, 1664, 1720, 1757, 1805, 1932, 1934, 2249; relations of, with Southern Africa, 1369, 1397, 1492, 1564, 1686, 1694, 1749, 1753-4, 1787, 1904, 1906, 2015, 2077
Groundnuts, 705
Guinea, 589, 629, 654, 698, 701, 724, 2025, 2325, 2539; foreign relations of, 107, 127, 263-4, 421, 456, 467-8, 483; invasion of, 31, 488, 1130; relations of, with France, 179, 1776-8, relations of, with PRC, 1752, 2001, 2068; relations of, with USA, 1752, 2213, 2539; relations of, with USSR, 1752, 1913, 2001, 2020, 2213, 2787, 2792
Guinea-Bissau, 369, 488, 1147, 1206, 1265, 2057, 2457; relations of, with Portugal, 2335, 2359, 2361

Haas, Ernest, 1111
Hadrami, Omar, 266
Hall, R.: works of, reviewed, 2008
Hammarskjold, Dag, 1174
High Commission Territories, 160. Also see Botswana; Lesotho; Swaziland
Horn of Africa, 363, 1731-2, 1851, 1926, 2465; conflicts in, 269, 489, 517, 539-40. 542, 545, 566, 571, 582, 613, 623, 628, 638-9, 641, 647, 1731-3, 1760, 1845, 1894, 1897, 1999, 2005; and the USA, 1732, 1833, 1851, 1894, 1979, 1999, 2160; and the USSR, 1731-2, 1814, 1823, 1833, 1851, 1859, 1876, 1894, 1908-9, 1931, 1979, 1999, 2040
Houphouët-Boigny, F., 212, 224, 271, 354, 1040
Human rights, 63, 920, 1187, 1192, 1202, 1301, 1335, 1345, 1812; and US foreign policy, 2074, 2190, 2277
Humanism, 437
Humphrey, H., 2111
Hungary, 139
Huntington, Samuel, 1110
Hutchison, A.: works of, reviewed, 1861

IBRD, 1156, 2152, 2511
ICJ, 1182, 1192, 1317, 1326, 1450; and Namibia case, 1159, 1180-2, 1184, 1192, 1203, 1215, 1217, 1219,

ICJ (continued), 1221, 1228-31, 1233, 1243, 1247, 1263, 1270, 1279, 1285, 1294, 1301-2, 1304, 1307, 1310, 1318, 1321, 1329, 1333, 1340, 1347, 1488, 1526, 1603, 1640; and South Africa, 1181, 1184, 1243, 1279, 1304, 1665; and Western Sahara case, 1152, 1163, 1167-8, 1179, 1208, 1338
ICVA, 2518
Idang, Gordon, J.: works of, reviewed, 119
Ideology, 67, 295, 330, 437, 460-2, 484, 533, 594, 629, 772, 781, 820, 922, 1048, 1087, 1598, 1815, 1951, 1960, 2296, 2380; in USSR foreign policy, 1772, 1803, 1888, 1896, 1937, 1947, 1961
Ifni, 1142
ILO, 1065, 1352, 1634
IMF, 2152
Imperialism, 75, 145, 366, 439, 615, 1775, 1960, 1982, 1993, 1997, 2024-5, 2043, 2064, 2108, 2234, 2250, 2268, 2342, 2430, 2595, 2628-9, 2718, 2220, 2831; and South Africa, 1522, 1530, 1555, 1557, 1596, 1601, 2221; and the OAU, 924, 1933
India, 295, 1853, 2388; relations of, with Africa, 66, 349, 1234, 2310, 2376, 2423, 2431; relations of, with South Africa, 2480
Indian Ocean, 181, 191, 253, 362-3, 1618, 1754, 1782, 1799, 1875, 1990, 2035, 2046; and littoral states, 161, 191, 253, 320, 482, 1367, 1471, 1767, 1979, 1985, 2148; military and strategic concerns in, 215, 1384, 1570, 1618, 1678, 1685, 1716, 1799, 1814, 1905, 2010, 2039, 2270; USA and, 1767, 1891, 1903, 1979, 1981, 2046, 2091, 2115, 2148, 2179, 2237, 2270; USSR and, 1814, 1875, 1881-2, 1891, 1903, 1905, 1979, 1981, 2010, 2039, 2237, 2270
Indians. See Asians
Indigenization, 2749
Indonesia, 295
Industrialization, 689, 830-1, 979, 1890, 2586
Institution-building, 2634
Interdependence, 87, 89, 1655, 1699, 2612, 2812
International Conference of African Studies (4th.), 47
International Conference on Economic Sanctions against South Africa, 1639
International Defence and Aid Fund, 1414
International division of labor, 2561, 2594, 2652, 2707
International integration, 11, 58, 78, 418, 695, 700, 703, 710, 739, 745, 756, 767, 784, 813, 819, 822, 831, 853, 866-7, 875, 883, 947, 979, 1010, 1078, 1094, 1187, 2557, 2601; economic, 731, 734, 747, 749, 752, 764, 777, 830, 858, 898, 969, 1007, 1220, 1419, 2765, 2712; political, 684, 704, 715, 752, 780, 865, 872, 899, 931-2, 971, 1007, 1049, 1069, 1648, 2534
International law, 56, 429, 505, 517, 520, 562, 591, 919, 1022, 1137-8, 1143, 1149, 1153-4, 1160, 1168, 1179, 1187, 1192, 1205, 1210, 1241, 1278, 1293, 1299, 1306, 1308, 1316, 1327, 1450, 1697; African contribution to, 1170, 1192-3, 1200, 1218; African views of, 1159, 1210, 1240, 1276, 1317, 1326; and de-colonization, 1162, 1169,

International law (continued), 1178, 1295, 1303, 1447; and international sanctions, 1438, 1443, 1452, 1487, 1515, 1729; and the Namibian situation, 1157, 1178, 1215, 1263, 1301, 1310, 1311, 1318, 1329, 1332, 1488, 1525, 1603, 2192; and the Nigerian Civil War, 1185, 1241, 1280, 1349-50; of boundaries, 495-6, 512, 664, 1353; of international organizations, 675, 737, 746, 839, 885, 1273, 2564, 2640, 2764; of the sea, 1170, 1177, 1240, 1296

International organizations, 3, 70, 669, 674-5, 688, 739, 743-4, 747, 752, 768, 771, 821, 832, 885, 896, 900, 913, 929, 944, 958, 968, 973, 994, 1010, 1047, 1057, 1059, 1134, 1150, 1187, 1224-5, 1322, 1341, 1379, 1410, 1426, 1447, 1610, 1648, 1657, 1670, 1679, 1699, 2090, 2368, 2562, 2602-3, 2641; bibliography of, 1013, 1072. Also see names of specific organizations

International trade, 58, 67, 671, 690, 705, 720, 765-6, 827, 848, 896, 969, 1198, 1268, 1389, 1528, 1574-5, 1983, 2020, 2060, 2075, 2295, 2307, 2393, 2419, 2431, 2445, 2464, 2494, 2497, 2517, 2521, 2530, 2549-50, 2583, 2587, 2635, 2654, 2693, 2705, 2739, 2753, 2757, 2759, 2775, 2783, 2788, 2792, 2816-7, 2832-3, 2838

Internationalism, 1568
Iran, 1999, 2441
Ireland, 2379
Irredentism, 493, 502, 582, 648
Islam, 228, 349, 445, 482, 570, 575, 589, 974, 978, 1066, 2302, 2540

Israel, 1731, 1999, 1213-4; aid programs of, 2333, 2377, 2386, 2391-3, 2400, 2451, 2497, 2512, 2556, 2599, 2642, 2644-5, 2655, 2665, 2728, 2739, 2759; and Entebbe raid, 102, 1153, 1316, 1327; economic relations of, with Africa, 2393, 2401, 2494, 2684, 2739, 2759; relations of, with Africa, 80, 120, 218-20, 250, 291, 329, 471, 984-5, 1175, 2300-1, 2304, 2307-9, 2311, 2318-22, 2333, 2341-2, 2350-3, 2357-8, 2367-8, 2377, 2382, 2384, 2386, 2391-3, 2395, 2397, 2400-2, 2413, 2415, 2418, 2425, 2428, 2446, 2449, 2451, 2459, 2461, 2463, 2467, 2474-5, 2479, 2481, 2739, 2825; relations of, with South Africa, 291, 1362, 1448, 1551, 1651, 1704, 1707, 2318, 2446, 2467, 2474; relations of, with Third World, 2350, 2352, 2392-3

Italy, 565, 1516, 2346, 2373
Ivory Coast, 537, 739, 864; foreign relations of, 124, 127, 212, 224, 271, 483, 726, 1405, 1913, 2111, 2642

Jansen, G. H.: works of, reviewed, 172
Japan, 1518, 2464, 2511, 2587, 2811
Jews: in South Africa, 1362, 1579, 1718, 2467

Kaunda, K., 142, 239, 351, 374
Keita, Modibo, 292
Kennedy, J. F., 2159, 2254
Kenya, 439, 588, 609, 763, 738-9, 1147, 2668; foreign relations of, 127, 216,

Kenya (continued), 272, 320, 373, 403-4, 723, 728, 778, 878, 1752, 1846, 2020, 2111, 2598, 2738, 2792; relations of, with Somolia, 269, 489, 507, 519, 536, 661, 952
Khama, Sir Seretse, 202, 2182
Kissinger, H.: Africa foreign policy of, 1446, 1541, 1587, 2078, 2082, 2089, 2133, 2183, 2188, 2210, 2259, 2278
Kitchen, H.: works of, reviewed, 2255

Labor movements, 1032, 1061, 1641, 1769, 2220-1, 2532
Lake. See River basin
Lake, A.: works of, reviewed, 2256
Lamizana, 226
Land-locked states, 9, 514-6, 606, 623, 1293, 1308, 1361, 1409, 1533, 2637. Also see names of specific states
Larkin, B.: works of, reviewed, 1793, 1883, 2028
Latin America, 32, 87, 744, 876, 1281, 1309, 1339, 1395, 1658, 1738, 1844, 1849, 1887, 2061, 2352, 2381, 2417, 2497, 2737
League of Arab States, 951
Legitimacy, 543, 1110-1
Legum, C.: works of, reviewed, 1424
Legum, M.: works of, reviewed, 1424
Legvold, R.: works of, reviewed, 2028
Leiss, A. C.: works of, reviewed, 1424
Leninism, 1989
Lesotho, 240, 298, 452, 481, 517, 636-7; foreign relations of, 160, 237, 240, 450, 1147, 1514, 1547, 1560, 1753, 2015, 2648; relations of, with

Southern Africa, 237, 240, 297-8, 448, 451-2, 476, 480, 517, 1413, 1419, 1512, 1524, 1545-6, 1563, 1641, 1643
Liberation Committee. See OAU, institutions of
Liberation movements, 91, 293-4, 316, 427, 988, 995, 1019, 1162, 1169, 1189, 1265, 1295, 1356, 1426, 1647, 1697-8, 1734; and the PRC, 1470, 1532, 1734, 1773, 1806, 1815, 2031; and the USSR, 1470, 1532, 1773, 1968, 1976
Liberia, 304, 812, 1304, 2094, 2111, 2266, 2689, 2692
Libya, 423, 552, 565, 2573
Lomé Convention. See EEC
Lonrho, 2546, 2629, 2793
Lonrho Agreement, 1016
Lumumba, Patrice, 289, 473
Lusaka Conference, 187, 236, 322
Lusaka Manifesto, 431-2, 1373
Lusophonie, 2398, 2457

Machel, Samora, 309
Machiavelli, 45
Maghreb. See North Africa
Makonnen, Ras, 1035
Malagasy, 243, 302, 1390, 1766
Malawi, 313, 351, 370, 375, 457, 2365, 2603, 2703; relations of, in Southern Africa, 183, 255, 267, 275, 313, 323, 351, 375, 426, 449, 457, 476, 549, 583, 588, 871, 1516, 1545, 1693
Mali, 494, 623, 698, 1892, 2001, 2068; foreign policy of, 190, 292, 483; relations of, with USSR, 1913, 2001, 2020, 2792
Mandate system, 1157, 1182-3, 1242, 1318, 1332, 1347, 1472, 1488
Marxism, 1965, 1989
Mauritania, 186, 305-6, 698, 724, 1766; and Western

Mauritania (continued),
  Sahara, 186, 231, 430,
  502, 544, 649
McGowan, P. : works of,
  reviewed, 2662, 2777
Mengistu, 539
Mennen, G. W., 2295
Mercenaries, 21, 45, 98, 2140
Middle East, 34, 70, 2458;
  relations of, with Africa,
  445, 2327, 2387; relations
  of, with USSR, 1779, 1795,
  1887, 1889, 2014. Also
  see Arab states
Middle East Crisis, 106,
  218-20, 250, 291, 296,
  471, 984-5, 1001, 1026,
  1213-4, 1252, 2333, 2352-
  3, 2355, 2393, 2428, 2451,
  2459
Middle powers, 50, 101, 473,
  478, 738-9, 1475-6, 1478-
  9, 1642, 1652, 2800
Migration, 498, 584; in
  Southern Africa, 1383,
  1388, 1705; in Western
  Africa, 490-1, 498, 537,
  607, 632, 696, 720
Military aid, 1748, 1984,
  1988, 2545, 2560, 2573,
  2600, 2614-6, 2645, 2651,
  2679, 2750, 2818
Minority rights, 1440
Mobutu, J., 194, 290, 393,
  2152
Monetary institutions and
  arrangements, 670, 818,
  843, 896, 1413, 1583,
  1673, 2504, 2559, 2563,
  2588, 2608, 2689
Monrovia group, 659
Morocco, 214, 423, 618; and
  Western Sahara, 231, 430,
  472, 502, 507, 544, 593,
  611, 649, 657-8, 661
Mozambique, 588, 995, 1169,
  1390, 1465, 1499, 1576,
  1860, 2039, 2057, 2129,
  2132, 2174, 2457, 2575;
  and UN, 1147, 1195, 1265,
  1351; foreign policy of,
  254, 265, 309, 360, 427,
  1645; relations of, in
  Southern Africa, 265, 1390,
  1407, 1431, 1463, 1520,
  1545, 1565, 1572, 1629-30,
  1633, 1641, 1643, 1659,
  1715; relations of, with
  Portugal, 2335, 2359, 2361
Mozambique Channel, 1390
Multinational corporations,
  227, 481, 822, 1556, 1624,
  1641-2, 2546, 2548, 2629-
  30, 2659, 2668, 2748, 2774,
  2801; and economic development, 2565, 2629, 2649
Muslim Africa, 2387
Mwalimu. See Nyerere, J.

Namibia, 1242, 1423, 1451, 1462,
  1467, 1499, 1503, 1525, 1563,
  1624, 1641, 1643, 1659, 1709,
  2153; and the ICJ, 1159,
  1180-2, 1184, 1192, 1203,
  1215, 1217, 1219, 1221,
  1228-31, 1233, 1243, 1247
  1263, 1270, 1279, 1285, 1294,
  1301-2, 1304, 1307, 1310,
  1318, 1321, 1329, 1333, 1340,
  1347, 1488, 1526, 1603, 1624,
  1640; and the UN, 1144,
  1147, 1161, 1164, 1182, 1184,
  1191, 1199, 1221, 1242-3,
  1246, 1279, 1301, 1311, 1313,
  1318, 1323, 1334, 1342, 1347,
  1488; international law
  and, 1157, 1178, 1183,
  1263, 1301, 1310-11, 1329,
  1332, 2192; US and, 1418,
  1550, 1566, 1709, 2073,
  2075, 2103, 2153, 2169,
  2192
Nasser, A., 248
National interest, 484, 1099,
  1689, 1989, 1951, 2239
National Liberation Council,
  Ghana, 246
Nationalism, 250, 389, 460,
  502, 543, 570, 628, 647,
  702, 849, 1073, 1404, 1426,
  1568, 1745, 1794, 2070,
  2165, 2173, 2359, 2742,
  2749

Nationalization, 2493, 2547, 2749
NATO, 57, 1468, 1569, 1647, 2016, 2027, 2131, 2172, 2214
Natural resources: as a factor in international relations, 1521, 1641, 1652, 2152, 2214, 2489, 2517, 2589, 2718, 2736, 2769
Negritude, 1120, 1140; Senghor's, views of, 319, 1115
Neo-capitalism, 2593-4
Neocolonialism, 13, 51, 75, 145, 148, 358, 370, 921, 924, 992-3, 2042, 2108, 2220, 2361, 2434, 2590, 2595, 2631, 2743, 2831; and the EEC, 2550, 2638; French, 1952, 2021, 2049, 2800; in Southern Africa, 1461, 1463, 1475, 1522, 1643, 2154; US, 2226, 2268, 2539
Neofunctionalism, 747
Neo-imperialism, 2760
Netherlands, 1516
Neutralism, 55, 58, 81, 139, 163, 212, 292, 389, 418-21, 1739, 1794, 2059, 2739
New International Economic Order, 2495, 2534
New Zealand, 2471
Niger, 225, 494, 623, 705, 726, 864
Niger River Commission, 902
Nigeria, 50, 51, 133, 499, 589, 653, 739, 1143, 2722-3, 2803; civil war in, 49, 125, 455, 528, 766, 916, 923, 955, 962, 1118, 1128, 1280, 1292, 1349-50, 1740-1, 1784, 1801, 1827, 1958, 1964, 2017, 2022, 2468, 2570; domestic politics in, 119, 121, 137, 153, 180, 188, 196, 232, 258-9, 277-8, 399, 400; foreign relations of, 108, 114, 116, 118, 119, 121-2, 125, 127-8, 131, 135-7, 141, 147-8, 153-4, 162, 171, 188, 196, 204, 211, 232-3, 258-9, 274, 277, 310-1, 324, 390, 394-7, 400, 402, 406, 408, 411-2, 422, 428, 453, 455, 475, 478, 493, 497, 604, 627, 705, 709, 732, 758, 765-6, 837, 1739, 1958, 2308, 2311, 2452, 2455, 2587, 2598, 2712; relief operations in, 2515, 2551, 2553-5, 2569, 2696-7, 2755, 2827, 2828; relations of, with France, 1801, 1820, 1958, 2647; relations of, with Great Britain, 132, 161, 278, 397, 399, 455, 1740-1, 1801; relations of, with USA, 134, 397, 455, 1784, 1827, 1964, 2166, 2202, 2238, 2263; relations of with USSR, 134, 377, 397, 401, 1784, 1801, 1827, 1913, 1951, 1961, 1964, 2017-8, 2020, 2787
Nigerian Society of International Affairs, 109
Nile Waters Agreement, 486, 505
Nixon, R., 2099, 2100, 2107, 2130, 2133, 2200
Nkrumah, Kwame, 62, 112, 248, 966, 1937, 2718; foreign policy views of, 117, 222, 238, 383-4, 464, 469, 780, 1063, 1099; views of, on Pan-Africanism, 715, 915, 1099, 1267
Non-alignment, 63, 163, 172, 184, 198, 205, 239, 249, 282, 321-2, 342, 355, 358, 378, 392, 398, 444, 464, 931, 2262, 2375, 2419, 2432, 2445; in Nigerian foreign policy, 125, 135, 141, 397
Non-interference: principles of, 1697
North Africa, 589, 691; international relations of, 445,

North Africa (continued), 483, 624, 667, 853, 885, 1969
North Korea, 2432
Northwestern University, 2537
Norway: and the Nigerian Civil War, 2452, 2515, 2551, 2696; relations of, with South Africa, 1662, 2469-70, 2472
Nuclear questions, 1274, 1404, 1498, 1527, 1617, 1677, 1712, 2281, 2283, 2340, 2360, 2390, 2597
Numeiri, 178, 477
Nyasaland. See Malawi
Nyerere, J., 1067-9, foreign policy views of, 365, 391-2, 429, 915, 1048, 2184

OAMCE, 760, 788, 855, 1354. Also see OCAM; UAM; UAMCE
OAU, 5, 18, 53, 63, 176, 235, 516, 594, 640, 689, 743, 853, 885, 912, 916, 920-1, 924, 930, 954, 958, 962, 967, 986, 992-3, 1008, 1016-7, 1025, 1034, 1043, 1052, 1059, 1071, 1076, 1089, 1091, 1093, 1102, 1110-1, 1116-7, 1124, 1129, 1131, 1135, 1192, 1754, 2305; and African foreign policy, 113, 136, 147, 155, 237, 309, 314, 391, 1006; and international law, 919, 948, 958, 1137-8, 1218; and recognition of states/governments, 922, 955, 983, 1006, 1018, 1101, 1104, 1127, 1392; and refugee problems, 520, 561-2, 662; and resolution of domestic conflict, 269, 923, 926, 928, 937, 955, 958, 983, 1023, 1073, 1098, 1112, 1128, 1345; and Southern Africa, 271, 959, 1101, 1106, 1119, 1559, 1580, 1629, 1647, 1651, 1698, 1753; and the Middle East conflict, 957, 984-5, 1001, 1026, 2428, 2451; and the Nigerian Civil War, 455, 923, 955, 958, 962, 1073, 1128; and the Rhodesian crisis, 958-9, 995, 1005, 1578; and the UN, 926-8, 954, 958, 986, 1002, 1014, 1029, 1036, 1075, 1112, 1139; bibliography of, 1055, 1060, 1072; Charter of, 934, 948, 958, 987; decolonization/liberation function of, 911, 925, 940, 945, 951, 954, 958-9, 962, 980, 986, 988, 995, 1005, 1015, 1019, 1033, 1088, 1093, 1103, 1106, 1119, 1136; economic/social function of, 926-7, 951, 956, 962, 1121; functions of, 1000, 1002, 1012, 1029, 1042, 1074, 1083, 1087, 1105, 1135; institutions of, 925, 946, 950, 976, 987, 1002, 1016, 1022, 1033, 1039, 1054, 1077, 1080, 1107, 1132; history of, 910, 924, 933, 941-3, 945, 948, 953, 956, 973, 997-8, 1009, 1011-2, 1027, 1029, 1041, 1056-7, 1074, 1081-2, 1084, 1087, 1090, 1092, 1096, 1101, 1113-4, 1123, 1132; problems in, 933, 997, 1050, 1085, 1087, 1101, 1121; relations of, with other international organizations, 1064-5, 1134, 2534; settlement of international disputes by, 522-3, 611, 628, 646, 655, 657-8, 923, 926, 928, 938, 951-2, 958, 960-3, 976-7, 983, 986, 1014, 1020-1, 1023, 1033, 1036, 1053, 1058, 1080, 1083, 1093, 1154
Obote, Milton, 886
O'Brien, Conor Cruise: works of, reviewed, 1216
OCAM, 682, 719, 724, 727, 730, 735, 760, 809, 842, 853-4, 867, 945. Also see UAM; OAMCE; UAMCE

OERS, 698, 714, 736, 742, 810
Ogunsanwo, A.: works of, reviewed, 1898
Olympio, Sylvanus E., 405
OPEC, 2607
Organization of African Trade Union Unity, 555
Osagyefo. See Nkrumah, K.
Outward policy. See South Africa: relations of, with African states
Oxfam, 2518

PAFMECA, 779, 1032. Also see PAFMECSA
PAFMECSA, 711, 779, 1032. Also see PAFMECA
Pakistan, 2381
Pan-African Congresses, 910, 1070, 1034. Also see Pan-Africanism
Pan-Africanism, 57, 63, 71, 77-8, 121, 149, 157, 270, 365, 715, 781, 790, 914, 918, 939-40, 947, 965, 974-5, 978, 1003, 1031-2, 1051, 1061, 1065, 1099, 1234, 1267, 1934, 2058, 2601; bibliography of, 91, 972, 981, 1009, 1072; definition of, 772, 835, 950, 989-91, 1028, 1035, 1037, 1044-5, 1074, 1097, 1108, 1140; goals of, 917, 931-2, 935, 1000, 1079, 1135; history of, 66, 711, 715, 835, 899, 900, 910, 915-7, 929, 934, 973, 982, 990-1, 999, 1000, 1004, 1010, 1030, 1032, 1034-5, 1037, 1047, 1059, 1062, 1070, 1086, 1090, 1097, 1100, 1109, 1118, 1126, 1130, 1135
Pan-Arabism, 111, 914
Pan Negroism, 469
Pariah state, 1490
Parsons, T., 1497
Patrice Lumumba Friendship University, 1818

Patron-client relations, 13, 201, 1768, 2592
People's Republic of China, 527, 1731-2, 1734, 1756, 1829, 1851, 1875, 1955, 1983, 2013, 2046, 2060-1, 2067, 2516; aid programs of, 1470, 1734, 1756, 1764, 1773, 1856, 1861, 1983, 2008, 2011, 2023, 2051, 2057, 2060, 2063, 2071, 2505, 2507, 2530, 2592, 2618, 2633, 2657, 2670, 2704, 2805, 2815-7, 2835; competition of, with USSR, 1734, 1773-4, 1839, 1852, 1874, 1886, 1910, 1924-5, 1956, 1983, 1989, 1991, 2000, 2033, 2064-5; relations of, with Africa, 1734-8, 1750-2, 1764, 1769, 1770-1, 1780, 1785, 1788, 1793, 1796, 1806, 1809, 1811, 1830-1, 1839, 1843-4, 1847, 1852, 1857, 1861-2, 1866, 1868-70, 1873-5, 1877-9, 1883, 1989-1900, 1912, 1916, 1919, 1924, 1929-30, 1941-4, 1953-5, 1962, 1975, 1977-8, 1983, 1989, 1991, 2000-1, 2008, 2013, 2028, 2030, 2044-5, 2051, 2054, 2057, 2061-6, 2068-70; relations of, with Southern Africa, 1404, 1431, 1468, 1470, 1486, 1549, 1734, 1773, 1815, 1910, 2031; relations of, with Tanzania, 205, 1756, 1856, 1983, 2008, 2011, 2023, 2036, 2060, 2063, 2068-9, 2071, 2505; relations of, with the Third World, 1735, 1751, 1861, 1869, 1919, 2036, 2061, 2064
Peripheral states, 739. Also see Dependency
Persian Gulf, 1795
Petroleum, 2346, 2517, 2534; and economic development, 2487, 2506, 2531, 2580, 2607, 2746; in Afro-Arab

Petroleum (continued), relations, 984, 1175, 1213, 1651, 2424, 2446, 2512, 2540, 2746; in Nigerian foreign policy, 324, 2732; in Southern African international relations, 1641, 1651, 2046
Peyman, H.: works of, reviewed, 2008
Phelps Stokes Foundation, 2080
Poland, 2737
Polisario, 266
Political economy, 4, 89, 442, 971, 1003, 1479, 1651-2, 2148, 2315, 2668, 2686, 2768-9
Pompidou, G., 1808, 2053
Portugal, 1322, 1798, 2155, 2214; relations of, with Africa, 488, 1846, 2337, 2366, 2403, 2414, 2439; relations of, with Southern Africa, 140, 255, 375, 549, 1358, 1364, 1372, 1457, 1483, 1531, 1533-5, 1629, 1645, 1667, 1684, 2429, 2433; relations of, with UN, 1147, 1173, 1196, 1254, 1265, 1291, 1351, 1667; relations of, with USA, 2075, 2095, 2120-30, 2155, 2174, 2204, 2206, 2284
Principe, 2297
Propaganda, 350, 1625, 1627, 1726, 1960
Public administration, 1091
Public corporations, 723, 878
Public opinion, 137, 196, 277, 1158, 1492, 2140, 2598
Public relations, 204

Rabat Conference, 228
Race, 203; as a factor in international relations, 37, 235, 327, 443, 517, 1331, 1441, 1529, 1558, 1653, 1658, 1661, 1757, 1872, 2372; as a factor in Sino-Soviet relations, 1839, 1873, 1991, 2057; as a factor in US foreign policy, 1873, 2138-9, 2197, 2260-1
Recognition: and the OAU, 922, 955, 983, 1018, 1101, 1127, 1392; and the UN, 1392, 1450; of governments, 528, 663, 922, 1241, 1265, 1353; of states, 663, 1149, 1206, 1241, 1353, 1391-2, 1450, 2447, 2732
Red Cross: in Nigerian Civil War, 49, 2551, 2697, 2755
Red Sea, 513, 539, 545, 558, 620, 1732, 1823, 1845, 2005
Refugees, 501, 511, 520, 531-2, 547-8, 553-4, 561-2, 580-1, 595, 603, 612, 621, 630-1, 650, 660; and international law, 520, 553, 562, 591, 650; in Southern Africa, 485, 532, 538, 567, 609, 631
Republic of China, 1490; aid programs of, 2538, 2676, 2709; relations of, with Africa, 2013, 2067, 2411
Republic of South Africa, 261, 532, 1157, 1303, 1322, 1356, 1379, 1391-2, 1396, 1429, 1447, 1499, 1506, 1568, 1612, 1634, 1647, 1657, 1670, 1679, 2601, 2610; aid program of, 1389, 1545-6, 1672, 1714; and the Commonwealth, 359, 1415, 1509, 1573, 1717, 1856, 2378; and the ICJ, 1181, 1184, 1243, 1279, 1304; and the OAU, 911, 959, 1106, 1753; and the Southern Oceans, 1367, 1471, 1569, 1618, 1663, 1678, 1685, 1716, 2148; and the UN, 1144, 1148, 1156, 1161, 1164, 1166, 1176, 1184, 1188, 1191, 1194, 1197, 1256, 1560, 1269, 1297-8, 1301, 1313, 1315, 1323, 1335-6, 1342, 1369, 1379, 1439-40, 1558, 1586, 1607, 1667, 1694, 1710, 1715, 1717, 1753;

Republic of South Africa (continued), arms sales to, 911, 1460, 1482, 1519, 1577, 1606, 1703, 1719, 1787, 1904, 1923, 2147, 2597, 2611, 2791; as a middle power, 50, 101, 439, 1395, 1475-6, 1478, 1555, 1642-3, 1652; as a nuclear power, 1404, 1498, 1527, 1617, 1677, 1712, 2281, 2283, 2340, 2360, 2390, 2597; domestic politics of, 1362, 1366, 1371, 1373, 1396, 1424, 1440, 1447, 1458, 1622, 1644, 1679, 1691, 1696, 1718, 1724, 1727, 2141, 2467; economic relations of, 375, 1389, 1403, 1449, 1481, 1518, 1544-5, 1548, 1550, 1552-4, 1556, 1583, 1597, 1626, 1631, 1638, 1662, 1673, 1678, 1690, 1701, 1717, 1904, 2015, 2236, 2745, 2756; foreign relations of, 162, 1358, 1360, 1369, 1371, 1381, 1395, 1404, 1410, 1424, 1428, 1431, 1439, 1441, 1445, 1449, 1453-4, 1458, 1475, 1484, 1486, 1490, 1492-6, 1507, 1510, 1521-2, 1530-1, 1540, 1553-4, 1562-4, 1577, 1579, 1592, 1599, 1604-5, 1607, 1613, 1620, 1627-8, 1636, 1644, 1647, 1658, 1665, 1679-85, 1687, 1694, 1700, 1706, 1771-2, 1717, 1726, 1798, 1983, 2027, 2046, 2147, 2315, 2372, 2421-2, 2457, 2464, 2471, 2480; relations of, in Southern Africa, 1360, 1372, 1375, 1381, 1404, 1413, 1419, 1425, 1431, 1463-4, 1472, 1483, 1495, 1502, 1511, 1517, 1522, 1524, 1531, 1533-5, 1540, 1542-3, 1545, 1560, 1563, 1570, 1518, 1601, 1629, 1632, 1640-3, 1645-6, 1676, 1683, 1708, 1710, 2097-8; relations of, with African states, 108, 123, 142, 146, 165, 271, 313, 330, 375, 455, 583, 639, 1040, 1101, 1106, 1119, 1359, 1369, 1373, 1377, 1389, 1393, 1405, 1416, 1428, 1432, 1434, 1442, 1444, 1464, 1476, 1481, 1486, 1495, 1513, 1537, 1540, 1544-6, 1559, 1563-4, 1580, 1582, 1588-90, 1594, 1596, 1598, 1622, 1626, 1637, 1653, 1674, 1676, 1679, 1683-4, 1686, 1693, 1696, 1714, 1725, 1730, 1749, 1787, 1818, 1846, 2097, 2458; relations of, with Angola, 309, 1444, 1489, 1545, 1694, 1723, 1863; relations of, with Botswana, 165, 199, 240, 286, 307, 451, 476, 1545, 1753, 2118; relations of, with Canada, 1528, 1577, 2303, 2312-3, 2315, 2378, 2756; relations of, with Great Britain, 245, 1369, 1397, 1564, 1626, 1637, 1686, 1694, 1749, 1753, 1787, 1904, 1923, 2015, 2155; relations of, with Israel, 291, 329, 1362, 1394, 1448, 1551, 1704, 1707, 2318, 2446, 2467, 2474; relations of, with Lesotho, 237, 240, 297-8, 448, 451-2, 476, 480, 1545-6, 1753; relations of, with Malawi, 183, 275, 313, 323, 426, 476, 583, 1693; relations of, with Rhodesia, 1358, 1531, 1545, 1637, 1664, 1688, 1753; relations of, with Swaziland, 413, 451, 476, 1545, 1743; relations of, with the USA, 1404, 1406, 1417, 1439, 1500, 1519, 1541, 1550, 1556, 1564, 1566, 1591, 1609, 1625, 1628, 1641-2, 1647, 1685, 1689, 1694,

Republic of South Africa (continued), 1703, 1712, 1723, 1749, 2027, 2074-5, 2086-7, 2097-8, 2107, 2110, 2113, 2126, 2128-9, 2139, 2141, 2147-8, 2153, 2155, 2161, 2168, 2180, 2215, 2221, 2228, 2234, 2236, 2240, 2245, 2259, 2269, 2276, 2279, 2281-3, 2513, 2745, 2778; relations of, with the USSR, 1381, 1404, 1431, 1486, 1517, 1521, 1613, 1628, 1663, 1880, 1974, 2029, 2046, 2153; relations of, with West Germany, 1606, 1712, 2027, 2340, 2360, 2390, 2442, 2597; relations of, with Zambia, 142, 221, 476, 1495, 1508, 1983; sanctions against, 1365, 1435, 1459-60, 1482, 1497, 1500, 1518-9, 1529, 1539, 1624, 1635, 1639, 1669, 1690, 1703, 2147, 2161, 2223, 2245, 2269, 2745, 2791; strategic/military considerations by, 1360, 1380-1, 1390, 1458, 1468, 1486, 1517, 1556, 1569, 1606, 1609, 1612-3, 1618, 1623, 1635, 1647, 1653, 1659, 1671, 1680, 1684, 1703, 2046, 2281, 2360, 2597

Republic of the Congo, 588, 619, 809, 2069

Rhodesia-Zimbabwe, 738, 1273, 1303, 1322, 1446, 1499, 1538, 1710; and the Commonwealth, 359, 1504, 1578, 1757; and the OAU, 958-9, 1005, 1578; and the UN, 1144, 1147-8, 1151, 1158, 1191, 1226, 1251, 1256, 1262, 1268, 1273, 1325, 1365, 1387, 1400, 1402, 1435, 1438, 1452, 1515-6, 1578, 1584, 1586, 1667, 1710, 1729; domestic politics in, 1382, 1426-7, 1571, 1578, 1584; economic development in, 1422, 1567, 1595, 1692, 1702; foreign relations of, 1358, 1364, 1421, 1427, 1460, 1462, 1474, 1516, 1519, 1531, 1545, 1571, 1619, 1676, 1826, 2153, 2249, 2454; relations of, with Africa, 138, 142, 375, 384, 414, 451, 871, 1505, 1516, 1584-5, 1664; relations of, with Great Britain, 1433, 1516, 1578, 1600, 1602, 1637, 1660, 1664, 1720-1, 1753, 1805, 1932, 1934, 2155, 2200, 2249; relations of, with Southern Africa, 1358, 1368, 1493, 1531, 1533, 1545, 1590, 1602, 1637, 1641, 1664, 1676, 1688; relations of, with USA, 331, 1460, 1473-4, 1516, 1519, 1541, 1574, 1587, 1608, 1722, 2075, 2081, 2112, 2140, 2151, 2153, 2155, 2168, 2191, 2200, 2234, 2242, 2245, 2249, 2256, 2276; sanctions against, 331, 1226, 1268 1363, 1365, 1382, 1387, 1400, 1402, 1422, 1435-6, 1443, 1452, 1456, 1487, 1515-6, 1519, 1567, 1574, 1593, 1595, 1600, 1608, 1669, 1692, 1699, 1702, 1710, 1720, 1728, 2081, 2112, 2168, 2200, 2217, 2222, 2234, 2242, 2245, 2454

Rio Mano Union, 812

River basins (as focus of international cooperation), 369, 733, 810, 839. <u>Also see</u> Chad, Basin Commission; Niger River Commission; OERS

Rogers, B.: works of, reviewed, 1445

Roman Catholic Church, 69

Rosenau, James, 474

Rubinstein, A. Z.: works of, reviewed, 1861

Rwanda, 531, 588, 623, 693, 724, 2051, 2678

Sahara, 796, 805, 909
Sahel, 559; drought in the, 42, 857, 2770
Sanctions, 1437, 1500, 2277; against Rhodesia, 331, 1226, 1268, 1363, 1365, 1382, 1387, 1422, 1435-6, 1443, 1456, 1459, 1487, 1515, 1519, 1567, 1574, 1593, 1595, 1600, 1608, 1669, 1692, 1699, 1702, 1710, 1720, 1728-9, 2081, 2166, 2223, 2245, 2454; against South Africa, 1365, 1435, 1435, 1459, 1482, 1497, 1500, 1518-9, 1529, 1539, 1577, 1624, 1635, 1639, 1669, 1690, 1703, 2147, 2161, 2223, 2245, 2269, 2745, 2791
Santiago Conference, 1240
Sao Tomé, 2297
SATO, 2210
Saudi Arabia, 545, 1999, 2465
Scandinavia, 2339; aid programs of, 1547, 2822; relations of, with Southern Africa, 1577, 2315, 2430, 2460
Scandinavian Institute of African Studies, 514
SCOA, 1988
Secession, 983, 1018, 1112, 1319, 1350
Selassie, Haile, 1095
Self-determination, 1018, 1218, 1222, 1245, 1274, 1292, 1303, 1332-4, 1440, 2070, 2103, 2132, 2159, 2629
Self-help, 1316
Self-reliance, 387, 1305, 2070, 2606, 2825
Senegal, 213, 214, 369, 447, 483, 535, 589, 1913, 2322, 2458, 2598; relations of, with France, 379, 1952, 1959, 2006, 2610, 2720;
relations of, with Gambia, 369, 651, 811, 850, 862-3, 900
Senegal River basin, 369.
Also see OERS
Senghor, Lamine: views of, on imperialism, 2628
Senghor, Leopold, 296, 319, 354, 369, 1069, 1115, 2458
Seychelles, 318, 2297
Shaba Conflict, 556
Shepherd, J.: works of, reviewed, 2584
Sierra Leone, 589, 812, 2563, 2742
Sierra Leone Development Company, 2630
Simonstown Agreement, 1787, 1904
Simulation, 1378, 1500
Slavophilism, 1120
Slonim, S.: works of, reviewed, 1423
Smith, Stewart: works of, reviewed, 2226
Socialism, 205-6, 415, 460, 1048
Somalia, 158, 588, 628, 661, 2005, 2111, 2567, 2694; relations of, with Affars and Issas, 602, 639, 647, 1851; relations of, with Ethiopia, 158, 269, 489, 507, 519, 524, 536, 545, 558, 576, 582, 647, 1854; relations of, with Kenya, 269, 489, 507, 519, 536, 582, 628, 647, 661, 952; relations of, with USSR, 1761, 1802, 1814, 1908, 1931, 2792
South Atlantic Ocean: and US policy, 2179, 2210, 2234; military and strategic concerns in, 1570, 1658, 1685, 1716, 1848, 2179.
Also see Cape route; Indian Ocean
South East Asia, 2693
South Korea, 1490, 2389, 2432,
South West Africa. See Namibia

Southwest Africa, 1875
Southern Africa, 220, 300,
  1169, 1404, 1419, 1514,
  1523, 1545, 1621, 1698,
  2154, 2233, 2369, 2801;
  and the OAU, 920, 954,
  962, 988, 1015, 1019,
  1088, 1093, 1103, 1119;
  and the UN, 1146, 1177,
  1189, 1204, 1259, 1271,
  1435; and the western
  states, 1461, 1557, 1628,
  1754, 1906, 1910, 2114,
  2155, 2433, 2437; Canada
  and, 1528, 2303, 2313,
  2328, 2336, 2365, 2383,
  2410, 2455; economic
  relations in, 300,
  375, 1357, 1401, 1405,
  1413, 1419, 1479, 1483,
  1524, 1542-3, 1565, 1632,
  1641, 1705; federation in,
  853, 860, 1467, 1561,
  1621, 1724; foreign
  investment in, 1412, 1445,
  1501, 1528, 1550, 1642;
  international relations
  of states in, 123, 146,
  200, 202, 265, 267, 375,
  444, 449, 517, 538, 680,
  1357, 1361, 1368, 1370,
  1372, 1374-6, 1384-6,
  1397-9, 1407-8, 1414, 1430,
  1435, 1446, 1457, 1459
  1469, 1476-7, 1479-80,
  1483-5, 1508, 1512, 1520,
  1531, 1533-4, 1536, 1538,
  1572, 1602, 1610, 1614-6,
  1629-30, 1645, 1647-52,
  1654-6, 1661, 1666-8,
  1675-6, 1684, 1695, 1713,
  1715, 2429; liberation
  movements in, 293-4, 316,
  1477, 1549, 1698, 2631;
  military and strategic
  considerations in, 1357,
  1364, 1420, 1499, 1517,
  1569, 1611, 1613, 1618,
  1633, 1686, 1985, 2210,
  2237, 2289; PRC and,
  1734, 1910; refugees and
  migration in, 485, 565,
  631, 1383, 1388, 1705;
  relations of African states
  with, 123, 155, 373, 408,
  412, 431-2, 444, 2485;
  Scandinavia and, 2430,
  2469-70, 2472; South Africa
  in, 1360, 1372, 1381, 1404,
  1425, 1431, 1463-4, 1472,
  1483, 1495, 1502, 1508,
  1511, 1517, 1531, 1533-5,
  1540, 1542-3, 1563, 1581,
  1601, 1629, 1632, 1641-3,
  1645, 1651-2, 1676, 1683-4,
  1710, 1713, 1715; US and,
  350, 1378, 1446, 1541,
  1576, 1628, 1675, 1935,
  1765, 2074, 2077, 2079,
  2082, 2099, 2100, 2114,
  2117-8, 2125, 2133-4,
  2143-4, 2153-5, 2161-2,
  2166, 2169, 2174-6, 2180-1,
  2186-7, 2195, 2199, 2203,
  2205, 2209-10, 2216, 2219,
  2223, 2230, 2233, 2237,
  2245, 2249, 2253, 2256-7,
  2261, 2275-6, 2282, 2284,
  2289, 2430; US economic
  relations with, 1462, 1576,
  2223, 2745; USSR and, 1428,
  1628, 1663, 1765, 1910, 1935,
  2039, 2153, 2233, 2249;
  Zambia in, 140, 142, 371-2,
  376, 442, 679, 685, 1378,
  1407, 1495, 1508
Southern African Customs Union,
  298, 677, 1524, 1560, 1570,
  1708
Southern hemisphere, 1395
Sovereignty, 1301, 2630
Soweto, 1537
Spain, 2348; and decolonization of Western Sahara,
  506, 544, 577, 2345, 2362,
  2416, 2420
Spanish Sahara. See Western
  Sahara
Sports: and the anti-apartheid
  campaign, 1529, 2471
Stabex, 2549, 2708. Also see
  EEC
Stalin, J., 1738, 1815
Statelessness, 1149

Statistics: collections of, on African states, 3, 53, 70, 72, 621, 2080, 2519, 2816, 2819
Status-field theory, 642
Stockwell, J.: works of, reviewed, 2288
Stratification: international, 433
Subimperialist state. See Middle powers
Sudan, 111, 115, 159, 178, 229, 244, 273, 280, 453, 477, 486, 505, 531, 588-9, 937, 1023, 1982, 2573, 2792
Suez Canal, 620
Suez Crisis, 1282, 1579
Supranationaliam, 1045, 1073
Swaziland, 160, 177, 450, 517, 1147, 1514, 1547, 1560, 1753, 2015, 2648; relations of, in Southern Africa, 413, 451, 476, 517, 1413, 1419, 1512, 1524, 1545, 1563, 1641, 1643, 2015
Sweden, 2719; aid programs of, 2063, 2562, 2719, 2738, 2743; relations of, with Southern Africa, 1516, 1552-4
Switzerland, 2399, 2639; aid programs of, 2639, 2656; relations of, with Southern Africa, 1516, 2421
Systems analysis, 25, 32, 102-3, 105, 152, 202, 265, 433, 438, 444, 640, 657, 668, 821, 1076, 1078, 1089, 1110-1, 1129, 1249-50, 1252, 1385, 1469, 1476, 1480, 1520, 1534, 1614, 1649-50, 1654-6, 1884, 2098

Taiwan. See Republic of China
Tanganyika. See Tanzania
Tanzam Railway, 207, 1409, 1856, 1983, 2008, 2011, 2023, 2071, 2505, 2618, 2835
Tanzania, 151, 253, 583, 588-9, 609, 763, 1207; aid programs in, 1756, 1856, 1983, 2008, 2011, 2022, 2060, 2063, 2071, 2505, 2602-3, 2618, 2644, 2706, 2738, 2792, 2835; foreign relations of, 127, 205-7, 268, 284, 295, 365, 380, 286-7, 391-2, 407, 415, 429, 433, 435, 459, 1101, 1104, 1409, 2365, 2423, 2455, 2521, 2610; relations of, with East Africa, 205, 728, 814, 881; relations of, with PRC, 205, 1756, 1856, 1983, 2008, 2011, 2022, 2036, 2060, 2063, 2068-9, 2071, 2505, 2618, 2835; relations of, with Southern Africa, 414, 583, 1407
Telli, Diallo, 1107
Third World, 41, 62, 65, 219, 374, 424, 464, 600, 954, 1008, 1148, 1197, 1203, 1213, 1256, 1328, 1408, 1840, 1894, 2350, 2352, 2392-3, 2417, 2443, 2445, 2496, 2565, 2628, 2656, 2710, 2729, 2737, 2757, 2781, 2790-1, 2801-3, 2817, 2829; and the EEC, 2508, 2533, 2536, 2551, 2593, 2638, 2664; PRC and, 1735, 1751, 1861, 1869, 1919, 1955, 1983, 2036, 2061, 2064, 2071; USA and, 2211, 2216; USSR and, 1790, 1818, 1824, 1861, 1886-9, 1967, 1993, 2020, 2573, 2651,
Tito, J., 2375
Togo, 190, 193, 405, 483, 492, 503, 726
Touré Sekou: foreign policy of, 179, 354, 467-8, 1778
Tourism, 1389, 2486, 2735, 2771
Transaction analysis, 679
Trans-Africanism, 2104

Transkei, 1149, 1391-2, 1450, 1537, 1676
Transnationalism, 1657, 2532
Transport, 9, 680, 692, 702, 753, 796, 840, 1361
Treaty for East African Co-operation, 725, 816
Treaty succession, 429, 1278
Tshombe, M., 2193
Tunesia, 423, 1842, 2792
Turkey, 565

UAM, 719, 760, 795, 845-7. Also see OCAM; OAMCE; UAMCE
UAMCE, 760. Also see OAMCE; OCAM; UAM
UDAO, 808
UDEAC, 675, 739, 824, 826-7, 858-9, 869, 897, 902-3, 916, 2780. Also see UEAC
UEAC, 697. Also see UDEAC
Uganda, 217, 241, 356, 364, 440, 458, 588, 597, 728, 763, 1101, 1127, 1853, 1872, 1933, 2319, 2367, 2386, 2602, 2792
Ujamaa, 206
UN, 19, 58, 62, 71, 110, 113, 139, 391, 418, 527, 544, 747, 817, 931, 1142, 1186, 1195, 1227, 1234, 1240, 1248, 1272, 1309, 1319, 1354, 1392, 1450, 1789, 2013, 2164, 2168, 2194, 2385, 2389, 2451, 2467, 2561; African role in, 78, 90, 94, 99, 474, 1116, 1141, 1165, 1171, 1175, 1188, 1192, 1209, 1218, 1237-9, 1249-50, 1277, 1283-4, 1287, 1324, 1331, 1345-6, 1829, 2512; analysis of voting in, 139, 464, 527, 1198, 1213-4, 1236, 1252-3, 1274, 1281, 1339, 1344, 1789, 1829, 2307, 2516; and conflict resolution, 926, 928, 960, 977, 1014, 1036, 1112, 1177; and decolonization, 1142, 1147, 1158, 1173, 1188, 1190-1, 1196, 1251, 1254, 1265, 1275, 1291, 1319, 1337, 1342, 1346, 1351, 2373; and development, 78, 926-7, 1144-5, 1150, 1271, 1288, 1305, 2604, 2613, 2626; and Namibia, 1144, 1147, 1161, 1164, 1182, 1184, 1199, 1222, 1230, 1243, 1246, 1251, 1279, 1294, 1301, 1311, 1313-4, 1318, 1325, 1334, 1342, 1347, 1440, 1472, 1488, 1624; and the OAU, 926, 954, 958, 1002, 1014, 1029, 1036, 1075, 1112, 1139; and Portugal, 1147, 1173, 1196, 1254, 1265, 1291, 1351, 1667; and refugees, 531, 553, 562, 630, 650; and Rhodesia, 1144, 1147-8, 1151, 1158, 1226, 1256, 1262, 1268, 1273, 1325, 1365, 1387, 1400, 1402, 1435, 1438, 1452, 1515-6, 1578, 1584, 1586, 1667, 1710, 1729; and sanctions, 1226, 1268, 1365, 1387, 1400, 1402, 1435, 1452, 1515-6, 1578, 1584, 1586, 1667, 1710, 1729, 2454; and South Africa, 1144, 1148, 1156, 1161, 1164, 1166, 1176, 1184, 1188, 1194, 1197, 1256, 1260, 1269, 1274, 1297-8, 1301, 1313, 1335-6, 1342, 1369, 1379, 1439, 1450, 1558, 1580, 1584, 1586, 1606-7, 1667, 1694, 1717, 1753; and Southern Africa, 138, 331, 1146, 1177, 1189, 1204, 1259, 1271, 1315, 1408, 1629, 1698, 1715; and trusteeship, 80, 1158, 1182, 1207, 1242, 1246, 1290; impact of, on Africa, 66, 926, 1171, 1212, 1216, 1271, 1288, 1346; institutions of, 1142, 1146-7, 1151, 1182, 1213, 1222,

UN (continued), 1297, 1320, 1325, 1440, 1450; operation in the Congo, 78, 367, 473, 1145, 1174, 1211, 1232, 1235, 1244, 1255, 1257, 1264, 1282, 1289, 1306, 1312, 1319-20, 1323, 1343, 1348, 1355, 1364, 2168, 2194, 2379
UNCTAD, 926, 2496, 2566, 2591, 2608, 2661, 2687, 2702
UNDP, 926, 2571, 2602
UNHCR, 531, 553, 926, 2518
UNICEF, 926, 1150
UNIDO, 926
Union of African States, 716, 847, 900; bibliography of, 717
Union of South Africa. See Republic of South Africa
University of Ibadan, 109
Upper Volta, 190, 226, 483, 494, 537, 589, 623, 726, 864
Uruguay, 1658
USA, 98, 348, 660, 1441, 1732, 1884, 1921, 2085, 2088, 2090, 2095, 2125, 2140, 2142, 2164, 2168, 2178, 2189, 2194, 2198, 2208, 2233, 2284, 2441; Afro-Americans in the foreign policy process of, 347, 2104-5, 2120, 2135, 2140, 2145, 2165, 2167, 2173, 2175, 2186 2207, 2211-2, 2246, 2284, 2286, 2293; aid programs of, 1462, 2075, 2119, 2152, 2157, 2243, 2251, 2295, 2297, 2299, 2511, 2516, 2522, 2560, 2579, 2600, 2692, 2700, 2725, 2750; and Southern Oceans, 1384, 1685, 1767, 1848, 1891, 1981, 2091, 2115, 2148, 2179, 2210, 2234, 2237, 2270; and the Horn of Africa, 545, 571, 1851, 1894, 1979, 1999, 2160; competition of, with PRC, 1734, 1764, 1844, 1910, 1919; competition of, with USSR, 1765, 1886, 1910, 1914, 1949, 1956, 2037, 2072, 2091, 2153, 2153, 2178, 2190, 2198, 2233, 2243, 2249-50, 2270, 2600; covert operations of, 2125, 2195, 2221, 2273, 2288; economic relations of, with Africa, 2075, 2132, 2166, 2247, 2258, 2280, 2517, 2539, 2581, 2654, 2682, 2724, 2745, 2776, 2778, 2788-9, 2820; economic relations of, with Southern Africa, 1406, 1418, 1445, 1460, 1462, 1501, 1519, 1548, 1550, 1556, 1566, 1575, 1608, 1641-2, 1703, 2144, 2147, 2281, 2513, 2745, 2778; military/strategic concerns of, 1460, 1519, 1703, 2131-2, 2147, 2159, 2172, 2179, 2214, 2237, 2239, 2258, 2270, 2281, 2289, 2560, 2600, 2750; race as a factor in the foreign policy of, 1873, 2138-9, 2197, 2260-1; relations of, with Africa, 53, 66, 71, 80, 90-1, 94, 349, 444, 521, 571, 604, 904, 973-4, 1234, 1257, 1283, 1331, 1333, 1576, 1752, 1783, 1791, 1804, 1809, 1817, 1843, 1919, 1953, 1970, 1997, 2003, 2012, 2073-99, 2218, 2220, 2517, 2560, 2589, 2654, 2820; relations of, with Angola, 57, 1538, 1576, 1863, 1917, 2048, 2088-9, 2195, 2115, 2121, 2125, 2129, 2132, 2158, 2163, 2198, 2203, 2262, 2273, 2278; relations of, with Namibia, 1294, 1418, 1462, 1550, 1566, 2073, 2075, 2103, 2153, 2168, 2192; relations of, with Nigeria, 134, 397, 1784, 1801, 1827, 1964, 2166,

USA (continued), 2202, 2238, 2263, 2274; relations of, with Portuguese Africa, 2075, 2095, 2129-30, 2155, 2174, 2204, 2206, 2214, 2284; relations of, with Rhodesia, 331, 1460, 1462, 1473-4, 1516, 1519, 1574, 1587, 1608, 1660, 1709, 1722, 2075, 2081, 2112, 2151-3, 2155, 2168, 2191, 2200, 2217, 2222, 2235, 2242, 2245, 2249, 2256, 2276; relations of, with South Africa, 1406, 1417, 1439, 1460, 1492, 1500, 1541, 1548, 1550, 1556, 1564, 1566, 1591, 1609, 1625, 1647, 1670, 1689, 1694, 1712, 1723, 1749, 2057, 2074, 2086-7, 2097-8, 2107, 2113, 2115, 2118, 2121, 2126, 2128, 2139, 2141, 2147-8, 2153, 2155, 2161, 2168, 2180, 2223, 2228, 2236, 2245, 2259, 2269, 2276, 2279, 2281, 2283, 2340; relations of, with Southern Africa, 350, 1378, 1404, 1445, 1462, 1519, 1576, 1628-9, 1642, 1675, 1765, 1935, 2074-5, 2077, 2079, 2082, 2098-9, 2100, 2111, 2114, 2117, 2125, 2129, 2132-4, 2143-4, 2154-5, 2163, 2166, 2169, 2174-6, 2180-2, 2186-7, 2195, 2199, 2203, 2205, 2209-10, 2216, 2219, 2223, 2226, 2230, 2233, 2237, 2240, 2245, 2253, 2256-7, 2261, 2275-6; relations of, with Zaire, 2111, 2125, 2149, 2156, 2156, 2164, 2166, 2178, 2193-4, 2198, 2252, 2287
USSR, 446, 1281, 1779, 1789, 1795, 1839, 1849, 1873, 1884, 1887, 1889, 1921, 1950, 1991, 1993, 2014; aid programs of, 1470, 1773, 1814, 1861, 1886, 1890, 1993, 2020, 2573, 2585, 2592, 2600, 2657, 2704, 2733, 2787, 2788-9, 2791, 2805, 2816-7, 2833; and Cuba, 1823, 1849, 2037, 2325, 2355, 2441; and detente, 1765, 1812-3, 1833, 1902, 1968, 2037, 2048; and Southern Africa, 1294, 1381, 1404, 1408, 1428, 1431, 1459, 1468, 1470, 1486, 1521, 1549, 1613, 1628-9, 1645, 1659, 1670, 1765, 1773, 1910, 1935, 1968, 1974, 1976, 2029, 2039, 2077, 2153, 2233, 2249; and the Horn of Africa, 545, 571, 1731-2, 1795, 1814, 1823, 1851, 1859, 1876, 1894, 1908-9, 1931, 1999, 2040; and the Indian Ocean, 1384, 1517, 1613, 1663, 1685, 1716, 1875, 1881-2, 1891, 1903, 1905, 1979, 1981, 1985, 2010, 2039, 2046, 2091, 2237, 2270; competition of, with PRC, 1734, 1751, 1773-4, 1839, 1844, 1852, 1874, 1910, 1919, 1924-5, 1956, 1983, 1989, 1991, 2000, 2033, 2045, 2064-5; competition of, with USA, 1765, 1886, 1903, 1910, 1914, 1949, 1956, 2016, 2024, 2037, 2072, 2153, 2178, 2190, 2198, 2233, 2249-50, 2600; economic relations of, with Africa, 55, 1889, 2020, 2583, 2592, 2751, 2757, 2775, 2788, 2791, 2816-7, 2833; naval strategy of, 1613, 1814, 1848, 1881-2, 2010, 2270, 2651; relations of, with African states, 53, 55, 66, 280, 349,521, 525, 575, 904, 973-4, 1234, 1257, 1281, 1464, 1549, 1742, 1752, 1769, 1770-2, 1779, 1783, 1788-9, 1791-2, 1794-5, 1803, 1809, 1811-3,

USSR (continued), 1824,
1833-4, 1843, 1847-50,
1855, 1857, 1861, 1871,
1877-8, 1880, 1883, 1885,
1887-9, 1895-6, 1901,
1907, 1911-14, 1916, 1922,
1936-7, 1941, 1943-8, 1951,
1953, 1956, 1961, 1965,
1970, 1975-6, 1994, 2001,
2003, 2012, 2017-8, 2020,
2028, 2065, 2178, 2190,
2198, 2250, 2324, 2366,
2422, 2573; relations of,
with Angola, 1538, 1747,
1827, 1841, 1853, 1902,
1917, 1967, 2002, 2019,
2037, 2048, 2158, 2198, 2278;
relations of, with Ethiopia,
55, 1823, 1832, 2055-6; relations of, with Ghana, 1762,
1824, 1913, 2001, 2018, 2020,
2573; relations of, with
Guinea, 1913, 2001, 2020,
2213; relations of, with
Nigeria, 134, 377, 397, 401,
1784, 1801, 1827, 1913, 1951,
1961, 1964, 2014, 2018, 2020;
relations of, with Somalia,
1761, 1802, 1814, 1931; relations, of with Third World,
1521, 1790, 1818, 1861, 1886-9, 1966, 1993, 2573; research
of, on Africa, 1786, 1877-8,
1948; role of ideology in
foreign policy, 1772, 1803,
1888, 1896, 1937, 1947,
1951, 1961
UTA, 692

Vance, Cyrus, 2184
Vietnam, 2022
Volta Dam, 2660
Voluntary organizations,
2076, 2259, 2743, 2781;
and development aid, 2518,
2701, 2734; and refugee
aid, 581, 630
Vorster, J., 1411, 1537, 1511,
1591

Waiyaki, Munyua, 216
WAMU, 675
Weinstein, W.: works of,
reviewed, 1861
West Africa, 125, 369, 633,
654, 659, 667, 680, 689,
908; economic regionalism
in, 671-2, 689, 691, 701,
713, 718, 720, 726, 776,
808, 832, 838-40, 843, 848,
857, 880, 902, 905-7, 2765;
migration in, 498, 607,
632, 696, 720; political
regionalism in, 847, 900,
902; regional organizations
in, 669, 885, 900, 902;
regionalism in, 668, 673,
678, 688, 709, 712, 716,
757, 774, 797, 825, 851,
856, 885, 902, 908; trade in,
671, 705, 720, 765-6, 848;
USSR and, 1913, 2028, 2788
West Africa Rice Development
Association, 901
West Germany, 1817, 2598,
2732; aid programs of,
2511, 2519, 2767; and
South African nuclear
power, 1712, 2340, 2360,
2390, 2597; relations of,
with Africa, 2316, 2326,
2357, 2436, 2482, 2622,
2732, 2761; relations of,
with Southern Africa, 1492,
1516, 2027, 2340, 2360,
2390, 2403, 2437, 2442,
2454, 2597; sale of arms
by, to South Africa, 1606,
2360, 2597
Western Sahara, 186, 231, 266,
472, 502, 506, 544, 564,
573, 577, 592, 611, 613,
649, 1142, 1147, 1245, 1871;
and the ICJ, 1152, 1163,
1167-8, 1179, 1208, 1338;
and Spain, 2345, 2362,
2416, 2420
Western states, 1257, 1832-3,
1903, 2046, 2237; relations
of, with Africa, 175, 195,
280, 327, 380, 464, 467,
482, 1281, 1832, 1844,

Western states (continued), 1846, 1877-8, 1943, 1960, 1983, 1985, 2024, 2108, 2152, 2297, 2464, 2672, 2812; relations of, with Southern Africa, 1360, 1369, 1449, 1453, 1486, 1521, 1613, 1618, 1647, 1657, 1680, 1694, 1910, 1983, 2114, 2147, 2215, 2237, 2289
Who's Who, 3, 70
World Bank. See IBRD
World Council of Churches, 1661
World Law Fund, 64

Yaoundé Agreement. See EEC
Yoruba, 632
Young, A., 2110
Yugoslavia: relations of, with Africa, 1770, 2375, 2443-5, 2453

Zaire, 289, 556, 693, 697, 724, 809, 817, 1343, 2485; crisis in, 58, 80, 121, 269-70, 287, 289, 367, 568, 962, 1098, 1128, 1174, 1319, 2718; foreign relations of, 101, 112, 127, 194, 289-91, 367, 393, 473, 1645, 1764, 1846, 1956, 2019, 2051, 2322, 2331, 2354, 2443; relations of, with USA, 2111, 2125, 2149, 2152, 2156, 2166, 2178, 2193, 2198, 2252, 2287. Also see UN, operation in the Congo
Zambia, 227, 239, 371, 410, 517, 551, 588, 609, 871, 1409, 2774; foreign relations of, 127, 140, 143-4, 164, 239, 251, 350, 372, 374, 410, 434, 436-8, 442, 551, 599, 679, 685, 2111, 2365, 2525, 2603, 2642; relations of, with Rhodesia, 142, 251, 1427, 1516, 1505, 1585, 1664; relations of, with Southern Africa, 37, 140, 142, 221, 371-2, 376, 442, 476, 517, 551, 679, 685, 1378, 1407, 1409, 1495, 1508, 1641, 1643, 1983
Zanzibar, 151, 763, 1147. Also see Tanzania
Zimbabwe. See Rhodesia-Zimbabwe
Zionism, 1362, 1691